Environmental Division 2018

Core Programming Area at the 2018 AIChE Annual Meeting

Pittsburgh, Pennsylvania, USA
28 October - 2 November 2018

ISBN: 978-1-5108-7624-8

Printed from e-media with permission by:

Curran Associates, Inc.
57 Morehouse Lane
Red Hook, NY 12571

Printed by Curran Associates, Inc. (2019)

For permission requests, please contact AIChE
at the address below.

AIChE
120 Wall Street, FL 23
New York, NY 10005-4020

Phone: (800) 242-4363
Fax: (203) 775-5177

www.aiche.org

Additional copies of this publication are available from:

Curran Associates, Inc.
57 Morehouse Lane
Red Hook, NY 12571 USA
Phone: 845-758-0400
Fax: 845-758-2633
Email: curran@proceedings.com
Web: www.proceedings.com

TABLE OF CONTENTS

Some abstracts in this book refer to images that were not made available by the publisher

530152 Advanced Oxidation Processes for the Treatment of Azole-Containing Industrial Wastewater

Abstract:

Advanced oxidation processes (AOPs) are based on the formation of hydroxyl radicals and have been explored to overcome the inability of conventional water treatment methods for removing a wide variety of organic contaminants. Because of their high effectiveness, AOPs have been used for the treatment of industrial effluents for over a decade.

In this study, the effectiveness of three types of AOPs including Fenton reaction, photo-Fenton process and electrical discharge plasma treatment have been compared for the treatment of an industrial wastewater containing azoles, a class of five-membered heterocyclic compounds with at least one nitrogen atom, hydrogen peroxide, and organic and inorganic co-contaminants. The key design and operational variables such as iron dosing, pH and temperature for the Fenton reaction, UV flux and intensity for the photo-Fenton process, and discharge power and reactor electrode configuration for the plasma treatment, were investigated. The fundamental reaction mechanisms were studied and degradation byproducts identified.

The results of this study offer recommendations on the suitability of the three AOPs for treating the azole-containing wastewater and provide insights regarding the economics of the processes and their scaleup.

531557 Cavitation and Immobilised Photo-Catalysis for Effluent Treatment: A Comparative Study of Individual and Combined Operations

Abstract:

Cavitation and Immobilised Photo-catalysis for Effluent Treatment: A Comparative Study of Individual and Combined Operations

Sebastien De-Nasri, Sanjay Nagarajan, Peter K. J. Robertson and Vivek. V. Ranade*

Multiphase Flows, Reactors & Intensification Group (mRIg)

School of Chemistry and Chemical Engineering

Queen's University Belfast

Belfast BT9 5AG

Email: V.Ranade@qub.ac.uk <>**Background**

Unsustainable strain on freshwater sources globally could be reduced if industries reuse their wastewater after treatment. Wastewater treatment (WWT) plants, which remove pollutants from wastewater, commonly use conventional physico-chemical processes to treat wastewater. These conventional processes are energy intensive, expensive or use environmentally unfavourable chemicals for treatment and hence an alternative has to be sought [1,2].

Advanced oxidation processes (AOPs) are a form of advanced treatment which are gaining attention as a potential cost-effective method of treating wastewater. AOPs are characterised by their in situ generation of hydroxyl radicals to mineralise organic pollutants. Various AOPs include Fenton, photo-Fenton, photocatalysis, cavitation, ozonation and electrochemical treatment. In Fenton and ozonation based processes, chemicals are added to induce hydroxyl radical production, whilst electrochemical utilises electricity to treat wastewater by inducing oxidation processes with electrodes. Cavitation and photocatalysis are both processes which have been extensively studied for water treatment, and recent studies show great potential with large scale operation [1,3,4].

Cavitation is the formation, growth and subsequent collapse of microbubbles in a liquid, and is of interest for WWT due to its ability to completely mineralise organic pollutants without the need of complex catalysts or extreme operating conditions. The two commonly studied types of cavitation are acoustic cavitation (AC) and hydrodynamic cavitation (HC). HC relies on cavity generation through induced pressure variations in a flowing liquid using a cavitating device whereas AC relies on the propagation of ultrasound in a liquid medium for cavity formation. HC can be easily scaled up for large scale operation due to its continuous nature and low energy requirement when compared to AC.

Photocatalysis is another emerging AOP which is a light driven chemical reaction capable of producing reactive radical species through irradiation of a semi-conductor. Key characteristics of photocatalysis are its ambient operating conditions, low operating costs, and complete mineralisation of pollutants without the formation of undesired intermediates, as such, this process has feasible applications in wastewater treatment[4–7]. Commonly, photocatalysts are suspended in solution to obtain an improved irradiated area to volume ratio to perform photocatalysis. However, the need to separate these nanoparticles is a major limiting step for scale-up. Conversely, immobilised photocatalyst (ImPC) systems that do not require a separation process could be used. Poor mass transfer in ImPC and poor diffusivity of OH radicals in bulk[8] are however issues faced by ImPC. These limitations can be partially overcome by operating it in combination with other AOPs, or as a continuous or recirculating batch systems to enhance mass transfer.

The extent of degradation of pollutants is dependent on the quantity of hydroxyl radicals produced, which is increased significantly when combinations of AOPs are combined[9]. HC is an effective AOP which is operated in continuous flow, this characteristic is ideal for coupling with ImPC process. By combining HC with ImPC, mass transfer limitations can be negated through the continuous flow, whilst the degradation efficiency of both processes is increased through increased production of hydroxyl radicals. Fig. 1 depicts such a coupled process.

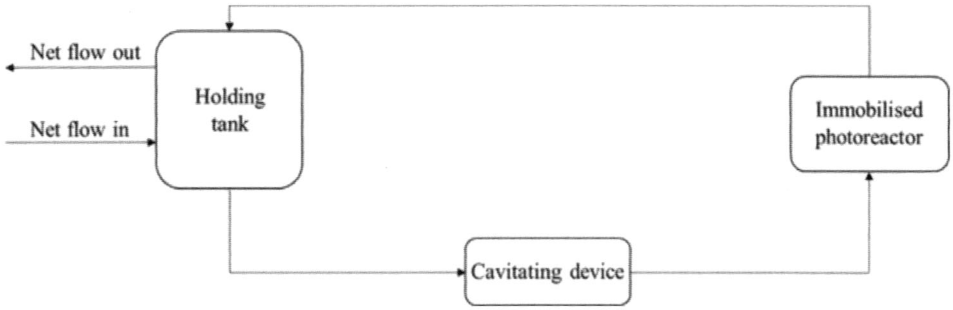

Figure 1 - schematic diagram of cavitation coupled with an immobilised photocatalyst system <>**Scope of present work**

Both HC and AC produces in situ OH radicals as a result of cavity collapse. Although methods of cavity generation are different, the chemistry of the reactions upon bubble collapse is equivalent in both cases. Hence, for the ease of experiments, this paper will focus on AC coupled with ImPC (sonophotocatalysis) which will then be used for translating it to a HC-ImPC system. Sonophotocatalysis has been extensively studied in the past for degrading a range of pollutants and proves to be an effective hybrid technique. However, the only combinations reported have used suspended photocatalysts with AC [10]. Due to the issues surrounding the scale-up of a suspended photocatalyst system, ImPC coupled with AC/HC is suggested in this work. By effectively degrading a model pollutant through ImPC and cavitation, an effective hybrid system can be employed which has neither mass transfer limitations nor a necessary separation process. The results will provide a sound basis for design of such hybrid systems for effluent treatment. <>**Experimental methods**

Acetone was used as the model pollutant for degradation experiments. The photocatalyst used was TiO_2 P25. The ImPC/cavitation system is described Fig. 1., the pollutant will be initially passed through a cavitating device (Sonics VCX 500 Ultrasonic processor) before being treated with photocatalysis. Given the nature of this system, treatment will consist of multiple passes occurring within a certain time. Individual processes (AC and ImPC) and their optimal operating parameters will be determined (amplitude, catalyst coating) through initial experiments. Sonophotocatalysis experiments will then be carried out and optimised. By comparing degradation data for individual and hybrid processes, the extent of synergy of the two processes can be quantitively compared and assessed against other hybrid systems in literature. <>**Results and discussion**

Processes were compared on a basis of acetone degradation. Acetone was selected as a model pollutant as it is a common industrial solvent, it also possesses a simple organic structure and has a well-documented reaction scheme with hydroxyl radicals, reducing the likelihood of unknown intermediates forming. Furthermore, sonophotocatalysis has been established as a process which can degrade a range of pollutants from a variety of industries, which does not need to be further established by investigating complex pollutants.

Initial experiments with AC have been carried out using 20kHz Sonics VCX 500 Ultrasonic processor, where the amplitude has been optimised (90%) for acetone solution (1000ppm) during 15 mins of sonication. Analysis was performed using a Cary 300 Scan, UV–vis Spectrophotometer at 263 nm, with a scan rate of 400 nm/min. Detailed results including optimised conditions for individual ImPC catalyst coating, and hybrid process will be presented at AIChE2018. <>**Summary and outlook**

This work focuses on comparing the performance of a batch recirculating system hosting an ultrasound/hydrodynamic cavitation reactor and an immobilised photoreactor in series. The performance of this sonophotocatalytic system is compared to the individual processes based on the degradation of a model pollutant - acetone. Additionally, the effect of operating parameters of cavitation reactors and weight of photocatalyst coating on the degradation performance will be tested and optimised. This comparison will allow the extent of synergy between the systems to be determined and would form the basis for developing a mathematical model to understand sonophotocatalytic degradation performance for a broader range of effluents. This approach and work presented is useful for future development of immobilised photocatalytic – hydrodynamic cavitation hybrid systems.

Keywords: AOP, Sonophotocatalysis, process intensification, immobilised photocatalyst

<>**References**

1. Ranade, V. V & Bhandari, V. M. *Chapter 1 - Industrial Wastewater Treatment, Recycling, and Reuse: An Overview. Industrial Wastewater Treatment, Recycling and Reuse* (2014). doi:http://dx.doi.org/10.1016/B978-0-08-099968-5.00001-5

2. Gupta, V. K., Ali, I., Saleh, T. A., Nayak, A. & Agarwal, S. Chemical treatment technologies for waste-water recycling—an overview. *RSC Adv.* **2,** 6380 (2012).

3. Carpenter, J. *et al.* Hydrodynamic cavitation: An emerging technology for the intensification of various chemical and physical processes in a chemical process industry. *Rev. Chem. Eng.* **33,** 433–468 (2017).

4. Chong, M. N., Jin, B., Chow, C. W. K. & Saint, C. Recent developments in photocatalytic water treatment technology: A review. *Water Res.* **44,** 2997–3027 (2010).

5. Tokode, O., Prabhu, R., Lawton, L. A. & Robertson, P. K. J. Controlled periodic illumination in semiconductor photocatalysis. *J. Photochem. Photobiol. A Chem.* **319–320,** 96–106 (2016).

6. Suryawanshi, P. G., Bhandari, V. M., Sorokhaibam, L. G., Ruparelia, J. P. & Ranade, V. V. Solvent degradation studies using hydrodynamic cavitation. *Environ. Prog. Sustain. Energy* **37,** (2017).

7. Bhandari, V. M., Sorokhaibam, L. G. & Ranade, V. V. Industrial wastewater treatment for fertilizer industry—A case study. *Desalin. Water Treat.* **57,** 27934–27944 (2016).

8. Choi, W., Kim, S., Cho, S., Yoo, H.-I. & Kim, M.-H. Photocatalytic reactivity and diffusing OH radicals in the reaction medium containing TiO2 particles. *Korean J. Chem. Eng.* **18,** 898–902 (2001).

9. Bethi, B. *et al.* Investigation of TiO2 photocatalyst performance for decolorization in the presence of hydrodynamic cavitation as hybrid AOP. *Ultrason. Sonochem.* **28,** 150–160 (2016).

10. Joseph, C. G., Li Puma, G., Bono, A. & Krishnaiah, D. Sonophotocatalysis in advanced oxidation process: A short review. *Ultrason. Sonochem.* **16,** 583–589 (2009).

532481 Degradation of Acesulfame Potassium By Ferrate(VI) and HCl-Activated Ferrate(VI) in Aqueous Solution

Abstract:

Abstract

The increasing discharge of emerging organic pollutants (e.g. pharmaceuticals and personal care products, and endocrine disrupting compounds) to the water bodies is receiving great attention due to their potential toxicity to humans and aquatic organisms. Many of these contaminants cannot be efficiently removed by conventional biological wastewater treatment processes indicated by their presence in influents and effluents of wastewater treatment plants, surface waters, and ground water across the globe, which ascertains the need to develop a more efficient water and wastewater treatment technologies. Artificial sugars, which are widely used as a low-calorie sugar supplements, are considered as emerging organic pollutants. Acesulfame potassium (ACK) is a compound extensively used as a sugar substitute in pharmaceuticals, and personal care and food products. Significantly, ACK has been detected in influents and effluents of several wastewater treatment plants worldwide at concentrations up to 80 μg/L. Hence, its removal from water is important.

The high-valent iron (oxidation state of +6), also known as ferrate(VI) (Fe(VI)), is receiving pronounced attention as a green chemical in a wide range of applications in different research fields, e.g. organic synthesis, water and wastewater treatment, and iron batteries. The oxidation, disinfection, and coagulation properties of Fe(VI) make it a multipurpose water-treatment chemical. Recently, we demonstrated the activation of Fe(VI) by acid (e.g. HCl and HNO_3) to significantly enhance the oxidative transformation of emerging pollutants in aqueous solution at slightly basic pH conditions, where Fe(VI) has decreased reactivity (E^0 (acidic) = +2.2 V vs. E^0 (basic) = +0.72 V). This study deals with the enhanced degradation of ACK by HCl-activated Fe(VI) at mild-alkaline environmental-relevant pH conditions.

Almost complete removal of ACK (>95%) by HCl-activated Fe(VI) has been achieved at Fe(VI) to ACK molar ratio of 8.0, and HCl to Fe(VI) molar ratio of 2.0. Comparatively, only ~65% degradation of ACK have been attained by Fe(VI) (nonactivated) at the same molar ratio of Fe(VI) to ACK of 8.0. Importantly, the Fe(VI)-decay time was much shorter in the case of HCl-activated Fe(VI) (5 minutes) compared to Fe(VI) (4 hours). In the case of activated Fe(VI), the pH of the ACK-Fe(VI) reaction mixture lowered by 0.6 pH units only, due to the addition of HCl (from an initial pH of 9.4 to 8.8). Further experiments at pH of 8.8 and 9.7, at a molar ratio of Fe(VI) to ACK of 2.0, without activation, showed that the effect of the pH is not significant at the studied pH range (i.e. 27% and 24% ACK removal efficiencies at pH of 8.8 and 9.7 respectively). The final pH (i.e. the Fe(III) formed by the reduction of Fe(VI) during the oxidation reaction was particulate enabling its removal by filtration) of the ACK-Fe(VI) and ACK-Fe(VI)-HCl reaction mixtures was similar (~7.7). Control experiments have also been done to rule out the effect of HCl and Fe(III) alone on the degradation of ACK, confirming that the enhanced oxidation is due to the activation of the chemical oxidant by HCl.

Results showed clearly the enhanced degradation of ACK by HCl-activated Fe(VI) at shorter time and environmental-relevant slightly-basic pH. Possible reason for the observed enhancement is the increased formation of iron-based reactive species, Fe(V) and Fe(IV), upon activation of Fe(VI) by HCl. It is known that Fe(V)/Fe(IV) species are 2-3 orders of magnitude more reactive than Fe(VI), thus causing enhanced oxidation of ACK. The effect of individual anions (i.e. Cl^-, HCO_3^-, and SO_4^{2-}), and cations (N^+, Ca^{2+}, and Mg^{2+}) usually present in wastewater on the oxidation of ACK by Fe(VI) and HCl-activated Fe(VI) has been also investigated. Moreover, the oxidation of ACK by Fe(VI) with and without activation has been also studied in the presence of organic matter to evaluate the applicability of this technology under water-treatment conditions. The mineralization of ACK was explored by dissolved organic carbon measurements.

533804 Advanced Oxidative Degradation of Benzoic Acid and 4-Hydroxy Benzoic Acid in Aqueous Phase â€“ a Comparative Study

Abstract:

Advanced Oxidative Degradation of benzoic acid and 4-hydroxy benzoic acid in aqueous phase – a comparative study.

Bhavna D Deshpande[a*], Dr.M.K.N.Yenkie[b], Dr.P.S.Agrawal[c]

a. Laxminarayan Institute of Technology.RTMNU,Nagpur.

b. Professor, LaxmiNarayan Institute of Technology.RTMNU,Nagpur.

c. Professor, LaxmiNarayan Institute of Technology.RTMNU,Nagpur.

*For correspondence - dnd.bhavna@gmail.com

ABSTRACT

Risks to environment and human health vary considerably depending upon the type and extend of exposure to hazardous chemicals. Environment protection agencies characterize risk on the basis of locally measured or predicted exposure scenarios[1]. Benzoic acid and 4-hydroxy benzoic acid are widely used in cosmetic, food preservatives and pharmaceutical industries[2]. Anthropogenic releases of such chemicals into the environment are primarily into water and soil. In the present study, advance oxidative processes namely UV, dark,solar assisted Fenton were investigated for the degradation in lab-scale photoreactor. The process being highly dependent on experimental parameters, degradation studies were carried out in order to establish optimum experimental conditions for the processes studied. The optimum degradation conditions for 4-HBA by the photo Fenton process are at pH 3.5, [Fe^{2+}] =7 ppm and [H_2O_2]/[4-HBA] molar ratio = 2.2, whereas the optimum degradation conditions for BA by the photo Fenton process are at pH 3.5, [Fe^{2+}] =7 ppm and [H_2O_2]/[4-HBA] molar ratio = 2.3.

Keywords: Advanced oxidation processes; 4-hydroxybenzoic acid; water treatment.

METHODOLOGY

AOPs work on the principal of production of hydroxy radical in situ, it serves as strong oxidizing agent to oxidize pollutants to nontoxic inorganic products[4][5][6]. Generally speaking, chemistry in AOPs could be essentially divided into three parts[3][7][8][9]:

- Formation of ·OH
- Initial attacks on target molecules by ·OH and their breakdown to fragments;
- Subsequent attacks by ·OH until ultimate mineralization.

Present work was carried out as follows:

1) Since both BA and 4HBA show strong absorption peaks in the UV region with high molar extinction coefficients, UV spectrophotometry was used for the quantitative determination of their concentration.

2) Optimization of experimental parameters like pH, Fe^{++} ,Fe^{+++}concentration and temperature.

3) The studied AOPs were also used to study its effectivity for the treatment of industrial waste effluent from dairy.

RESULTS

Table 1: Optimisation of amount of H_2O_2 for the oxidative degradation after 90 minutes.

AOP studied	Substrate	H_2O_2 ppm	% degragation	AOP studied	Substrate	H_2O_2 ppm	% degragation
H_2O_2+Fe^{2+} (UVFenton)	BA	504	45.980	H_2O_2+Fe^{2+} (UVFenton)	4HBA	400	62.010
		560	57.879			450	54.246
		616	39.230			500	69.868

Table 2: Optimisation of amount of Fe^{2+}/Fe^{+3} for the oxidative degradation after 90 minutes.

Substrate						
Benzoic acid	Fe^{2+} in ppm	10	8	7	5.5	4
	% Degradation	29	34	57.8	37.312	49.22
	Fe^{3+} in ppm	10	8	7	5.5	4
	%	34	41.67	69.39	67.529	44.596

	Degradation					
4HBA	Fe^{2+} in ppm	10	8	7	5.5	4
	% Degradation	44.760	52.123	69.868	63.201	60.270
	Fe^{3+} in ppm	10	8	7	5.5	4
	% Degradation	32.250	49.015	47.863	55.638	45.450

Table 3.Optimization of Fe^{2+}/Fe^{3+} for solar Fenton after 90 min

BA Solar Fenton	Fe^{2+} (4ppm)	% Degradation 76.80	Fe^{3+} (4ppm)	% Degradation 72.86
4HBA Solar Fenton	Fe^{2+} (4ppm)	% Degradation 95.25	Fe^{3+} (4ppm)	% Degradation 83.55

Amid the various processes dark Fenton, photo Fenton, dark Fenton like, photo Fenton like and solar degradation, studies show that solar is most effective and efficient process when applies to 4 hydroxybenzoic acid system as compared to benzoic acid. This shows that substituted benzoic acid is easily degraded as compared to benzoic acid itself. Batch systems were studied in the lab. Vis- UV spectrophotometer (Labman) was use to study absorbance at specific interval. Absorbance of benzoic acid and hydroxy benzoic acid was observed at 272nm and 245nm respectively. As H_2O_2 shows self-degrading, its degradation was studied spectrophotometrically using sodium titanium oxalate at 420nm.

It is observed, that with the increase in Fe ion concentration, % degradation first increases then decreases, this can be due to the reduced transparency of the solution, when excess of Fe ions get deposited in the reactor which thereby requires a process of removal. Fenton process shows best results at acidic pH (2.5 -3.5). Therefore, optimization is required. All the batch scale experiments were carried out at room temperature. It is observed with the increase in temperature, % degradation increases. In the beginning of reaction, some colored intermediates are perceived which may lead to the increase in absorption spectra, such intermediate is shot lived and soon degrade, which in turn depends upon nature of pollute and the condition applied.

AOPs are welcoming new trends, modified AOPs are more efficient and cost effective. A lot of research on solar base system are ongoing. Use of non-metallic elements are doped to enhance photocatalytic activity[10]. To increase the working pH range complex like Fe-EDTA[11], Fe oxalate[12], Fe-EDDS[13], are introduced in waste water for quicker and complete degradation.

REFERENCES

[1] NHS Choices, "Health A-Z - Conditions and treatments," *www.nhs.uk*, 2016. [Online]. Available: http://www.nhs.uk/Conditions/Pages/hub.aspx.

[2] "BENZOIC ACID AND SODIUM BENZOATE." .

[3] C. L. P. S. Zanta and C. A. Martínez-Huitle, "Degradation of 2-hydroxybenzoic acid by advanced oxidation processes," *Brazilian J. Chem. Eng.*, vol. 26, no. 3, pp. 503–513, 2009.

[4] W. Gernjak, T. Krutzler, A. Glaser, and S. Malato, "Photo-Fenton treatment of water containing natural phenolic compounds Photo-Fenton treatment of water containing natural phenolic pollutants," vol. 50, no. FEBRUARY, pp. 71–78, 2003.

[5] W. H. Glaze, "Reaction products of ozone: A review," *Environ. Health Perspect.*, vol. Vol. 69, pp. 151–157, 1986.

[6] W. H. Glaze, J.-W. Kang, and D. H. Chapin, "The Chemistry of Water Treatment Processes Involving Ozone, Hydrogen Peroxide and Ultraviolet Radiation," *OZONE Sci. Eng.*, vol. 9, pp. 335–352, 2016.

[7] V. K. Saharan, D. V. Pinjari, P. R. Gogate, and A. B. Pandit, "Advanced Oxidation Technologies for Wastewater Treatment: An Overview," in *Industrial Wastewater Treatment, Recycling and Reuse*, 2014, pp. 141–191.

[8] M. Cheng *et al.*, "Hydroxyl radicals based advanced oxidation processes (AOPs) for remediation of soils contaminated with organic compounds: A review," *Chemical Engineering Journal*, vol. 284. pp. 582–598, 2016.

[9] Y. Deng and R. Zhao, "Advanced Oxidation Processes (AOPs) in Wastewater Treatment.," *J. Ind. Eng. Chem.*, vol. 1, pp. 167–176, 2015.

[10] M. M. Huber *et al.*, "Oxidation of pharmaceuticals during ozonation of municipal wastewater effluents: A pilot study," *Environ. Sci. Technol.*, vol. 39, no. 11, pp. 4290–4299, 2005.

[11] G. Ghiselli, W. F. Jardim, M. I. Litter, and H. D. Mansilla, "Destruction of EDTA using Fenton and photo-Fenton-like reactions under UV-A irradiation," *J. Photochem. Photobiol. A Chem.*, vol. 167, no. 1, pp. 59–67, 2004.

[12] J. Jeong and J. Yoon, "pH effect on OH radical production in photo/ferrioxalate system," *Water Res.*, vol. 39, no. 13, pp. 2893–2900, 2005.

[13] L. Clarizia, D. Russo, I. Di Somma, R. Marotta, and R. Andreozzi, "Homogeneous photo-Fenton processes at near neutral pH: A review," *Appl. Catal. B Environ.*, vol. 209, pp. 358–371, 2017.

"AOP Performance at Wastewater Treatment Plants: Recent Developments"

by:

Jason Heberling, President, Heberling Environmental Consultants (HEC), Birmingham, AL 35234
Yusuf Adewuyi, Professor, Department of Chemical, Biological and Bio Engineering, NC A&T State University, Greensboro, NC 27411
Ahmed S. Mahmoud. Sanitary and Environmental Institute (SEI), Housing and Building National Research Center (HBRC), Giza, Egypt
Mohamed Mostafa, Lecturer, Faculty of Engineering and Technology, Badr University in Cairo (BUC), Cairo, Egypt
Robert W. Peters, Professor, Department of Civil, Construction, and Environmental Engineering, University of Alabama at Birmingham, Birmingham, AL 35294

ABSTRACT:

Advanced Oxidation Processes (AOPs) are chemical treatment techniques used to remove contaminants from water and wastewater using hydroxyl radical reactions. AOPs are efficient methods for removing organic contamination not degradable by means of biological processes. AOPs are a set of treatment processes involving the production of very reactive oxygen species (hydroxyl radicals (•OH) able to destroy a wide range of organic compounds). This paper examines selected recent studies involving AOP's application for wastewater treatment. Discussions of selected reviews and summaries of results from recent reports in the technical literature are provided below. Brief descriptions of the most popular AOPs are also provided. In addition, nano zero-valent iron systems (nZVI) for degrading organic contaminants are briefly examined. Performance data is provided where available or relevant. Of particular interest are treatments for emerging contaminants. Multiple studies involving remediation of wastewater contaminated by emerging contaminants by AOP's and nZVI systems are reviewed.

Introduction

The two main techniques of reducing toxins in wastewaters are chemical neutralization and photodegradation. Advanced Oxidation Processes (AOPs) utilize one or both of these approaches in tandem. AOPs were first defined in 1987 as 'near ambient temperature and pressure water treatment processes which involve the generation of hydroxyl radicals in sufficient quantity to effect water purification' [Glaze *et al.*, 1987; Munter *et al.*, 2001]. AOP's are generally expensive and often applied as pre-treatment in conjunction with more conventional methods [Cesaro *et al.*, 2013]. AOP treatments are divided into two categories:

1) Non-Photochemical Methods:

Ozone, Catalytic Ozone, Peroxone, Fenton systems, Ferrate

2) Photochemical Methods

Ultraviolet radiation (UV), Ozone-UV Radiation, Ozone – Hydrogen Peroxide – UV Radiation, Photocatalytic ozonation, Photo Fenton/Fenton-like Systems, Photocatalytic Oxidation (UV/TiO_2), Electron Beam Irradiation, Ultrasound/sonication

Recent research and application for municipal wastewater will be briefly discussed for each of these methods.

Emerging contaminants

Emerging contaminants are a broad group of chemicals that are extremely difficult to treat and are often unregulated, even when toxicity is well documented. Emerging contaminants endanger aquatic life and they are extremely resistant to biological degradation due to their stability [Roth *et al.*, 2018]. Emerging contaminants are primarily synthetic organic chemicals that have been recently detected in natural environments [Ahmed *et al.*, 2017]. Emerging contaminants are often derivatives of manufactured products which adversely affect aquatic environments, surface waters, drinking water, and soils. Endocrine disrupting chemicals (EDCs) and pharmaceutical and personal care products (PPCPs) are two groups of emerging contaminants that have garnered significant interest in the past decade [Fast *et al.*, 2017]. AOP's, and nZVI, have frequently been used to treat water tainted with emerging contaminants and recent studies involving the treatment of emerging contaminants are discussed throughout this paper.

AOP's

Ozone

Ozone (O_3) is an unstable chemical that decomposes spontaneously producing hydroxyl free radicals. Ozone's instability requires generation at the point of application for effectiveness [Cesaro *et al.*, 2013]. Due to power requirements and cost, ozonation has been deemed less effective than other AOPs (including ozonation used in tandem with other processes) in the reduction of wastewater micropollutants [Prieto-Rodriguez *et al.*, 2013]. Ozone has shown as much as an 8% increase in pharmaceuticals destruction when used a pre-treatment with other methods such as UV and Hydrogen Peroxide. The ozone pre-treatment also reduced energy consumption by 32% while increasing pharmaceutical deactivation [Nilsson, 2015]. One study showed significant reduction in chemical oxygen demand (COD), specific ultraviolet absorbance at 254 nm, total suspended solids (TSS), and biological oxygen demand (BOD_5). However most recent studies concentrate on ozone used in tandem with other disinfection approaches or ozone with a catalyst [Bettaller *et al.*, 2005].

Catalytic Ozonation

Catalysts introduced into the ozonation process have shown greater efficiency in the degradation of many contaminants compared with ozone alone. Transition metals are commonly used for catalytic ozonation [Wang and Xu, 2012]. In one study, a synthesized nanocomposite was used for catalytic ozonation of phenol in a batch reactor. The results showed that this catalyst significantly improved the degradation of phenol compared to that of ozonation without the catalyst [Heidarizad and Sengor, 2017]. Photocatalysts are light reactive chemicals which convert photons into chemical energy. These photocatalytic ozonation process has shown success in breaking down several dozen organic contaminants [Xiao *et al.*, 2017]. One particular concern is the durability of photocatalysts in wastewater application. A commonly used photocatalyst, titanium dioxide (TiO_2) was studied for its durability in a 2017 study. The study findings showed that the photocatalysts could be reused at least three times without significant loss in their activity [Mecha *et al.*, 2017]. A titanium dioxide – tungsten trioxide photocatalyst was recently studied in the degradation of emerging contaminants. Significant removal of caffeine, metoprolol, and ibuprofen (up to 88%) were achieved using photocatalytic ozonation with the titanium dioxide – tungsten trioxide catalyst when compared to more conventional catalysts [Rey *et al.*, 2014].

Peroxone (Ozone / Hydrogen Peroxide) – E. Peroxone

Peroxone systems produce hydroxyl radicals from the decomposition of ozone enhanced by hydrogen peroxide [Deng and Zhao, 2015]. Peroxone, as a combination of ozone and hydrogen peroxide, is a new and advanced oxidation process that can be used for the treatment of polluted soils, groundwater and wastewater. Peroxone processes have been widely used for the elimination of biological refractory compounds in municipal wastewater effluents [Liu *et al.*, 2015]. In a 2017 study a number of micropollutants and bromates were targeted for destruction by a peroxone system finding similarities between with the removal rates of ozonation. These micropollutants were categorized into three groups: non-reactive compounds (i.e. amidotrizoate), moderately reactive compounds (i.e. metoprolol, acesulfame potassium, bezafibrate, and benzotriazole), and highly reactive compounds (i.e. carbamazepine and diclofenac). However, the peroxone system showed promise in the suppression of bromates [Phattarapattamawong *et al.*, 2018]. A hybrid electro-peroxone (E-peroxone) system, developed in 2013, showed success in degrading carbamazepine (CBZ), an emerging contaminant derived from pain and epileptic medication [Yang *et al.*, 2018; Wang *et al.*, 2018]. E-peroxone systems have shown more success in the breakdown of emerging contaminants and the suppression of bromates as compared to ozonation and conventional peroxone systems. In a recent study, the E-peroxone process reduced the reaction time and electrical energy consumption required to remove 90% of all spiked pharmaceuticals from the secondary effluents as compared to ozonation [Yao *et al.*, 2015]. Abatement of ozone-resistant emerging contaminants such as ibuprofen, clofibric acid, and chloramphenicol were enhanced using the E-peroxone process [Wang *et al.*, 2018]. Results from a 2015 study show that the E-peroxone process increased total organic carbon (TOC) elimination rate (up to 5.5 %) when compared to conventional ozonation [Wang *et al.*, 2015]. TOC's react with chlorine to form regulated and potentially carcinogenic disinfection byproducts, which are known carcinogens..

Fenton Systems / Photo-Fenton Process/ Fenton-like Systems

The most commonly used metal to activate hydrogen peroxide is iron, specifically Fe^{2+}. Ferrous ions are produced in tandem with hydroxyl radicals to deactivate contaminants. This process is known as the Fenton process [Deng and Zhao, 2015]. The Fenton process has been applied to highly loaded, refractory, toxic and discolored wastewaters with success. One drawback to the Fenton process is the excessive ferric sludge generated. A 2014 study found success in using recycled ferric sludge as a source of ferrous ions needed for the Fenton process, thereby reducing generated waste and making the process more economically and environmentally attractive [Bolabajev *et al.*, 2014].

A photo-Fenton process is a solar activated technology which inactivates microorganisms by photoactivated iron species and the direct action of solar photons [Garc a-Fern ndez, 2018]. Photo-Fenton systems have also shown promise in detoxifying wastewater with heavy metal contamination. This treatment is considered environmentally friendly and particularly effective in areas without access to sewage wastewater treatment facilities [Barwala and Rubina, 2017].

The Fenton process has glaring disadvantages such as high operating cost, limited optimum pH range, and (as alluded to earlier) large sludge production. The concentration range of the iron ion is 50–80 ppm for batch processes, which exceeds most regulatory requirements worldwide.

Fenton-like systems are alternative methods which replace Fe^{2+} with other catalyst such as Fe^{3+}, Cu^{2+}/Cu^{+}, schorl, pyrite, and nano zero-valent iron. Fenton-like processes comprise two categories: heterogeneous and homogeneous. Heterogeneous Fenton-like processes replace Fe^{2+} in the Fenton reagent with a solid catalyst. Homogeneous Fenton-like processes combine metal ion(s)/metal ion-organic ligand complexes and hydrogen peroxide (H_2O_2). [Wang *et al.*, 2016]. One study examined the applicability of Fenton-like system in the destruction of pharmaceuticals, specifically steroid hormones. It was determined that the Fenton-like system (using Fe^{3+} as the catalyst) was an economical and sound tertiary treatment for the contaminants studied [Ifelebuegu and Ezenwa, 2011]. Fenton like processes have been successfully used to treat water contaminated from by-products of traditional Chinese medicine production [Su *et al.*, 2016].

Ferrate

Ferrate treatment uses Fe(VI) to oxidize micropollutants and remove phosphate by formation of ferric phosphates [Lee, 2009]. Previous studies have shown effectiveness of ferrate(VI) in removing a broad range of water pollutants including sewage organic substances, phosphate, toxic inorganic toxins, algae, micro-organisms, and emerging pollutants [Li, 2017]. Ferrate treatment resulted in selective oxidation of phenol, olefin, amine and aniline moieties. In another study selected pharmaceuticals were spiked into effluent samples. Ferrate showed promise in the removal of suspended solids and COD. Ferrate, however, failed to react with triclocarban, three androgens, seven acidic pharmaceuticals, two neutral pharmaceuticals and erythromycin [Jiang, 2014]. Ferrate has also shown promise for removing formaldehyde [Talaiekhozani *et al.*, 2016].

Ultraviolet Radiation

Ultraviolet irradiation (UV) is produced by mercury arc lamps or light-emitting diodes to breakdown contaminants and micro-organisms. UV has been used for several applications including destruction of contaminants and removal of color. One of the most popular uses of UV is the inactivation of micro-organisms. A comprehensive, two-year study in Canada examined UV's ability to destroy viruses in a wastewater stream. Some removal rates were as high as 50% from pre- to post-treatment [Qiu *et al.*, 2018]. UV light-emitting diodes (UV-LED) have shown success in micro-organism inactivation [Song *et al.*, 2018]. UV is more commonly used in tandem with other processes or catalysts.

Ozone – UV (O_3/UV); *Ozone – Hydrogen Peroxide – UV* (O_3/H_2O_2/UV)

Ozone is expensive to generate and is often used in concert with ultraviolet radiation (UV). UV photons activate ozone molecules generating hydroxyl free radicals [Cesaro *et al.*, 2013]. Hydrogen Peroxide (H_2O_2) is often used in tandem with UV and ozone to enhance hydroxyl free radical production. A combination O_3/UV system was compared with an ozone and a UV system in removing organic load, measured as chemical oxygen demand (COD). The removal efficiency of the combined process was higher than the sum of the removal efficiencies of ozone and UV light treatments separately [Bustos-Terrones *et al.*, 2016]. Another 2016 study also found improved efficiency in the oxidation of different parabens (such as methylparaben, ethylparaben, propylparaben, and butylparaben) with the pairing of UV and ozone systems compared to an ozone system without UV [Cuerda-Correa *et al.*, 2016].

A study early in 2018 examined an O_3/H_2O_2/UV system for treating contaminants in gray water. This O_3/H_2O_2/UV system was particularly effective in reducing COD [Hassanshahi and Karimi-Jashni, 2016]. Another recent study looked at bisphenol degradation with a O_3/H_2O_2/UV system. It was discovered that the addition of UV to an ozone or ozone/peroxide had little effect in the breakdown of bisphenols [Mehrabani-Zeinabad *et al.*, 2016].

Photocatalytic Oxidation (UV/TiO_2)

Dating to the early 1970's, photocatalytic oxidation (UV/TiO_2) involves irradiation with near-UV light of a semiconducting catalyst (generally TiO_2) which is easily photo-excited to form electron-donating and electron-accepting sites. This permits induction of oxidation–reduction reactions. When the absorbed UV photons have an energy larger than the energy gap (between the valence and the conducting bands) of the semiconductor, electron-hole pairs are formed. These pairs either recombine or migrate to the semiconductor surface and then react with chemical species adsorbed on the surface. UV/TiO_2 has been widely applied to persistent organic pollutants (POPs). Also, this process is effective on pathogenic biologic pollutants, including viruses, bacteria, and mold [Oturan and Aaron, 2014]. In a recent study, destruction of chlorophenol was significantly improved with UV/TiO_2 as compared to TiO_2 without irradiation. TOC removal improved by 30% [Tian *et al.*, 2018].

Electron Beam Irradiation

Free radicals can be generated by the excitation of molecules using high-energy electrons. These high energy electrons can be produced by gamma ray sources or, more commonly, through electron beam accelerators. Electron beam irradiation has been applied on the industrial scale for decades. Only recently have application for municipal wastewater been explored [Oturan and Aaron, 2014]. In an accelerator, a current produces a stream of electrons which are accelerated using an electric field generated by an applied voltage [Munter *et al.*, 2001]. A study using EB showed a 90% reduction in the number of coliforms during irradiation. However, COD and BOD reduction reached about 50% in another study [Emami-Meibodi *et al.*, 2016].

Ultrasound/Sonication

Sonochemistry is a promising oxidation process for wastewater treatment. Cavitation forms micro-bubbles that grow to a critical size then implode. Temperatures at the collapsing bubble interface reach about 5,000° K and pressures near 500 atmospheres with the bulk solution remaining near ambient conditions. In these highly reactive conditions, hydroxyl radicals and hydrogen ions, which are very effective in degrading organic compounds, are generated. If organic compounds are present in the treated water, these compounds are rapidly destroyed either directly or by reactions with free radicals. The intensity of cavity implosion can be controlled by process parameters such as ultrasonic frequency, ultrasonic intensity per unit volume of liquid medium, static pressure, choice of ambient gas, and addition of oxidants (e.g. H_2O_2, O_3). Ultrasound was used successfully in the removal of inorganic sulfide in a simulated wastewater [Matei and Scarpete, 2016]. Ultrasound has had mixed results when applied to various emerging contaminants. In a study from 2014 sonolytic degradation efficiencies of diclofenac and ibuprofen were 80% and 10%, respectively. The sonolytic degradation efficiency of ciprofloxacin was less than 10%. As with many AOP's, ultrasound showed greater success when coupled with other treatment methods [Yap *et al.*, 2018].

Nano Zero-Valent Iron (nZVI) *Systems*

Nanoscale zero-valent iron (nZVI) particles are typically 5 – 40 nm sized FeO/Fe-oxide particles that rapidly transform contaminants to non-toxic by-products. These particles are able to effectively eliminate or neutralize certain recalcitrant pollutants in water. A study was performed in 2014 to determine the effectiveness of nZVI in removing steroidal estrogen, an emerging contaminant. Reductions in overall estrogen concentration was significant at all the tested concentrations (2 to 6 g/L) and that the removal of estrogens did not increase significantly with exposure time. Maximum estrogen removal was primarily achieved already within the first hour of interaction with nZVI particles [Jarosova *et al.*, 2015]. Another application for nZVI is the removal of nitrate and phosphate from wastewater. A study described in 2017 showed that the effectiveness in removal of nitrate and phosphate by nZVI can be increased by 12% to 38% with the addition of activated carbon [Khalil *et al.*, 2017].

Summary

There are virtually unlimited applications for AOPs and nZVI for wastewater treatment. AOPs and nZVI methods also show great promise in the treatment of wastewater contaminated with emerging contaminants. Ultrasound performed well in the deactivation of pharmaceuticals according to recent research. In recent years ozone has shown effectiveness in deactivating pharmaceuticals and reducing COD, BOD, and suspended solids, while reducing energy costs. Ferrate has also showed promise in the reduction of COD as has UV. UV, particularly UV with ozone or peroxide, showed effectiveness in treating pollutants in graywater and the breakdown of bisphenols. In a 2013 study, catalytic ozone showed contaminant removal of up to 88% compared to conventional methods. Research within the last half decade showed that peroxone and E. peroxone improved ozone effectiveness in contaminant destruction while reducing cost compared to ozone alone. Recent studies involving Fenton processes have shown promise in the treatment of metal ions and pharmaceuticals. Contemporary studies have shown success in the deactivation of biological pathogens by UV/TiO_2. Electron beam irradiation showed effectiveness in the reduction of coliforms. Nanoscale zero-valent iron systems were effective in estrogen removal according to the findings of a five-year old study. When multiple methods are applied in tandem removal efficiencies improve significantly. Also, when multiple methods are used simultaneously, power demands decrease for the same level of contaminant destruction. What this implies is that there is a virtually unlimited number of process combinations which are more efficient in reducing contaminants and use less energy. To suggest that there is plenty of opportunity for research in AOP's and nZVI in the destruction of wastewater contaminants, particularly emerging contaminants, is a phenomenal understatement.

References

Ahmed, M. B., J. L. Zhou, H. H. Ngo, W. Guo, N. S. Thomaidis, and Jiang Xu, 2017. "Progress in the Biological and Chemical Treatment Technologies for Emerging Contaminant Removal from Wastewater: A Critical Review", ***Journal of Hazardous Materials***, ***323***: 274–298.

Barwala, A., and R. Chaudhary, 2017. "Effectiveness of Solar Photo-Fenton Process for Simultaneous Detoxification of Heavy Metals and Disinfection in Municipal Wastewater by Using Response Surface Method", ***Environmental Progress and Sustainable Energy, 36*** (2): 448–459.

Bataller, M., E. Veliz, L. A. Fernandez, C. Hernandez, I. Fernandez, C. Alverez, and E. Sanchez, 2005. "Secondary Effluent Treatment of Ozone", ***IOA* 17th *World Ozone Congress***, Strasbourgh, France, (August 22–25).

Bolobajev J., E. Kattel, M. Viisimaa, A. Goi, M. Trapido, T. Tenno, and N. Dulova, 2014. "Reuse of Ferric sludge as an Iron Source for the Fenton-based Process in Wastewater Treatment", ***Chemical Engineering Journal, 255***: 8–13.

Bustos-Terrones, B., J.G. Rangel-Peraza, A. Sanhouse, E.R. Bandala, and L.G. Torres, 2016. "Degradation of Organic Matter from Wastewater Using Advanced Primary Treatment by O_3 and O_3/UV in a Pilot Plant", ***Physics and Chemistry of the Earth*, *91***: 61–67.

Cesaro, A., V. Naddeo, and V. Belgiorno, 2013. "Wastewater Treatment by Combination of Advanced Oxidation Processes and Conventional Biological Systems," ***J. Biorem. Biodeg., 4***: 8–15.

Cuerda-Correa, E. M., J. R .Dom nguez, M. J. Mu oz-Pe a, and T. Gonza ez, 2016. "Degradation of Parabens in Different Aqueous Matrices by Several O_3-Derived Advanced Oxidation Processes", ***Ind. Eng. Chem. Res.*, *55***: 5161–5172.

Deng, Y., and R. Zhao, 2015. "Advanced Oxidation Processes (AOPs) in Wastewater Treatment", ***Curr. Pollution Rep., 1***: 167–176.

Emami-Meibodi, M., M. R. Parsaeian, R. Amraei, M. Banaei, F. Anvari, S.M.R. Tahami, B. Vakhshoor, A. Mehdizadeh, N. Fallah Nejad, S.P. Shirmardi, S.J. Mostafavi, and S.M.J. Mousavi, 2016. "An Experimental Investigation of Wastewater Treatment Using Electron Beam Irradiation", ***Radiation Physics and Chemistry*, *125***: 82–87.

Fast, S.A., V. G. Gude, D. D. Truax, J. Martin, and B. S. Magbanua, 2017. "A Critical Evaluation of Advanced Oxidation Processes for Emerging Contaminants Removal", ***Environ. Process.*, *4***: 283–302.

Garc a-Fern ndez, I., 2018. "Inactivation of E. coli and E. faecalis by Solar Photo-Fenton with EDDS Complex at Neutral pH in Municipal Wastewater Effluents", ***Journal of Hazardous Materials***, https://doi.org/10.1016/j.jhazmat.2018.07.037.

Glaze, W.H., J.W. Kang, and D.H. Chapin, 1987. "The Chemistry of Water Treatment Processes Involving Ozone, Hydrogen Peroxide and Ultraviolet Radiation", ***Ozone: Science & Engineering*, *9***: 335–352.

Hassanshahi, N., and A. Karimi-Jashni, 2018. "Comparison of Photo-Fenton, O_3/H_2O_2/UV and Photocatalytic Processes for the Treatment of Gray Water", ***Ecotoxicology and Environmental Safety*, *161***: 683–690.

Heidarizad, M., and S.S. Sengor, 2017. "Graphene Oxide/Magnesium Oxide Nanocomposite: A Novel Catalyst for Ozonation of Phenol from Wastewater", ***World Environmental and Water Resources Congress***, Sacramento, CA, (May 21–25).

Ifelebuegu, A.O., and C.P. Ezenwa, 2011. "Removal of Endocrine Disrupting Chemicals in Wastewater Treatment by Fenton-Like Oxidation", ***Water Air Soil Pollut.*, *217***: 213–220.

Jarosova, B., J. Filip, K. Hilscherova, J. Tucek, Z. Simek, J. P. Giesy, R. Zboril, and L. Blaha, 2015. "Can Zero-Valent Iron Nanoparticles Remove Waterborne Estrogens?", ***Journal of Environmental Management*, *150***: 387-392.

Jiang, J., 2014. "Advances in the Development and Application of Ferrate (VI) for Water and Wastewater Treatment", ***J. Chem. Technol. Biotechnol.*, *89***: 165–177.

Khalil, A.M.E., O. Eljamal, T.W.M. Amen, Y. Sugihara, and N. Matsunaga, 2017. "Optimized Nano-Scale Zero-valent Iron Supported on Treated Activated Carbon for Enhanced Nitrate and Phosphate Removal from Water", ***Chemical Engineering Journal*, *309***: 349–365

Lee,Y., S.G. Zimmermann, A.T. Kieu, and U.V. Gunten, 2009. "Ferrate (Fe (VI)) Application for Municipal Wastewater Treatment: A Novel Process for Simultaneous Micropollutant Oxidation and Phosphate Removal", ***Environ. Sci. Technol.*, *43***: 3831–3838.

Li, N., 2017. "Ferrate as a New Treatment Chemical for Removal of Effluent Organic Matter (EfOM) and Emerging Micro-Pollutants in Treated Municipal Wastewater for Water Reuse", Theses, Dissertations and Culminating Projects. 37, Montclair State University, Montclair, NJ.

Liu, Y., J. Jiang, J. Ma, Y. Yang, C. Luo, X. Huangfu, and Z. Guo, 2015. "Role of the Propagation Reactions on the Hydroxyl Radical Formation in Ozonation and Peroxone (Ozone/Hydrogen Peroxide) Processes", ***Water Research*, *68***: 750–758.

Matei, N., and D. Scarpete, 2016. "Assessment of Inorganic Sulfide Removal from Simulated Wastewater by Ultrasound", ***International Journal of Trend in Research and Development*, *3*** (4): 2394-9333.

Mecha, A.C., M.S. Onyango, A.i Ochieng, and M.N.B. Momba, 2017. "Ultraviolet and Solar Photocatalytic Ozonation of Municipal Wastewater: Catalyst Reuse, Energy Requirements and Toxicity Assessment", ***Chemosphere*, *186***: 669–676.

Mehrabani-Zeinabad, M., G. Achari, and C.H. Langford, 2016. "Degradation of Bisphenol S Using O_3 and/or H_2O_2 with UV in a Flow-through Reactor", ***J. Environ. Eng.*, *142*** (8): 06016004 1–4.

Munter, R., S. Preis, J. Kallas, M. Trapido, and Y. Veressina, 2001. "Advanced Oxidation Processes (AOPs): Water Treatment Technology for the Twenty-first Century", ***Kemia-Kemi*, *28***: 5.

Nillson, F., 2015. "Application of Ozone in Wastewater Treatment for Mitigation of Filamentous Bulking Sludge & Reduction of Pharmaceutical Discharge", Ph.D.Dissertation, Lund University. Lund, Sweden.

Oturan, M.A., and J. Aaron, 2014. "Advanced Oxidation Processes in Water/Wastewater Treatment: Principles and Applications. A Review", ***Critical Reviews in Environmental Science and Technology*, *44***: 2577–2641.

Phattarapattamawong, S., A.M. Kaiser, E. Saracevic, H.P. Schaar, and J. Krampe, 2018. "Optimization of Ozonation and Peroxone Process for Simultaneous Control of Micropollutants and Bromate in Wastewater", ***Water Sci. Technol.*, *2***: 404–411.

Prieto-Rodriguez, L., I. Oller, N. Klamerth, A. Aguera, E.M., Rodrıiuez, and S. Malato, 2013. "Application of Solar AOPs and Ozonation for Elimination of Micropollutants in Municipal Wastewater Treatment Plant Effluents", ***Water Research***, ***47:*** 1521–1528.

Qiu, Y., Q. Li, B.E. Lee, N.J. Ruecker, N.F. Neumann, N.J. Ashbolt, and X. Pang, 2018. "Inactivation of Human Infectious Viruses at Two Full-Scale Wastewater Treatment Plants in Canada", ***Water Research***, ***147***: 73–81, doi: https://doi.org/10.1016/j.watres.2018.09.057.

Rey, A., P. Garc a-Munoz, M.D. Hern ndez-Alonso, E. Mena, S. Garc a-Rodr guez, and F.J. Beltr n, 2014. "WO_3–TiO_2 Based Catalysts for the Simulated Solar Radiation Assisted Photocatalytic Ozonation of Emerging Contaminants in a Municipal Wastewater Treatment Plant Effluent", ***Applied Catalysis B: Environmental***, ***154–155***: 274–284.

Talaiekhozani, T., M. Salari, M.R. Talaei, M. Bagheri, and Z. Eskandari, 2016. "Formaldehyde Removal from Wastewater and Air by Using UV, Ferrate(VI) and UV/ferrate(VI)", ***Journal of Environmental Management***, ***184***: 204–209.

Tian, M., S.S. Thind, J.S. Dondapati, X. Li, and A. Chen, 2018. "Electrochemical Oxidation of 4-Chlorophenol for Wastewater Treatment Using Highly Active UV Treated TiO_2 Nanotubes", ***Chemosphere***, ***209***: 182–190.

Song, K., F. Taghipour, and M. Mohseni, 2018. "Microorganisms Inactivation by Continuous and Pulsed Irradiation of Ultraviolet Light-emitting Diodes (UV-LEDs)", ***Chemical Engineering Journal***, ***343***: 362–370.

Su, C., W. Li, Y. Lu, M. Chen, and Z. Huang, 2016. "Effect of Heterogeneous Fenton-like Pre-treatment on Anaerobic Granular Sludge Performance and Microbial Community for the Treatment of Traditional Chinese Medicine Wastewater", ***Journal of Hazardous Materials***, ***314***: 51–58.

United States Environmental Protection Agency (USEPA), 2009. "Assessing the Use and Application of Zero-Valent Iron Nanoparticle Technology for Remediation at Contaminated Sites", Office of Solid Waste and Emergency Response Office of Superfund Remediation and Technology Innovation, U.S. Environmental Protection Agency, Washington, D.C.

Wang, H., S. Yuan, J. Zhan, Y. Wang, G. Yu, S. Deng, J. Huang, and B. Wang, 2015. "Mechanisms of Enhanced Total Organic Carbon Elimination from Oxalic Acid Solutions by Electro-Peroxone Process", ***Water Research***, ***80***: 20–29.

Wang, J.L., and L.J. Xu, 2012. "Advanced Oxidation Processes for Wastewater Treatment: Formation of Hydroxyl Radical and Application", ***Critical Reviews in Environmental Science and Technology***, ***42*** (3): 251–325.

Wanga, N., T. Zhengb, G. Zhangb, and P. Wang, 2016. "A Review on Fenton-like Processes for Organic Wastewater Treatment", ***Journal of Environmental Chemical Engineering***, ***4***: 762–787.

Wang, Y., G. Yu, S. Deng, J. Huang, and B. Wang, 2018. "The Electro-Peroxone Process for the Abatement of Emerging Contaminants: Mechanisms, Recent Advances, and Prospects", ***Chemosphere***, ***208***: 640–654.

Xiao, J., Y. Xie, and H. Cao, 2015. "Organic Pollutants Removal in Wastewater by Heterogeneous Photocatalytic Ozonation", ***Chemosphere*, *121***: 1–17.

Yanga B., J. Deng, G. Yu, S. Deng, J. Lia, C. Zhu, Q. Zhuo, H. Duan, and T. Guo, 2018. "Effective Degradation of Carbamazepine Using a Novel Electro-Peroxone Process Involving Simultaneous Electrochemical Generation of Ozone and Hydrogen Peroxide", ***Electrochemistry Communications*, *86***: 26–29.

Yao, W., X. Wang, H. Yang, G. Yu, S. Deng, J. Huang, B. Wang, and Y. Wang, 2016. "Removal of Pharmaceuticals from Secondary Effluents by an Electro-Peroxone Process", ***Water Research*, *88***: 826–835.

Yap, H.C., Y.L. Pang., S. Lim, A.Z. Abdullah, H.C. Ong, and C.H. Wu, 2018. "A Comprehensive Review on State-of-the-Art Photo-, Sono-, and Sonophotocatalytic Treatments to Degrade Emerging Contaminants", ***Int. J. Environ. Sci. Technol.***,1–28, https://doi.org/10.1007/s13762-018-1961-y.

Table 1. AOP Application/performance in treating municipal wastewaters.

Study Reference	Treatment technique	Pilot/lab/full scale study	Contaminants	Results/Findings
Nillson 2015	Ozone	Pilot	Pharmaceuticals	8% increase in contaminant reduction with 32% energy reduction
Heidarizad and Sengor, 2017	Catalytic Ozonation	Laboratory	Phenols	Reduction of Phenols improved over ozonation
Mecha *et al.*, 2017	Catalytic Ozonation	Laboratory/full	Phenols	photocatalysts can be reused at least three times without significant loss of activity
Rey *et al.*, 2014	Catalytic Ozonation	Laboratory	caffeine, metoprolol, and ibuprofen	88% reduction of contaminants achieved
Liu *et al.*, 2015	Peroxone	Full	Biological refractory compounds	Showed improvement over ozone alone
Yang *et al.*, 2018	E-peroxone	Full	CBZ	99.8% removal of CBZ achieved
Yao *et al.*, 2015	E-peroxone	Laboratory	pharmaceuticals	Energy reduction of 90% achieved compared to ozonation
Wang *et al.*, 2015	E-peroxone	Laboratory	TOC	TOC reduction 5.5% greater than ozonation
Barwala and Rubina, 2017	Photo-Fenton	Full	Heavy metals	Removals between 36-100% depending on metal
Talaiekhozani *et al.*, 2016	Ferrate	Full	Formaldehyde	Nearly 100% removal in 35 minutes
Qiu *et al.*, 2018	UV	Laboratory	Viruses	50% destruction in wastewater stream

Emami-Meibodi *et al.*, 2016	Electron Beam Irradiation	Pilot	Coliforms, COD, BOD	90%reduction in coliforms – 50% reductions in COD BOD
Yap *et al.*, 2018	Ultrasound	Laboratory	Diclofenac, ibuprofen, and ciprofloxacin	Diclofenac reduction 80% - ibuprofen and ciprofloxacin reduction 10%.
Jarosova *et al.*, 2015	nZVI	Laboratory	Estrogen	Near 100% removal in one hour.
Khalil *et al.*, 2017	nZVI	Pilot	Phosphates	12-38% phosphates reduction with activated carbon

538665 A Multi-Dimensional Strategy for Treatment of Raw Landfill Leachate Wastewater Using Electroperoxone

Abstract:

Population explosion and rapid urbanization have led to rampant increase in waste generation and environmental pollution. Solid waste generated by cities across the globe is expected to become 2.2 billion tons per year by 2025. Landfilling has become the preferred strategy for disposal of solid wastes. However, mismanaged landfill leachate is capable of polluting the surrounding surface and ground water, resulting in potential health hazards. Moreover, the presence of dissolved organics in ground water causes ineffective chlorination and carcinogenic disinfection by-products. Hence, landfill leachate needs appropriate and sufficient treatment before its discharge into water bodies. Unfortunately, conventional treatment technologies are unable to treat raw (i.e., not pre-treated) leachate to a satisfactory level. Electroperoxone, being a promising advanced oxidation process, is herein investigated in the treatment of raw leachate wastewater from Buck Run landfill, Mississippi. We critically compare the performance of electroperoxone with electrolysis and ozonation, for the removal of humic acid, a model persistent organic pollutant and a major component of leachate. Secondly, a comprehensive treatment of the leachate was carried out using relatively inexpensive Reticulated Vitreous Carbon (RVC) cathode and Platinum-coated Titanium anode. Optimization of H_2O_2 generation under varying aeration rates, applied current, electrode surface area and solution pH was performed. ·OH radical estimation in all three processes showed that the maximum radical generation was in electroperoxone. This experimental system offers a multi-pronged attack and tackles (i) Total Organic Carbon (TOC), (ii) Color, (iii) Disinfection, (iv) Organic suspended solids, (v) Turbidity and (vi) Toxicity. We obtained a TOC degradation of 72%, 81.9% turbidity removal, and 84% suspended solids removal within 480 min, 99.72% decolorization within 240 min, and 99.93% disinfection within 180 min. Mechanisms responsible for the above efficient performances were also investigated. Electroperoxone thus proves to be a highly promising technology for the effective removal of such persistent contaminants in complex real wastewater systems.

524478 Aqueous Organics Degradation in Visible Light and Ozone Integrated Process with WO_3 catalysts

Abstract:

Aqueous organics degradation in Visible Light and Ozone Integrated Process with WO_3 Catalysts

Yongbing Xie[1], Jin Yang[2], Hongbin Cao[1,2]*

[1] *Institute of Process Engineering, Chinese Academy of Sciences, Beijing, 100190, China*

[2] *School of Chemical Engineering & Technology, Tianjin University, Tianjin, 300072, China*

**Corresponding author: hbcao@ipe.ac.cn*

Semiconductors can trigger a super synergy between photocatalysis and ozonation, and the combined process is very effective in organics mineralization[1]. Ultraviolet light (UV) was mostly applied in this combined process with TiO_2 catalyst, but this cannot use solar energy. It is very meaningful to develop a visible light sensitive catalyst to adopt the green solar energy for wastewater treatment. In this paper, three tungsten oxides (WO_3) with different phases and exposed facets were synthesized and applied in the visible light photocatalytic ozonation process.

The three materials were synthesized with hydrothermal and calcination methods, and named with H-100 (hexagonal phase with [100] facet exposed), M-002 and M-100 (monoclinic phase with [002] and [100] facet exposed), respectively. The activities of the materials were evaluated upon oxalic acid degradation and cephalexin mineralization, and M-100 showed the highest activity (Fig. 1a).

THE GRAPHIC ON THIS PAGE WAS NOT AVAILABLE ON THE E-MEDIA VERSION OF THIS PAPER.

Fig 1. Oxalic acid degradation (a), band structures (b) and charge separation of different WO_3 materials (c).

To explain this, different optical and electronic properties were characterized. The band gaps (Eg) were calculated from UV-Vis DRS, and the valence bands (VB) were obtained by valence band XPS, and the results are finally shown in Fig. 1b. More negative CB position indicates the electrons with stronger reducing ability and thus favors ozone reduction[2]. M-100 with higher CB position is favor to reduce ozone molecule to generate reactive radicals. Electrochemical Impedance Spectroscopy (EIS) results indicated that the H-100 material has the highest charge transfer ability (Fig 1c), this is very important in photocatalysis but not in photocatalytic ozonation. Other operating parameters of this process were also optimized to achieve a better performance.

References

1. J.D. Xiao, Y.B. Xie and H.B. Cao, *Chemosphere*, 2015, *121*, 1-17.
2. J.D. Xiao, Y.B. Xie, F. Nawaz, S. Jin, F. Duan, M.J. Li and H.B. Cao, *Applied Catalysis B: Environmental*, 2016, 181, 420-428.

525757 Application of Advanced Oxidation Process for the Treatment of Hydrofracked Water

Abstract:

APPLICATION OF ADVANCED OXIDATION PROCESS FOR THE TREATMENT OF HYDROFRACKED WATER

Shikha Sinha, Debashis Roy, Sudarsan Neogi, Sirshendu De

Indian Institute of Technology, Kharagpur

Kharagpur-721302

One of the key technologies in the exploration of shale gas, an important unconventional natural gas, is hydraulic fracturing. During hydraulic fracturing, the fracturing fluid, consisting mainly of water mixed with proppant (sand) and chemicals (such as surfactants, chelating agents, friction reducers, scale inhibitors and biocides) is injected at high pressure (450-800 bar) into the producing formation, creating fissures that allows natural gas to be released from rock pores. About 70% of this fluid is returned to the surface after the fracturing process. It is commonly known as shale gas drilling flowback fluid [1],[2]. These fluids typically have high dissolved salt concentration (e.g., 10,000 to 300,000 mg/L TDS), and contain various metals, dissolved and suspended solids, radionuclide and organic compounds. To minimize the adverse affect on environment, and to reduce the production cost and to conserve fresh water, development of efficient on-site treatment technologies is required. The primary consideration for produced water treatment to a quality suitable for discharge or for external reuse is reducing the Total Dissolved Solids (TDS) concentration. Significant pretreatment is required to reduce fouling and scaling potential before produced water may be desalinated using membrane or thermal technologies.

The conventional primary treatment for wastewater consists of a coagulation-flocculation process which is only able to remove suspended solids, whereas soluble organic compounds remain unaffected. Advanced Oxidation Process (AOP) is a secondary treatment process based on generation of highly oxidative and non selective reactive species (free radicals) mainly hydroxyl radicals (oxidation potential of 2.8eV, next only to fluorine) facilitating the organics removal. AOPs have been successful in treating the non-biodegradable pollutants such as herbicides, landfill leachate and wastewater originating from textile industries. However, use of AOP for hydrofracked water treatment is scant in literature. AOPs commonly include various processes like Fenton process, photo-assisted Fenton process, photo catalysis, Ozone/UV, H_2O_2/UV process etc. Fenton process is one of the widely used AOPs because of its efficiency, easy operation, relatively cheap cost and rapid oxidation kinetics. The hydroxyl radicals generated by reaction between Fenton's reagent (ferrous sulfate and hydrogen peroxide) is responsible for organic substrate oxidation in the Fenton reactions. Besides, the ferric ions (Fe^{3+}) produced during the reaction can also aid in coagulation by charge neutralization making Fenton process suitable for both coagulation and oxidation [3]. The ideal solution pH for Fenton process should be between the ranges of 3.0-4.5. For pH 3.0, the reaction is slow due to the formation of complex ions; also, for pH 4.5, the generation of ferric hydroxo complexes reduces the production rate of OH· ions, thus reducing the oxidizing efficiency. In photo-assisted Fenton process, irradiation with UV-VIS light accelerates the rate of degradation of organic pollutant with Fenton reagent, which causes photolysis of Fe^{3+} to regenerate Fe^{2+}. However, the use of UV makes it an energy intensive and costly process. Photo catalysis makes use of a semiconductor metal oxide catalyst, titanium dioxide (TiO_2) being the most promising one. But separation of titanium dioxide nanoparticles from the sample is difficult and not suitable for further tertiary treatment (membrane separation) of hydrofracked water. The use of oxidation treatments based on O_3/UV and H_2O_2/UV systems requires the use of suitable UV sources and of appropriate photochemical reactors. Hence Fenton's process without UV irradiation, though a conventional process, is chosen among other AOPs.

In this work, performance of primary and secondary water treatment processes to treat hydrofracked water collected from Forward base, ONGC, Ankleshwar, Gujarat, India were investigated. Coagulation with Potash alum (K_2SO_4, $Al_2(SO_4)_3$, $24H_2O$), an inorganic coagulant was studied. Coagulant dose was varied between 10 to 500 mg/L. Variation of Total Organic Carbon (TOC) with varying coagulant doses was studied. It was found that 100 mg/l of Potash alum was giving the maximum organics removal with 52% removal efficiency. Fenton process was carried out using ferrous sulfate heptahydrate ($FeSO_4.7H_20$) and hydrogen peroxide (H_2O_2, 50 vol%). pH was maintained between 3-3.5 to avoid the formation of ferric-hydroxo complexes which inhibits the production of hydroxyl radical. Ferrous sulfate concentration was varied from 0.001mol/L to 0.002 mol/L while hydrogen peroxide concentration was varied from 0.1 mol/L to 0.2 mol/L, varying the reaction time from 15 to 60 minutes. The organic removal efficiency was found to be maximum for 0.002 mol/L ferrous sulfate concentration and 0.2 mol/L hydrogen peroxide concentration. The TOC removal efficiency with 0.002M ferrous sulfate and 0.2M hydrogen peroxide at pH 3-3.5 was found to be 97.7%. Graphical representation of the experiments conducted is shown below in figures 1, 2 and 3. Figure 1 shows that organics removal is maximum for 100 mg/L of Potash alum dose. Figure 2 shows that for a given iron concentration, rate of dissolved organics removal increases with increase in peroxide concentration (maximum removal for peroxide of 0.2M) due to rapid reaction and increase in generation of hydroxyl radicals. Figure 3 shows the variation of iron concentration for 0.2M peroxide concentration. It shows that with increase in iron concentration organics removal rate also increases because of increase in generation of hydroxyl radical (maximum removal for 0.002M iron sulfate). Figures show that the removal efficiency was observed to be more for Fenton process where both ferrous and ferric ions act as coagulant by charge neutralization. Hence, Fenton process is aiding both in coagulation and oxidation. We found that the use of Fenton process eliminates the need of primary treatment such as coagulation. The hydrofracked water can be directly treated by using Fenton process (secondary treatment) for organics and suspended solids removal eliminating the need for primary treatment.

Keywords: Hydrofracking, Flowback water, Coagulation, Fenton process, Total Organic Carbon

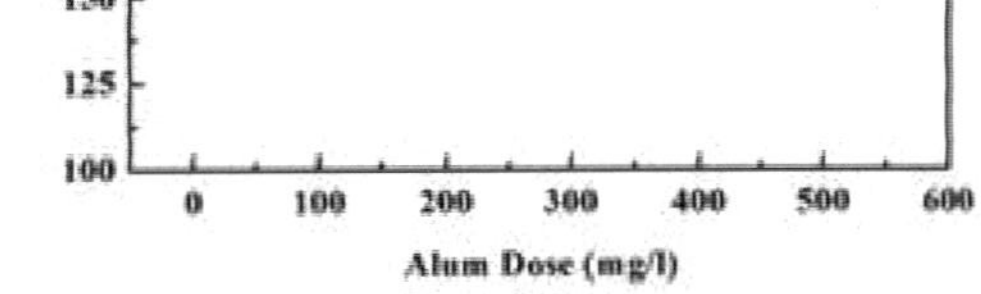

Figure 1. Variation of TOC with Potash alum dose (mg/L).

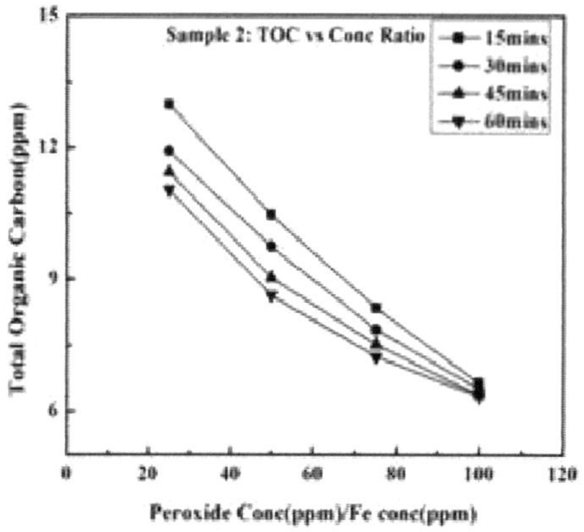

Figure 2. Variation of TOC with Concentration ratio (Peroxide conc./ 0.002M Iron conc.)

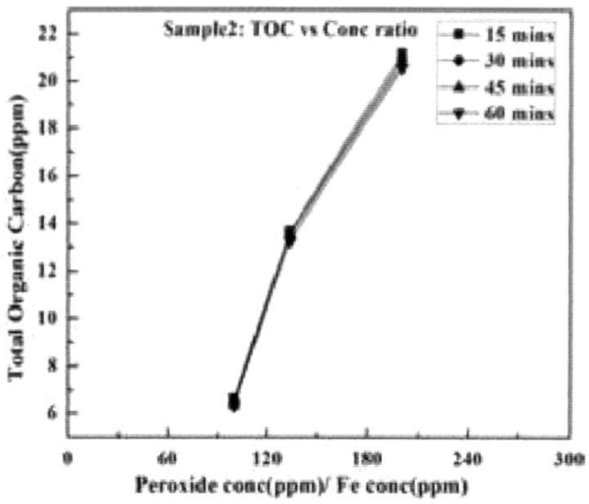

Figure 3. Variation of TOC with Concentration ratio (0.2M Peroxide conc. / Iron conc.)

References:

[1] *G. Chen, Zhouwei Wang et al.*, "Treatment of shale gas drilling flowback fluids (SGDFs) by forward osmosis: Membrane fouling and mitigation," *Desalination*, Vol. 366, pp. 113–120, June 2015.

[2] *R. D. Vidic, S. L. Brantley et al.*, "Impact of shale gas development on regional water quality," *Science*, Vol. 340, Issue no. 6134, May 2013.

[3] *P. Banerjee, Sunando Dasgupta et al.*, "Kinetic Study of Advanced Oxidation of Eosin Dye by Fenton's Reagent" *Internal Journal of Chemical Reactor Engineering*, Vol. 6, January 2008.

539217 Low Temperature NO and Hydrocarbon Trapping over Pd-Exchanged Zeolite Passive NO_x Adsorbers

Abstract:

Low temperature NO and hydrocarbon trapping over Pd-exchanged zeolite passive NOx adsorbers

Sam Malamis, Michael P. Harold*

Dept. of Chemical & Biomolecular Engineering, University of Houston, Houston, Texas

*corresponding author e-mail: mpharold@central.uh.edu

Introduction

Automotive catalysts including TWC, LNT, and SCR catalysts are all highly effective in reducing NO_x and HC emissions above 200Â°C [1]. Most engines, however, require a warm-up period during which the catalyst undergoes a â€œcold startâ€ environment. During this period of up to 200s, large amounts of pollutants such as NOx, N_2O, CO and Hydrocarbons slip through the catalyst into the atmosphere [2]. To mitigate this occurrence, it is beneficial to develop a passive NOx adsorber, or PNA, that can store NOx and hydrocarbons at these low temperatures, and then to release them at higher temperatures to improve the effectiveness of the aftertreatment system [3].

Few comprehensive studies have been performed on this topic, and only recently has there been a movement toward understanding the chemistry behind these low temperature traps. To date, a wide variety of materials has been explored, including palladium on ceria-zirconia (Pd-CZO) [4], CeO_2-M_2O_3 with supported Pt or Pd [5], and Palladium on various zeolites including BEA, MFI, CHA [3]. While it has been reported that Pd-CZO is a potentially effective NO_x trap at low temperatures, it also strongly absorbs SO_2, which inhibits NO_x storage [3,4]. This study focuses on the use of zeolites as PNA materials in order to trap both NOx and HC during the cold start period, with the potential to act as a partial oxidation catalyst upstream of the SCR or LNT. The study mainly explores the effect of temperature and various feed gases on NOx and HC trapping, as well as exploring the development of a kinetic model for HC trapping.

Materials and methods

Experiments were performed using 2wt% Pd-BEA zeolite supported on a cordierite monolith. A feed gas was flowed over the catalyst in a vertical quartz-tube reactor, and the effluent species concentrations were measured using FT-IR. The catalyst was first pre-treated in 5% O_2(bal. Ar) for 30 min at 550ºC. To determine the amount of NOx stored on the catalyst during an experiment, a feed gas of 2% O_2 and 400ppm NO was flowed over the catalyst for 5 min at a fixed feed temperature, either 50ºC, 80ºC or 150ºC. After 5 min., the NO was turned off and the reactor was maintained at the adsorption temperature until the concentration of NO dropped to less than 2ppm. To measure the species desorbing after adsorbing on the catalyst surface, a temperature programmed desorption (TPD) was performed at 20ºC/min from the adsorption temperature to 550ºC. The surface was then pretreated to re-oxidize the Pd to PdO to ensure proper behavior during subsequent experiments [6]. To measure the effect of hydrocarbons on catalyst performance, experiments similar to those described above were performed, using a feed with either 800ppm C_3H_6 and 2% O_2 or a mixture of 400ppm NO, 800ppm C_3H_6 and 2% O_2.

Results

It was found that using a feed of only NO and O_2, Pd-BEA stored between 0.4 and 1.0 mol NOx/gcat*10^4, but released it too early during the TPD (i.e. at a temperature <200ºC). It was also found that as the adsorption temperature increased, the amount of NOx stored and desorbed decreased. Experiments using only propylene in the feed indicate a very strong propensity for the catalyst to trap hydrocarbons at low temperatures (1.5-6.0 mol C_3H_6/gcat*10^4), more so than in trapping NOx. Adding C_3H_6 to the NO feed revealed a competitive adsorption effect of both NOx and C_3H_6, as the stored quantities of these species was significantly lower than the pure feeds. While the total amount of stored NOx decreased with the addition of C_3H_6, our studies show that the majority of NOx was released above 200ºC, which is a promising result.

References

1. Twigg M: **Progress and future challenges in controlling automotive exhaust gas emissions.** *Applied Catalyst B* 2007, **70**:2-15.
2. Wilenmann M, Favez J, Alvarez R: **Cold-start emissions of modern passenger cars at different low ambient temperatures and their evolution over vehicle legislation categories.** *Atmospheric environment* 2009, **43**:241-2429.
3. Chen HY *et al.*: **Low temperature NO storage of zeolite supported Pd or low temperature diesel engine emission control.** *Catal. Lett.* 2016.
4. Theis JR, Lambert C: **The effects of CO, C_2H_4, and H_2O on the NO_x storage performance of low temperature NO_x adsorbers for diesel applications.** *SAE International* 2017, **10.**
5. Jones S *et al.*: **CeO_2**
   2016 **CLEERS Issue.**

527203 Ion-Exchanged Zeolites for Hydrocarbon Traps and Passive NO_X Adsorption Applications

Abstract:

Regulations on vehicle emissions are becoming more stringent worldwide due to increasing concerns of the impact of air pollution on the environment and public health [1]. The current emission control technologies involve the use of three-way and diesel oxidation catalysts in gasoline and diesel vehicles, respectively. However, in most cases these catalysts are relatively ineffective until they reach their operating temperature (~200°C) [2]. In this context, foremost attention is required for low temperatures associated with engine cold starts that take place within the first 1-2 minutes of a driving cycle and emit around 80% of the polluting gases into the environment [3]. A very promising strategy to mitigate cold start hydrocarbon (HC) emissions is to a HC trap (HCT) that can adsorb the HC emissions at low temperatures and then oxidize the stored HCs to carbon dioxide (CO_2) and water (H_2O) at higher temperatures. To treat the NOx emissions during the cold start period, a passive NOx adsorber (PNA) can adsorb the NOx at low temperatures [4]. Zeolites were evaluated using ethylene (C_2H_4), toluene (C_7H_8) as model HC compounds and NOx in simulated exhaust conditions relevant to vehicle cold starts [5]. The specific objective was to investigate in detail trends observed between C_2H_4, C_7H_8 and NOx trapping-release performance, zeolite framework structure (SSZ-13 vs. ZSM-5 vs. BEA), exchanged cation (e.g. H^+ vs. Ag^+ vs Pd^{2+}), and gas composition. Our results show that Ag/SSZ-13 and Pd/SSZ-13 are able to adsorb C_2H_4 and NOx, respectively. Toluene is not adsorbed by either Ag or Pd/SSZ-13, because its kinetic diameter is larger than the pore size of SSZ-13. Ethylene and NOx are desorbed at temperatures higher than the operating temperature of conventional catalysts. Specifically, C_2H_4 and NOx release is observed at temperatures > 200 °C. Larger pore ion-exchanged zeolites, e.g. Ag/ZSM-5, are able to adsorb both C_2H_4 and C_7H_8. However, cracking of toluene to CO is observed during desorption over Ag/ZSM-5. The adsorption/desorption properties of ion-exchanged Ag/SSZ-13 are retained after hydrothermal aging at 800 °C for 10h. Moreover, the NOx adsorption efficiency of Pd/SSZ-13 is improved after aging under the same conditions. These results indicate that Ag and Pd ion-exchanged SSZ-13 zeolites are suitable as potential HCT and PNA materials.

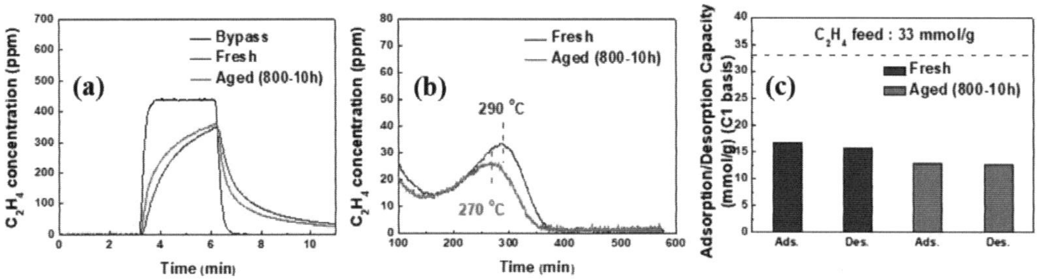

Figure 1. (a) C_2H_4 adsorption and (b) desorption profile over fresh and hydrothermally aged (800°C-10h) Ag/SSZ-13, and (c) comparison of its adsorption/desorption capacity.

[1] A. Westermann, B. Azambre, M. Chebbi, A. Koch, Microporous and Mesoporous Mater. 230 (2016) 76−88.

[2] M.J Hazlett, W.S Epling, Catal. Today 267 (2016) 157−166.

[3] I. Daldoul, S. Auger, P. Picard, B. Nohair, S. Kaliaguine, Can. J. Chem. Eng. 94 (2016) 931−937

[4] Y. Zheng, L. Kovarik, M.H. Engelhard, Y. Wang, Y. Wang, F. Gao, J. Szanyi, J. phys. Chem. C 121 (2017) 15793−15803.

[5] J. Lee, E.A. Kyriakidou, *submitted 2018.*

533950 Excellent Hydrothermal Stability of Composite Catalyst Based on Two Cu-CHA Type Molecular Sieves

Abstract:

The hybrid catalysts consisting of Cu-SSZ-13 and Cu-SAPO-34 are designed to investigate the synergistic effect on the hydrothermal stability of Cu-SSZ-13. The Cu-SSZ-13 and Cu-SAPO-34 synthesized with one-pot method were physically mixed, and the weight ratio of the Cu-SSZ-13/Cu-SAPO-34 were 9:1, 7:3 and 1:1. Both the single catalyst and the composite were hydrothermal treated in 10 vol. % H_2O/air at 750 °C for 16 h. NH_3-SCR activity shows that the composite sample with the weight ratio of 1/1 exhibits a significantly enhanced hydrothermal stability compared with the single-component catalysts under the same test condition. Textural characterization results show that hydrothermal treatment leads to the collapse of Cu-SSZ-13 framework, while negligibly degenerate effect on framework of the composite is observed. The structure collapse of aged Cu-SSZ-13 is ascribed to the dealumination. However, Cu-SAPO-34 in the composite impedes the dealumination of Cu-SSZ-13 during hydrothermal aging treatment as proved by the ^{27}Al MAS NMR and NH_3-TPD results. EPR results reveal that a larger content of isolated Cu^{2+} remains in the composite after hydrothermal treatment compared with that in the aged Cu-SSZ-13, which are consistent with the H_2-TPR results. ^{27}Al NMR and ^{31}P NMR results show that extra-framework P atoms in SAPO-34 transfer to certain sites and connect with the partially bonded framework Al of SSZ-13 during the hydrothermal process, maintaining the framework of Cu-SSZ-13. A large amount of Brønsted acid sites and Cu^{2+} species remained contribute to the excellent SCR performance of the aged composite. This work provides a new simple way to improve the hydrothermal stability of Cu-SSZ-13 and indicates that the SAPO-34 may provide the fragment to stabilize Cu-SSZ-13 during the aging process.

538039 Effects of Surface Species and Dispersion of CeO_2 Supported Transition Metal Oxide Catalysts for NO Reduction By CO Reaction

Abstract:

Nitrogen oxides (NO_x) are a main contributor to global air pollution [1]. During the past few decades, much attention has been paid to eliminating NO_x pollution and technologies such as NH_3-SCR and NO_x storage and reduction have been developed for NO_x reduction [2]. As of now, most commercial catalytic converters use platinum group metals (PGMs) and zeolites for NO_x abatement [3]. Although these catalysts are highly active in NO_x reduction reactions, NO_x decomposition over PGMs and zeolites has some practical issues such as high cost and high light-off temperature [4]. However, transition metal oxide catalysts were found to also be active in NO_x reduction reactions. By far, transition metal oxide catalysts have been investigated for many reactions including NO_x decomposition reaction by CO, which is a primary reaction in three-way catalysis (TWC).

In this study, series of CeO_2 supported CoO_x, FeO_x, or NiO_x catalysts were prepared by incipient wetness impregnation method for NO reduction by CO. Catalytic activity and catalyst physicochemical property were investigated by FTIR, BET, and Raman spectroscopy. No diffraction peaks due to crystalline CoO or Co_3O_4 were observed in XRD patterns under to 10 wt%, while nanocrystaline Co_3O_4 peaks at ~610cm^{-1} was observed in the Raman spectra with 5 wt% sample, indicating that 5-6 wt% CoO_x/CeO_2 contains monolayer coverage. Similarly, 1 wt% NiO_x/CeO_2 and 10 wt% FeO_x/CeO_2 also showed monolayer coverage. It was observed that both surface species and dispersion are closely related to the catalytic activity in NO reduction by CO. For example, for CoO_x/CeO_2 catalysts, monolayer and above monolayer coverage catalysts showed 80~90% CO and NO conversion at 300~400°C with high N_2 selectivity (>90%), whereas sub-monolayer coverage catalysts had significantly low activity.

532853 Fast Lean-Rich Cycling for Enhanced NO_x Conversion on $Pt/CeO_2/Al_2O_3$

Abstract:

Fast Lean-Rich Cycling for Enhanced NOx Conversion on $Pt/CeO_2/Al_2O_3$

Zhiyu Zhou, Michael P. Harold*, Dan Luss**

Department of Chemical and Biomolecular Engineering, University of Houston, Houston, TX 77204

* *mharold@uh.edu,* ** *dluss@central.uh.edu*

The more stringent emission standards pose a challenge for the NOx ($NO+NO_2$) abatement. A potential route to meet these standards is through fast lean-rich cycling on NOx storage and reduction (NSR) catalyst. The Di-Air (Diesel NOx aftertreatment by Adsorbed Intermediate Reductants), introduced by Toyota researchers [1] involves rapid fuel injection into the exhaust to a NSR converter. Their pioneering work revealed superior deNOx performance at high temperature [1]. There is an active debate about the underlying mechanism for NSR at high cycle frequency. Previous studies have advanced mechanisms, including reaction between hydrocarbon intermediates and adsorbed NOx [2], enhanced utilization of fast NOx storage sites [3], and NO decomposition on reduced ceria [4]. Ceria has been studied as an oxygen storage component [5] as well as a low temperature NOx storage component [6] in automobile catalytic converters for many years. Previous study showed that ceria enhances the deNOx performance of the Di-Air system at both low [7] (<300°C) and high [4] (>550°C) temperatures.

In order to isolate the role of ceria, we are evaluating the performance of both CeO_2/Al_2O_3 and $Pt/CeO_2/Al_2O_3$ washcoated monolith catalysts in a bench-scale flow reactor system over a wide range of operating conditions, including cycle time, feed temperature and composition, and reductant type. A lean to rich feed time ratio 6 to 1 of was fixed to maintain the same reductant penalty (reductant to oxidant ratio). The lean/rich cycling conditions included lean/rich times (in seconds) of 90/15, 60/10, 30/5 and 6/1. Several sets of feed conditions were applied to study the favorable and adverse conditions. 300 ppm NO was added to both rich and lean phases. 6.21% H_2 or 0.69% C_3H_6 was only added to rich phase and 0.5% or 5% O_2 was only added to lean phase.

Figure 1 shows the cycle-averaged NOx conversion for four cycle times from 150°C to 600°C using two sets of feeds. With a cycle-averaged lean feed, faster cycling of lean/rich promotes NOx conversion below 450°C. However, above 450°C, different cycling times have almost negligible impact on deNOx performance and NOx conversion drops to a low level (~15%). In contrast, with a cycle-averaged rich feed, the enhancement of NOx conversion due to faster cycling exists for the whole temperature range (150°C ~ 600°C). Above 450°C, NOx conversion obtained under a fixed cycling time remains an asymptotic value but the asymptotic value is elevated by faster cycling.

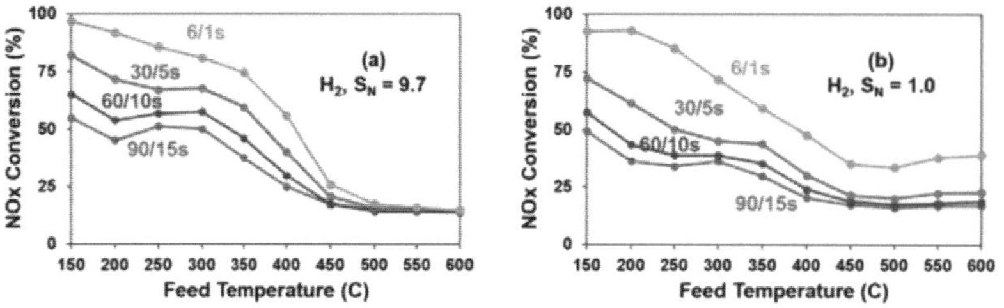

Figure 1. Cycle-averaged NO conversion as a function of feed temperature and lean/rich switching frequency. [Conditions: lean/rich switching frequency: 90/15s, 60/10s, 30/5s, 6/1s; rich feed: 6.21% H_2, 300ppm NO; lean feed: (a) 300ppm NO, 5% O_2; (b) 300ppm NO, 0.5% O_2].

Figure 2 shows the cycle-averaged NO conversion for a fixed cycle of 6/1s with H_2 or C_3H_6 as the reductant. With either cycle-averaged lean or cycle-averaged stoichiometric feed, H_2 is the better reductant than C_3H_6 below 400°C. Above 400°C, C_3H_6 surpasses H_2 when a cycle-averaged stoichiometric feed is applied. However, when a cycle-averaged lean feed is applied, the promotional impact from C_3H_6 disappears and NOx conversion diminishes to the asymptotic value (~15%) above 500°C.

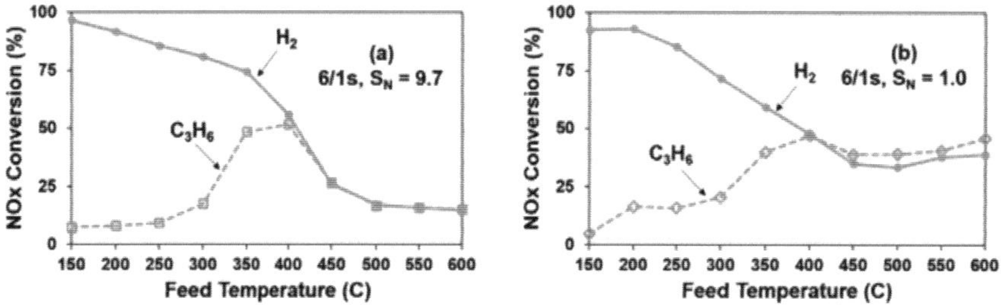

Figure 2. Cycle-averaged NO conversion as a function of feed temperature with fastest cycling frequency and H_2 or C_3H_6 as reductant. [Conditions: lean/rich switching frequency: 6/1s; rich feed: 6.21% H_2, or 0.69% C_3H_6, 300ppm NO; lean feed: (a) 300ppm NO, 5% O_2; (b) 300ppm NO, 0.5% O_2].

These and other data will be presented which provide insight into the underlying mechanism and performance features of ceria-containing NSR catalysts for enhanced NOx abatement for lean burn engines. The results which also provide critical data for model development needed for the development of improved catalyst formulations and operating strategies.

Reference

1. Y. Bisaiji et al., *SAE Int. J. Fuels Lubr.* 5 (2012) 380-388
2. Y. Bisaiji et al., *SAE Int. J. Fuels Lubr.* 5 (2012) 1310-1316
3. A. Ting et al., *Chem. Eng. J.*, 326 (2017) 419-435
4. Y. Wang et al., *Top. Catal.*, 59 (2016), 854-860
5. T., Montini et al., *Chem. Rev.*, 116 (2016), 5987-6041
6. Y., Ji et al., *Catal. Lett.* 110 (2006), 29-37
7. Y. Zheng et al., *Catal. Today*, 276 (2016), 192-201.

537120 Degrading Organic Compounds in Simulated Produced Water By Creating Hydroxyl Radicals Catalytically

Abstract:

Currently, most produced water is disposed of by downhole injection, however, there is limited availability of disposal wells and growing concern over potential environmental contamination from produced water to groundwater wells. Few treatment technologies address the removal of organic compounds present in produced water which promote hazardous bacterial growth. In this work, we explored the potential application of alumina-supported palladium catalysts to degrade organic compounds using low-cost chemical additives. We find that a 1 wt% Pd supported on Al_2O_3 ("Pd/Al_2O_3") catalyst can degrade phenol in DI water at room temperature and atmospheric pressure when in the presence of oxygen and formic acid or hydroxylamine. To understand this result, we carried out control experiments using formic acid or hydroxylamine, O_2, and catalyst in DI water without phenol. We observed hydrogen peroxide (H_2O_2) generation rates of 3360 and 8745 mmol-H_2O_2Â g_{metal}^{-1} h^{-1} using formic acid and hydroxylamine, respectively. Additionally, another control experiment showed no degradation of phenol with added H_2O_2 and Pd/Al_2O_3, and we further hypothesized that radicals directly generated O_2 and hydroxylamine or formic acid over the Pd/Al_2O_3 were key to phenol degradation. Using a terephthalic acid-based fluorescence assay, we monitored â€¢OH radical production and estimated its generation rate to be 5.3 mmol g_{metal}^{-1} h^{-1}.

532373 Swellable Organically Modified Silica (SOMS): A Novel Support for Pd Catalyzed Hydrodechlorination of Trichloroethylene in Aqueous Phase

Abstract:

Trichloroethylene (TCE) is a major groundwater contaminant which is strictly regulated by U.S. E.P.A. The maximum allowable concentration of TCE in water is 5 ppb, which demands complete removal of TCE from groundwater[1]. As of today, several techniques have been explored for degradation of TCE, among which, the catalytic hydrodechlorination (HDC) appears to be promising[2, 3]. Nonetheless, the challenges that need to be overcome in the use of catalytic HDC of TCE include slow kinetics, product inhibition and deactivation due to anionic species such as chloride and sulfites.

Our studies focus on the application of swellable organically modified silica (SOMS) as a novel catalyst support for HDC of TCE. SOMS belongs to the class of bridged polysilsesquioxanes, which are hybrid organic-inorganic materials, synthesized by sol-gel technique. A unique quality of SOMS is that it swells to almost 3 times itâ€™s dry volume, in presence of organic solvents. Apart from this, properties such as hydrophobicity, mesoporosity, high surface area and absorptivity, also make SOMS an attractive choice for a catalyst support[4]. Furthermore, catalysts involving SOMS as the support were synthesized by impregnation of Pd in its pores[5]. This study aims at investigating the effects of swelling and hydrophobicity of SOMS on the kinetics of HDC of TCE. The deactivation resistance of SOMS towards the anionic species in groundwater is also being studied.

Various techniques such as, near ambient pressure X-ray photoelectron spectroscopy (NAP-XPS), in-situ X-ray absorption near edge structure (XANES), in-situ diffuse reflectance infrared Fourier transform spectroscopy (DRIFTS) and laser Raman spectroscopy were used to investigate different characteristics of Pd/SOMS and their role in HDC of TCE. Additionally, techniques such as extended X-ray absorption fine structure (EXAFS), inductively coupled plasma optical emission spectrometry (ICP-OES), transmission electron microcopy (TEM) and mass spectrometry were employed to determine the deactivation resistance of Pd/SOMS. For this study, $Pd/Al_2O_3$was used as the basis of comparison. The results indicate that SOMS has great potential to serve as a catalyst support for HDC of TCE.

[1] National Primary Drinking Water Regulations: EPA 816-F-09-004, in, United States Environmental Protection Agency, 2009.

[2] C.G. Schreier, M. Reinhard, Catalytic hydrodehalogenation of chlorinated ethylenes using palladium and hydrogen for the treatment of contaminated water, Chemosphere, 31 (1995) 13.

[3] H.H. Russell, J.E. Matthews, W.S. Guy, TCE removal from contaminated soil and groundwater, EPA Environmental Engineering Sourcebook, (1992).

[4] C.M. Burkett, L.A. Underwood, R.S. Volzer, J.A. Baughman, P.L. Edmiston, Organicâ€“Inorganic Hybrid Materials that Rapidly Swell in Non-Polar Liquids: Nanoscale Morphology and Swelling Mechanism, Chemistry of Materials, 20 (2008) 10.

[5] H. Sohn, G. Celik, S. Gunduz, S.L. Dean, E. Painting, P.L. Edmiston, U.S. Ozkan, Hydrodechlorination of trichloroethylene over Pd supported on swellable organically-modified silica (SOMS), Applied Catalysis B: Environmental, 203 (2017) 641-653.

527934 Treatment of OIL Produced Water Using Advanced Oxidative Processes: Heterogeneous-Photocatalysis and Photo-Fenton

Abstract:

TREATMENT OF OIL PRODUCED WATER USING ADVANCED OXIDATIVE PROCESSES: HETEROGENEOUS PHOTOCATALYSIS AND PHOTO-FENTON.

SILVA PC[1], FERRAZ NP[1], PERPETUO EA[1], ASENCIOS YJO[1]*

[1] Instituto do Mar, Universidade Federal de São Paulo, Campus Baixada Santista yvan.jesus@unifesp.br

The treatment of contaminants present in the oil-produced water must be carried out before the disposal according to the legislation. This paper presents the use of Advanced Oxidative Processes in the degradation of phenol in sea water, and in treatment of a real sample of oil-produced water (OPW). Two catalysts were used: TiO_2 for photocatalysis-heterogeneous, and $FeSO_4.7H_2O$ for the Photo-Fenton process. The pH variation of the effluent (in acid, neutral and alkaline) and the addition of traces of hydrogen peroxide, were evaluated. The results obtained show an efficiency in the removal of the phenolic compound in sea water in about 96%, under an hour of experiment, having the best behavior when the solution is in pH 3. Regarding to the real sample of OPW, it was observed that larger reductions of the absorption bands relative to organic matter, in both processes proposed in pH 3.

INTRODUCTION

Water that reaches the surface during oil extraction is called oil-produced water (OPW) and is defined as a mixture of salts, suspended solids, heavy metals, microorganisms and oil droplets dispersed in the water. The treatment of OPW is necessary due to the high volumes of this water that are generated during the production of crude oil specially in Offshore processes. When OPW is not treated causes problems to the environment and to the equipment. In this work, phenol was used as the model molecule (representing the soluble aromatic fraction of the crude oil) in seawater. The choice of sea water was made in order to simulate the conditions of the water produced, since the high content of salts is similar to OPW, but in different concentrations. In addition, the composition of organic compounds in the OPW has benzene, toluene, ethylbenzene and xylene (BTEX), naphthalene, phenanthrene, dibenzothiophene (NFD), polyaromatic hydrocarbons (HPA), phenols, gases and heavy metals [Neff, 2002].

2. METHODOLOGY

Characterization: The crystalline phases of the catalysts used in this study were analyzed in a Rigaku miniflex diffractometer (30Kv-10mA). The compounds formed were identified by comparison with the JCPDS - International Center of Difraction Data.

The phenol concentration during the processes was monitored through: High Performance Liquid Chromatography (HPLC) and by colorimetric Method (EPA 9065). In the case of OPW, the photodegradation tests were accompanied by UV/Vis Spectrophometry in the scanning mode.

Catalytic test: Phenol degradation was performed with two different catalysts: TiO_2 (for Photocatalysis-Heterogeneous, Synth 99.99% P.A.) and $FeSO_4.7H_2O$ (for Photo-Fenton, Synth 99.99% P.A.). The tests were carried out using a solution of 43 ppm phenol prepared in sea water (previously filtered). In the case of Photocatalysis-heterogeneous the TiO_2 dosage was $1g.L^{-1}$ and H_2O_2 (in the proportion 0.3% v/v) for 25 mL of effluent. In the case of the Fenton process 4.2 mg of $FeSO_4.7H_2O$ (for 25 mL of effluent) was used, with a ratio of catalyst/H_2O_2 of 1:10. For both processes, the pHs of the phenol solution were varied in: 3, 7 and 10. UV radiation (254 nm) was used for the catalytic degradation assays. The same procedures were applied treatment of OPW extracted from the Rio Grande do Norte Basin, Brazil.

3. RESULTS AND DISCUSSION

In the X-ray diffraction analysis, it was possible to identify a single crystalline phase present in the $FeSO_4.7H_2O$ sample, having unit cell parameters indicating pure monoclinic phase (JCPDS 25-0409) with a = 7.624 Å, b = 7.648 Å and c = 7,123 Å. The TiO_2 diffractogram presents purely the Anatase phase (JCPDS 21-1272), with unit cell parameters of the tetrahedral phase of $a = b = 3.73$ Å and $c = 9.37$ Å. Diffusse reflectance spectroscopy (DRS) analyzes showed that band-gap energy for each material was 2.87 eV (TiO_2) and 1.98 eV ($FeSO_4.7H_2O$), respectively.

Photocatalytic degradation of Phenol in seawater by Photocatalysis-heterogeneous with TiO_2

In the test performed without addition of H_2O_2, the more favorable condition was at initial pH 3, where a decrease in 19 ppm of phenol from the initial solution was reported. On the other hand, no significant removal was observed when the initial pH was 7 and 10, indicating that the pH strongly influences in this process. Similar results were seen by [Seftel et al. 2014; Moraes 2003] which studied the removal of phenol from distilled water. In the tests in presence of traces of H_2O_2, it was observed that in pH 3 and 7 there were almost the total reduction in the concentration of phenol reaching 5.8 ppm and 2,7 ppm (as final concentration), respectively. In addition, it was observed by chromatographic analysis the marked formation of an intermediate organic compound similar to that already present in sea water, this could be associated to the formation of organic acids. Curiously, a slight increase in the final pH of the solution was observed after the photocatalytic test carried out at pHs 3 and 7, which can be explained by the formation of organic acids in the ionic form (formed by the degradation of phenol example: acetates), and which are in equilibrium. This could be related to the higher formation of intermediate compounds observed in the chromatogram at these pHs. The tests also showed that the reaction reaches equilibrium in 1 hour exposure to UV light.

Photocatalytic degradation of Phenol in sea water by Photo-Fenton with $FeSO_4.7H_2O$

In these assays it was noted that the different pHs influenced the process similarly to that reported in the photocatalysis-heterogeneous. The pH 3 and 7 assays obtained a complete removal of the phenol, quantified at 0 ppm by HPLC. The process was less efficient at pH 10. This is a fact that proves the Photo-Fenton is favored at acidic pH, due to the stability of the Fe ions in acid pH and its interaction with hydrogen peroxide. Wang et al.[2014] found an 85% reduction of phenol content for Photo-Fenton tests performed at pH 3; the similar finding was reported by Kuo et al. [1992] and Kiwi and Nadtochenko [2000]; according to them the hydroxyl radical is the most active species in the degradation of phenol at this condition. There was, however, a very marked formation of organic acids (intermediates compound) produced during Photo-fenton process, which have similar chemical behavior as those compounds present in sea-water. The kinetics assay demonstrated that 1 hour was sufficient to achieve equilibrium in the assays at three differrent pHs.

Photocatalysis-heterogeneous and Photo-fenton with a real sample of OPW

In this spectrophotometric analyzes the decharacterization of the absorption band of dissolved organic compounds, after the photocatalysis, indicates the occurrence of degradation processes [Tambani 2011]. The spectral scan of OPW has a single wide band below 400 nm with maximum intensity around 250 nm. Considering that pHs 3 and 7 are of more interest, the photocatalysis-heterogeneous and photo-fenton tests were performed at these pHs for 3 hours. The results showed a marked decharacterization of the spectral scan of OPW, in comparison to the same sample before photocatalysis; thus indicating degradation of part of the organic compounds dissolved in OPW. The decharacterization of the water profile produced was more marked in the Photo-fenton assays indicating a greater efficiency of this process in comparison to heterogeneous photocatalysis.

4. CONCLUSIONS

In the photocatalysis-heterogeneous and Photo-fenton processes for degradation of phenol in sea water, it was found that phenol removal is favored at low pHs (3 or 7). The degradation efficiency for phenol reached about 96% for Photo-fenton, which is very high considering the high content of salts in sea water. In the tests with a real sample of OPW it was identified that for the two catalytic processes, the best working pH is in acidic of neutral pH (3 or 7). Both processes are good for Treatment of OPW, Photo-fenton performed better that heterogeneous-phootcatalysis, however it forms a sludge at the final of the reaction.

5. ACKNOWLEDGMENTS

The authors thank the São Paulo Research Foundation (FAPESP) for the financial support (process Nº: 2014/24940-5), and to CEPEMA of Poli-USP for the analyzes by Liquid Chromatography.

6. REFERENCES Neff J.M. *Bioaccumulation in Marine Organisms Effect of Contaminants From Oil Well Produced Water,* (Ed.), Amsterdam, Elsevier Science, (2002) E.M.Seftel. M.C.Puscasu, M.Mertens, P.Cool, G.Carja. Assemblies of Nanoparticles of CeO_2 –Znti – Ldhs And Their Derived Mixed Oxides As Novel Photocatalytic System For Phenol Degradation. Applied Catalysis B: Environmental, 150 (2014) 157-166. J.E.F. Moraes, *Aplicação do Processo Foto-Fenton na Degradação De Efluentes Industriais Contendo Poluentes Orgânicos. Tese de Doutorado. Universidade de São Paulo*, Brasil, 2003. Wang C., Zhang S., Zhang Z., Zeng M., Yuji S. Optimization and Interpretation of Fenton and UV/Fenton Processes for Degradation of Syringyl Lignin. J Environ Anal Chem 1: (2014) 115-119. doi:10.4172/JREAC.1000115 J. Kiwi, V. Nadtochenko. Mechanism and kinetics of the oh-radical intervention during fenton oxidation in the presence of a significant amount of radical Scavenger (cl–). *Environmental science & technology*, 34 (2000) 2162-2168. P.C. Tambani, *Estudo da Degradação do Fenol e Sais Intermediários Pelo Processo UV/H_2O_2, Universidade de São Paulo*, Brasil, 2011, 115.

537593 Electrochemical Anthraquinone Process Enabled By Phase Transfer Catalysis

Abstract:

Currently, hydrogen peroxide is produced via the *anthraquinone process*, using hydrogen as terminal reductant for two-electron reduction of oxygen to H_2O_2. This is in order to ensure selectivity for two-electron reduction as well as to avoid mixing hydrogen and oxygen within the explosive concentration limits. However, this process is difficult to downscale because of the reliance on the use of hydrogen generally produced on-site by steam reforming. Electrochemical delivery of two protons and two electrons can be envisioned as an alternative, however there is currently no commercial production of hydrogen peroxide using electricity and very few small scale demonstrations.

Existing methods for the *direct* electrochemical production of hydrogen peroxide rely on precious metal catalysis, using palladium or platinum, as well as toxic alloying metals such as mercury. Furthermore, the production is limited by both solubility of oxygen in water, and the possibility of a buildup of hydrogen peroxide short-circuiting the device. Furthermore, a general existing problem with direct electrolysis is that H_2O_2 is generated in some electrolyte which must be removed, as well as the instability of peroxides in such catholytes.

We have developed a method for the electrolytic reduction of quinones to form hydroquinones in aqueous solutions; in effect an electrochemical analog of the anthraquinone process. Where precious metal oxygen evolution catalyst is avoided by the use of a bipolar membrane, allowing quinone reduction at acidic pH and oxygen evolution at basic pH using an iron-nickel (FeNi) water oxidation catalyst. The peroxide forming step is decoupled from the initial reduction by phase transfer catalysis, using a tetrabutylammonium salt and a water immiscible phase to shuttle the quinone between an electrolyte medium and an oxygenated salt-free water flow in which the product can be formed with high efficiency. A device utilizing this method was fabricated and the continuous production of hydrogen peroxide was demonstrated with a Faradaic efficiency of >90% at concentrations of > 2000 ppm at neutral pH.

542555 Important Hard and Soft Skills Difficult to Teach in the Classroom

Abstract:

There are many important skills and experiences imparted through working on international humanitarian engineering projects. Understanding your customer, dealing with other cultures, safety, dealing with uncertainty, and thinking on your feet are a few of them. Students can benefit greatly from taking on this kind of work either in implementing full projects or doing the conceptual work in a capstone design class.

542581 A Student Perspective: What I Learned from Working on an International Project

Abstract:

Students volunteer to work on international projects for many different reasons. They take on many different roles in planning, design, execution, communications, project management, and research; and they each develop different skills, have different experiences and are impacted differently. Undergraduate students will discuss why *they* volunteered, what *they* have done, and what *they* have gotten out of this work professionally and personally.

542554 What Does Sustainability Really Mean in These Projects?

Abstract:

To make a project successful partnering with a small community in the developing world requires many important attributes besides technical accuracy. In order to sustain the benefits over the long term, the project team must consider and master many other aspects of sustainability in addition to environmental sustainability. Financial and cultural are two of the many types of sustainability that will be discussed.

542556 What Is EWB-USA and Why Might You Want to Work with Them?

Abstract:

EWB-USA is organization of volunteer students and professionals. Student and professional chapters can be found on campuses and cities across the country. EWB-USA has very specific standards in terms of how partnerships with local communities are maintained and projects executed. The EWB-USA philosophy is tightly focused on executing projects that will actually deliver value over the long-haul, and the mandatory work processes take into account the unique characteristics of student volunteers.

542585 How International Projects Might Help with ABET Accreditation

Abstract:

Team humanitarian engineering projects align with many ABET Criterion 3 Student Outcomes. Incorporating the projects into a programâ€™s assessment structure is difficult when the projects are extra-curricular. We will present solutions from different schools that allow the projects to assist accreditation.

542580 Panel Discussion -- Making Successful International Projects Happen

Abstract:

There are many challenges to implementing humanitarian engineering projects in a university setting. Where do the ideas / projects come from? How can these projects be turned into meaningful capstone design classes? How can they be fit into existing curriculum and academic environment? How do you motivate students, faculty, administration? How can faculty motivated to make these projects happen collaborate across different universities? This panel is intended to be a forum for all those in the room to define challenges and offer solutions.

539584 Pittsburgh: Urban Agriculture and the Food-Energy-Water Nexus

Abstract:

Cities rely on external supply chains for food, energy and water. Urban areas are growing and the dependence on reliable supply chains for these resources grows. The interdependence between the infrastructure and connections of providing these resources has become increasingly significant. Urban agriculture provides an opportunity to address Nexus challenges. Urban agriculture benefits include increasing food resilience and improving stormwater management. Ancillary benefits can also include the potential to reduce the amounts of materials sent to landfills (through composting), the reduction urban heat-island effect, and reducing wastewater treatment loads â€“ all of which come with energy reduction benefits.

Pittsburgh lost a substantial portion of its population during the transition away from its industrial roots as the â€˜Steel City.â€™ A legacy of the population decrease includes empty and abandoned lots (e.g. over 8,000), aging infrastructure, and areas of decline that have created food deserts and challenges associated with the interdependence of food, energy and water. Pittsburgh has implemented multiple initiatives to address this challenge that will realize the benefits from urban agriculture (e.g. a 23 acre urban farm; community driven revitalizations that include energy efficient housing, stormwater infrastructure, rainwater management, and agriculture spaces).

This analysis examines urban agriculture strategic options, the potential benefits, and the challenges. Options include agriculture options (e.g. geoponic, hydroponic, aeroponic systems), water management options (e.g. water reuse), waste management (e.g. biochar as soil enhancer and carbon storage), and energy option (e.g. distributed generation).

538328 Quantifying Virtual Phosphorus Flows in Interstate Food Trade: Implications for Environmental Sustainability

Abstract:

Phosphorus(P) is one of the essential nutrients for agriculture. It is primarily supplied to crops as inorganic phosphorous fertilizer or though manure application. On farm use of phosphorus efficiency is very low, ranging between 10-15%. Rest of the applied P gets bonded in the soil which may subsequently become available for crop uptake. Current agronomic P recommendations for farmers exclude the accumulated P due to unknown timescales of release. Sustainable management of P is an important topic due to the limited availability of inorganic P sources and its downstream potential to cause eutrophication. Additionally, production of phosphorus fertilizers is highly energy intensive with limited potential of waste stream recovery due to techno-economic constraints. Therefore, an accurate understanding of regional P application and efficiency, combined with interregional food trade is necessary to understand the complexity of P cycle. We combine data on current phosphorus application practices with recommended and existing soil P levels to assess sustainability of P management on farm. Additionally, we use P application data with food trade to quantify phosphorus embodied in the inter-regional trade. This work will enable to identify trends and hotspots in US agriculture P application practices for management of food-water nexus.

We develop a food trade and associated virtual phosphorus model for the U.S. by evaluating over 60 crop items covering commodity groups of cereal grains, fruits, nuts, and vegetables. Regional phosphorus mass balances are constructed using datasets from US Department of Agricultureâ€™s Natural Resource Conservation Service, International Plant Nutrition Institute, and Census of Agriculture. The interstate food trade network is formulated from Freight Analysis Framework and translated to virtual phosphorus network by using crop yields and phosphate fertilizer application rates from Census of Agriculture. Preliminary results indicate that phosphorus fertilizer application differ considerably for crop ranging from 2-10 lb P per ton crop for grains to 10-68 lb P per ton of crop for nuts. To estimate whether calculated P application rates are excessive or necessary for specific soil type, we compare it with recommended P levels in each state. In 2012, 28 states had to recommended levels, including states with highly eutrophic waterbodies such as Ohio, Florida, California, Texas, and Maine. With addition of total P fertilizer across crops in the given year, the soil phosphorus for these states increases by a fraction of 2-4 compared to recommended levels. From food trade perspective, we find that 13 billion US tons of food and 5.3 million US tons of embodied P were transferred across the U.S. Specifically, grains dominate food and total embodied P transfers within the U.S., indicating the importance of considering scale of production. The implications of the study findings will be discussed in the context of balancing on-farm conservation policies for simultaneously enriching soil with nutrients and avoiding excess run-offs.

533778 Towards a Two-Level Superstructure Optimization Framework for Land Use Based on Food-Energy-Water Nexus

Abstract:

Land use optimization is the result of competitions between different land types [1]. The main reasons of these competitions are the quantitative constraints of land scales and corresponding Food-Energy-Water (FEW) resources under multiple conflicting objectives [2, 3]. Despite FEW resources play critical important roles for sustaining and improving human life, increasing production demands and sustainable concerns intensify these competitions due to their limitation. How to coordinate these competitions is a complex task for reasonable land allocation based on limited FEW resources while promoting efficiencies and sustainability [4, 5]. A nexus thinking by considering the FEW flow and interactions in multi-objective land use systems has become apparently necessary for efficient use of FEW resources and trade-off decisions [6, 7]. However, challenges arise in representing quantitative FEW-Nexus that supports land-use decisions while encountering conflict objectives, limited data, and multiple scales [8, 9].

To address these challenges, we develop a multi-objective optimization framework by considering total profit, food production, resources use, and environmental penalty as multiple objectives for land use systems. The proposed framework suggests a two-level superstructure optimization method: (1) In the first level, models of all the production units are developed by data-driven modeling and global optimization methods based on limited realistic data. FEW flow among them are quantified and interlinked to construct interval models, which can be represented as interval superstructures. The small-scale MINLP problems can be solved efficiently due to the limited combinations of land units in the interval subsystems; (2) In the second level, multiple interval models with optimal land and FEW allocations are used to construct the extended systematic network and represented as a large-scale superstructure, which can be solved as a MILP problem. A series of FEW indices are provided for decision-makers to analysis nexus in the system, carry out quantitative assessment based on different objectives, and achieve trade-off solutions. The framework is illustrated by a case study on the crop-livestock system, which shows that it can provide multiple land allocation solutions and predicts corresponding yields for land units within the systems under flexible scales. Computational results from interval models indicate that we achieve optimal solutions in the interval subsystems, and valuable production models for yield prediction with high confidence. The performance of these models can be improved by increasing feedback data [10]. The systematic network provides an efficient method for extending the land allocation solutions to multiple land scales. For multiple objectives, the proposed FEW index can be applied to select strategies for optimal land allocation that maximizing food productivity and minimizing water and energy consumptions.

Reference

[1] Memmah, M. M., Lescourret, F., Yao, X., & Lavigne, C. (2015). Metaheuristics for agricultural land use optimization. A review. Agronomy for sustainable development, 35(3), 975-998.

[2] Bergstrom, J. C., Goetz, S. J., & Shortle, J. S. (2013). Land use problems and conflicts: Causes, consequences and solutions. Routledge.

[3] Keairns, D. L., Darton, R. C., & Irabien, A. (2016). The energy-water-food nexus. Annual review of chemical and biomolecular engineering, 7, 239-262.

[4] Miralles-Wilhelm, F. (2016). Development and application of integrative modeling tools in support of food-energy-water nexus planningâ€"a research agenda. Journal of Environmental Studies and Sciences, 6(1), 3-10.

[5] Simpson, G., & Berchner, M. (2017). Water-energy nexus-Measuring integration: towards a water-energy-food nexus index. Water Wheel, 16(1), 22-23.

[6] Garcia, D. J., & You, F. (2016). The water-energy-food nexus and process systems engineering: a new focus. Computers & Chemical Engineering, 91, 49-67.

[7] Ringler, C., Bhaduri, A., & Lawford, R. (2013). The nexus across water, energy, land and food (WELF): potential for improved resource use efficiency?. Current Opinion in Environmental Sustainability, 5(6), 617-624.

[8] McCarl, B. A., Yang, Y., Schwabe, K., Engel, B. A., Mondal, A. H., Ringler, C., & Pistikopoulos, E. N. (2017). Model Use in WEF Nexus Analysis: a Review of Issues. Current Sustainable/Renewable Energy Reports, 4(3), 144-152.

[9] McCarl, B. A., Yang, Y., Srinivasan, R., Pistikopoulos, E. N., & Mohtar, R. H. (2017). Data for WEF Nexus Analysis: a Review of Issues. Current Sustainable/Renewable Energy Reports, 4(3), 137-143.

[10] Nie, Y., Avraamidou, S., Li, J., Xiao, X., & Pistikopoulos, E. N. (2018). Land use modeling and optimization based on food-energy-water nexus: a case study on crop-livestock systems. 13th International Symposium on Process Systems Engineering; Elsevier; Accepted.

537263 Land Availability, Utilization, and Intensification for a Solar Powered Economy

Abstract:

Among all forms of renewable energy, solar energy is the most promising one with sustainable availability and abundance to meet human beingâ€™s energy need. However, solar energy is a diffuse resource and harvesting it requires vast tracts of land. While most studies estimated that land availability is not a major challenge for the feasibility of a total solar economy, a few recent detailed analyses have raised the problem in land availability in some densely populated countries, using actual solar farm production data and considering other factors such as storage requirement and intermittency of solar availability. However, according to their estimation, countries with relatively low population density, such as the United States, have sufficient land for solar farms to meet all energy demand. In this study, we perform a detailed land requirement analysis in each state of the US for a solar powered future and propose land intensification strategies to resolve possible land use challenges.

Land requirement analysis in this study takes into account several issues that are usually ignored. First, efficiencies of major energy conversion steps from current primary energy to end use are different from those when solar energy is deployed. Second, available land is limited to only part of the urban area (rooftops) and miscellaneous land. Farmland, forest and other reserved land such as national and state parks cannot be utilized for solar plant construction. In addition, we prefer local photons for local use for consideration of minimizing transmission loss and energy security.

Our land requirement analysis shows that for many states there is insufficient currently available land and that other land areas will need to be used, which is contrary to previous report. We propose that farmland is the best choice for land use intensification as it minimizes additional environmental impact to existing habitat and propose several photovoltaic systems for this purpose. We then estimate the power output of these systems and estimate the amount of farmland for each state that is needed to meet energy needs. By land intensification, the energy demand of each state in the US can be locally satisfied.

529950 Microbial Community Profile Versus Water Quality in Urban Watersheds

Abstract:

Microbial community profiles can serve as an indicator for unbran water quality. The aim of this study was to gain insight into bacterial diversity and infer water quality based on selected operational taxonomic units (OTUs). Here we report bacterial communities of 12 watersheds sampled at different frequencies from year 2013 to 2016, were characterized by sequencing the V4 to V5 regions of 16S rRNA genes using 454 pyrosequencing and paired-end sequencing platforms. An nMDS plot of all samples revealed that bacterial communities of the same watershed tended to cluster together and a few watersheds shared similar communities. Effects of temporal variation were more pronounced on bacterial communities within watersheds ($P < 0.01$; 2-Factor PERMANOVA) than spatial variation. Cyanobacteria of *Microcystis* and *Cylindrospermopsis* dominated certain watersheds that tended to have algal blooms. Pairwise correlation of gut and sewage-associated OTUs revealed a strong positive correlation among *Enterobacter*, *Klebsiella* and *Escherichia* and *Clostridium sensu stricto 8*, suggesting that presence of these taxa as possible indicators of impairment. A comparison of the abundances of 13 potential antibiotic resistant bacteria, revealed four watersheds with samples that correlated with higher abundances of potential antibiotic resistant bacteria compared to the other watersheds. Thus microbial community determined by high-throughput sequencing in conjunction with qPCR can be used to indicate quality of water in urban watersheds.

528536 Mechanisms Whereby Microbes Promote Intermediate Soil Moisture Content

Abstract:

The function of microbial communities is inextricably with the soil microstructure. At the pore scale, water content regulates hydraulic connectivity and thereby modulates microbesâ€™ access to aqueous and gaseous substrates. Furthermore, soil bacteria near plant roots produce extracellular polymeric substances (EPS), a polymeric substance that helps modulate soil moisture and promote plant growth. EPS regulates soil moisture (i) by serving as hydrogel, swelling during wet conditions and remaining hydrated during dry conditions, (ii) by promoting soil particle aggregates via capillary forces, (iii) by altering soil surface properties through creating water repellent surfaces. EPS-mediated moisture retention behavior in soil is complicated but can be understood by systematically controlling the microscale environment. Here, we show the effect of physical microstructure on EPS-mediated moisture retention. We employ physically complex emulated soil micromodels as well as micromodels with uniform macropore or micropore geometries. EPS was collected from stationary-phase *Sinorhizobium meliloti* cultures and suspended at different concentrations in growth media salts or in artificial groundwater. Experimental results showed that the 0.25× EPS solution dried eight times slower than deionized water, and the 1× EPS solution dried 16 times slower than deionized water in identical soil micromodels. Also, the residual saturation of 0.25× and 1× EPS solutions were 6 ± 1% and 35 ± 21%, compared with a residual saturation of 0 for deionized water. We observed no effect on rate or extent of moisture loss for the micromodels with a uniform macropore geometry. Studies with uniform micropore geometry confirmed that higher EPS concentrations were associated with a reduce rate and increased variance of moisture loss at the pore scale. In this study, we provide a systematic method to evaluate the performance of EPS solutions in modulating soil moisture, and show how EPS concentration decreases the rate of water loss and also amplifies the variability of moisture content at the pore scale. Longer term, we anticipate that the impact of this work will be to aid in the development of sustainable agriculture biotechnology for increased food production using less water.

528379 Nitrogen Efficient Fertilizer Materials

Abstract:

Nitrogen cycle management is a critical challenge in modern engineering. NH_3 synthesis from N_2 consumes several percent of total world energy while only a fraction of the synthesized fertilizer nitrogen, such as urea, is absorbed by the plants. Majority of nitrogen initially confined in urea is decomposed and released in the environment as gaseous NH_3 or soil and water mobile nitrate, NO_3^-, resulting in low sustainability nitrogen fixation process. Furthermore, for a complete plant nutrient all of the nutrients (N,P,K,Ca,Mg,S) need to be supplied in one solid granulated or liquid mixture. Aqueous liquid fertilizers are expensive to transport since their dissolved nutrient content is governed by the solubility limits.

Molecular crystals of urea with inorganic acids (H_3PO_4 and H_2SO_4) as well as solid salts ($CaSO_4$, $MgSO_4$, $Ca(H_2PO_4)$ and $Mg(H_2PO_4)$) present a convenient form of high nutrient content fertilizers that can be inexpensively transported to the remote locations and applied directly or via fertigation. Importantly, these urea inorganic acid adducts have been shown to result in a much more efficient nitrogen uptake by plants combined with significantly lower nitrogen losses. The work will be presented that will explore design of these fertilizers that contribute to the enhanced sustainability of the chemical processing. In particular, mechanochemical synthesis methods that avoid use of liquid solvents that would otherwise necessitate evaporation or result in high aqueous solution transportation costs will be presented. Since dissolution of the solid urea adducts for fertigation requires in depth knowledge of the corresponding phase diagrams, these data will also be presented. Finally, molecular structure effects of the urea adducts on the improved nitrogen cycle management will be discussed as well as the synthesis of industrial and anthropogenic waste derived nitrogen efficient molecular fertilizers.

530806 On the Role of Fe^{2+} and O_2 in Oxidative Degradation of Aqueous Monoethanolamine (MEA)

Abstract:

The most suitable technology for post-combustion CO_2 capture appears to be chemical absorption using an aqueous amine-based solvent.[1,2] Monoethanolamine (MEA) has been widely studied and is typically considered the benchmark solvent, and it exhibits several advantages such as relatively fast absorption rates, low viscosity and volatility, and high solubility.[1–8] However, its widespread implementation remains hindered by costs associated with the parasitic energy consumption during regeneration, as well as problems associated with corrosion and solvent degradation.[1–8] Costs incurred from degradation include loss of solvent, the energy requirements for the reclamation process, and the corrosivity and toxicity of the degradation products.[9,10]

Oxidative degradation is thought to occur due to the presence of O_2 from the flue gas (~ 3-4%) and metal ions from stainless steel corrosion (Fe^{2+}, Cr^{3+}, Ni^{2+}, Mn^{2+}) in the absorber, but the underlying molecular mechanisms remain unclear.[11,12] It has been proposed that oxidative degradation is initiated by an electron or hydrogen abstraction mechanism, whereby a radical abstracts either an electron from the lone pair of N or hydrogen from the amine, respectively.[11,13,14] The radical could be a transition metal complex, such as solvated Fe^{2+}, or reactive oxygen species (ROS) formed from dissolved O_2.[15–17] Reported oxidative degradation products in aqueous MEA include ammonia/ammonium, aldehydes, formate/formic acid, glycolate/glycolic acid, oxalate/oxalic acid, acetone, formamides, and imidazolidones.[11,13,14,18,19] Few theoretical studies have been undertaken to elucidate the mechanisms underlying oxidative degradation,[20,21] while an improved fundamental understanding could aid in the development of inhibitors.

In this work, we elucidate the role of Fe^{2+} and O_2 in the oxidative degradation of MEA in aqueous solution using a theoretical approach. First, we evaluate the energetics of redox reactions and relative stabilities of intermediates from Fe^{2+} and O_2 using static QM calculations. Thereafter, AIMD simulations are used to demonstrate elementary reaction steps and involved during oxidative degradation of MEA assuming Fe^{2+} and O_2 are in aqueous solution.

Our analysis clearly demonstrates that Fe^{2+} can transfer 2 electrons to O_2 in aqueous amine solution to form ferryl (Fe^{4+}) hydroxide complexes and reactive oxygen species such as superoxide ($O_2^{-\bullet}$) and hydroxyl radical ($^{\bullet}OH$), which are short-lived intermediates that easily abstract hydrogen from the C-H, N-H and O-H bonds in MEA. These results clearly demonstrate that oxidative degradation by MEA is initiated by hydrogen abstraction, rather than electron abstraction. After hydrogen abstraction, MEA can dissociate to formaldehyde and imine radicals, which can further participate in reactions with H_2O and hydroxyl radical to form acetic acid, ammonia, and possible precursors to glycolic and oxalic acids. The basicity of aqueous amines enhances the electron transfer between Fe^{2+} and O_2 due to the electron donation to Fe^{2+}/Fe^{3+} from OH^- ligands, as described in previous studies[22,23] and predicted by our static QM calculations. This study highlights how first-principles studies can play a significant role in assessing the reaction mechanisms and intermediates underlying degradation, while it would be difficult to experimentally isolate and characterize short-lived reactive species involved.

References

(1) Rochelle, G. T. Amine Scrubbing for CO_2 Capture. *Science* **2009**, *325* (5948), 1652–1654.

(2) Figueroa, J. D.; Fout, T.; Plasynski, S.; McIlvried, H.; Srivastava, R. D. Advances in CO_2 Capture technology—The U.S. Department of Energy's Carbon Sequestration Program. *Int. J. Greenh. Gas Control* **2008**, *2* (1), 9–20.

(3) U.S. Department of Energy. *Carbon Capture Technology Program Plan*; 2013.

(4) D'Alessandro, D. M.; Smit, B.; Long, J. R. Carbon Dioxide Capture: Prospects for New Materials. *Angew. Chem. Int. Ed. Engl.* **2010**, *49* (35), 6058–6082.

(5) Strazisar, B. R.; Anderson, R. R.; White, C. M. Degradation Pathways for Monoethanolamine in a CO_2 Capture Facility. *Energy and Fuels* **2003**, *17* (4), 1034–1039.

(6) Abu-Zahra, M. R. M.; Niederer, J. P. M.; Feron, P. H. M.; Versteeg, G. F. CO_2 Capture from Power Plants. Part II. A Parametric Study of the Economical Performance Based on Mono-Ethanolamine. *Int. J. Greenh. Gas Control* **2007**, *1* (2), 135–142.

(7) Abu-Zahra, M. R. M.; Schneiders, L. H. J.; Niederer, J. P. M.; Feron, P. H. M.; Versteeg, G. F. CO_2 Capture from Power Plants. Part I. A Parametric Study of the Technical Performance Based on Monoethanolamine. *Int. J. Greenh. Gas Control* **2007**, *1* (1), 37–46.

(8) Stowe, H. M.; Hwang, G. S. Fundamental Understanding of CO_2 Capture and Regeneration in Aqueous Amines from First-Principles Studies: Recent Progress and Remaining Challenges. *Ind. Eng. Chem. Res.* **2017**, *56* (24), 6887–6899.

(9) Goff, G. S. Oxidative Degradation of Aqueous Monoethanolamine in CO_2 Capture Processes : Iron and Copper Catalysis , Inhibition , and O_2 Mass Transfer. **2005**, 283.

(10) Chi, S.; Rochelle, G. Oxidative Degradation of Monoethanolamine. *Ind. Eng. Chem. Res.* **2002**, *41* (17), 4178–4186.

(11) Gouedard, C.; Picq, D.; Launay, F.; Carrette, P. L. Amine Degradation in CO_2 Capture. I. A Review. *Int. J. Greenh. Gas Control* **2012**, *10*, 244–270.

(12) Lepaumier, H.; Picq, D.; Carrette, P.-L. New Amines for CO_2 Capture. II. Oxidative Degradation Mechanisms. *Ind. Eng. Chem. Res.* **2009**, *48* (20), 9068–9075.

(13) Goff, G. S.; Rochelle, G. T. Monoethanolamine Degradation: O_2 Mass Transfer Effects under CO_2 Capture Conditions. *Ind. Eng. Chem. Res.* **2004**, *43* (20), 6400–6408.

(14) Fredriksen, S. B.; Jens, K. J. Oxidative Degradation of Aqueous Amine Solutions of MEA, AMP, MDEA, PZ: A Review. *Energy Procedia* **2013**, *37* (1876), 1770–1777.

(15) Koppenol, W. H.; Stanbury, D. M.; Bounds, P. L. Electrode Potentials of Partially Reduced Oxygen Species, from Dioxygen to Water. *Free Radic. Biol. Med.* **2010**, *49* (3), 317–322.

(16) Koppenol, W. H. The Haber-Weiss Cycle – 70 Years Later. *Redox Rep.* **2001**, *6* (4), 229–234.

(17) Wood, P. M. The Potential Diagram for Oxygen at pH 7. *Biochem. J.* **1988**, *253* (1), 287–289.

(18) Da Silva, E. F.; Lepaumier, H.; Grimstvedt, A.; Vevelstad, S. J.; Einbu, A.; Vernstad, K.; Svendsen, H. F.; Zahlsen, K. Understanding 2-Ethanolamine Degradation in Postcombustion CO_2 Capture. *Ind. Eng. Chem. Res.* **2012**, *51* (41), 13329–13338.

(19) Vevelstad, S. J.; Johansen, M. T.; Knuutila, H.; Svendsen, H. F. Extensive Dataset for Oxidative Degradation of Ethanolamine at 55-75 °C and Oxygen Concentrations from 6 to 98%. *Int. J. Greenh. Gas Control* **2016**, *50*, 158–178.

(20) Vevelstad, S. J.; Eide-Haugmo, I.; Da Silva, E. F.; Svendsen, H. F. Degradation of MEA; A Theoretical Study. *Energy Procedia* **2011**, *4*, 1608–1615.

(21) Gupta, M.; Vevelstad, S. J.; Svendsen, H. F. Mechanisms and Reaction Pathways in MEA Degradation; A Computational Study. *Energy Procedia* **2014**, *63*, 1115–1121.

(22) Stumm, W.; Morgan, J. J. Kinetics of Redox Processes. In *Aquatic Chemistry: Chemical Equilibria and Rates in Natural Waters*; John Wiley & Sons, Inc.: New York, NY, 1996; pp 672–725.

(23) Morgan, B.; Lahav, O. The Effect of pH on the Kinetics of Spontaneous Fe(II) Oxidation by O2 in Aqueous Solution - Basic Principles and a Simple Heuristic Description. *Chemosphere* **2007**, *68* (11), 2080–2084.

533450 Amine-Functionalized Hierarchical Zeolites for Carbon Dioxide Capture

Abstract:

Amine-Functionalized Hierarchical Zeolites for Carbon Dioxide Capture

Maryam Khaleel*, Anna Tuneu Pujolras, Issam Ismail, Georgios Karanikolos, Lourdes F. Vega, Fèlix Llovell.

*E-mail: maryam.khaleel@ku.ac.ae

Abstract

One of the greatest challenges we are facing in the 21st century is our ability to provide sustainable energy sources to meet the demands for quality of life and economic growth. As most of the energy today is provided from fossil fuels, huge amounts of CO_2 and other greenhouse gases are emitted to the atmosphere, with negative effects on the environment. Depletion of fossil fuels and their associated environmental pollution lead scientists to the pursuit of renewable energy sources and clean energy. However, reducing dependency on fossil fuels is a major challenge for most economically advanced countries. To make fossil fuels clean, efficient, reliable and scalable technologies for CO_2 capture from large sources should be developed to control CO_2 emission in the short-medium term. Although some technologies for carbon capture are available in the market they still suffer from some limitations to be fully implemented at large scale.

Conventional CO_2 removal using aqueous alkanolamine solutions is energy intensive. This lead researchers to investigate various solid adsorbents for selective CO_2 removal at low pressure ranges while requiring less energy for regeneration compared to alkanoamine solutions [1]. Various porous solids were tested for post-combustion CO_2 capture including activated carbon [2], zeolites [3] and metal–organic frameworks (MOFs) [4]. Solid materials grafted or impregnated with amines were also studied to further increase the CO_2 binding strength and hence CO_2 uptake [5, 6]. Of particular interest are amine grafted zeolites that combine strong CO_2 capturing zeolite micropores with the chemical binding of CO_2 with amines [7]. Hierarchical zeolites, combining zeolite micropores and mesoporous domains, show increased surface area and pore volume compared to conventional zeolites. Functionalizing the mesoporous domains in such materials with amines can result in high CO_2 uptake because both the active sites of the zeolite and the amine groups in the mesopore space can simultaneously capture CO_2. However, there are very few studies of amine grafted hierarchical zeolites and the latter were prepared using expensive organics [7].

In this work we study the performance of amine grafted hierarchical zeolites, prepared from inorganic (i.e. inexpensive) routes, in CO_2 capture. A series of zeolites with varying textural properties are synthesized and grafted with different amines and their performance is tested for CO_2 adsorption. Synthesized materials are characterized using X-ray Diffraction, Nitrogen Physisorption and Scanning Electron Microscopy. CO_2 uptake properties are evaluated by equilibrium gas uptake measurements to study the effect of hierarchy on the performance of the materials. Thermogravimetric Analysis is used to estimate the amount of amine grafted on zeolite samples and Infrared Spectroscopy is used to study the functionality of the materials.

References

[1] S. Choi, J. H. Drese, C. W. Jones, ChemSusChem 2009, 2, 796 – 854.

[2] Lu, C.; Bai, H.; Wu, B.; Su, F.; Hwang, J. F. Energy Fuels 2008, 22 (5), 3050–3056.

[3] Chatti, R.; Bansiwal, A. K.; Thote, J. A.; Kumar, V.; Jadhav, P.; Lokhande, S. K.; Biniwale, R. B.; Labhsetwar, N. K.; Rayalu, S. S. Microporous Mesoporous Mater. 2009, 121 (1-3), 84–89.

[4] J. A. Mason, T. M. McDonald, T.-H. Bae, J. E. Bachman, K. Sumida, J. J. Dutton, S. S. Kaye, J. R. Long, J. Am. Chem. Soc. 2015, 137, 4787 – 4803.

[5] T. M. McDonald, J. A. Mason, X. Kong, E. D. Bloch, D. Gygi, A. Dani, V. Crocella, F. Giordanino, S. O. Odoh, W. S. Drisdell, B. Vlaisavljevich, A. L. Dzubak, R. Poloni, S. K. Schnell, N. Planas, K. Lee, T. Pascal, L. F. Wan, D. Prendergast, J. B. Neaton, B. Smit, J. B. Kortright, L. Gagliardi, S. Bordiga, J. A. Reimer, J. R. Long, Nature 2015, 519, 303 – 308.

[6] F. Su, C. Lu, S.-C. Kuo, W. Zeng, Energy Fuels 2010, 24, 1441–1448

[7] T. H. Nguyen, S. Kim, M. Yoon, T.-H. Bae, ChemSusChem 2016, 9, 455 – 461.

Improving Full-Scale Models of New Carbon Capture Technologies with Uncertainty Quantification

Christopher S. Russell[a,b], Sham K. Bhat[a], Joel D. Kress[a], Larry L. Baxter[b,c], Charles J. Freeman[d], Yuan Jiang[d], Joshua C. Morgan[e]

[a]Los Alamos National Lab, Los Alamos, New Mexico 87545

[b]Department of Chemical Engineering, Brigham Young University, Provo, Utah 84602

[c]Sustainable Energy Solutions LLC, Orem, Utah 84058

[d]Pacific Northwest National Lab, Richland, Washington 99352

[e]National Energy Technology Lab, Pittsburgh, Pennsylvania 10940

Abstract

Carbon capture technologies for combustion power plants aim to remove CO_2 from flue gas. One such technology is the novel CO_2-Binding Organic Liquid (CO_2BOL) process developed at the Pacific Northwest National Laboratory. The process utilizes their anhydrous CO_2BOL solvent in place of amine mixtures to reduce the energy penalty. A full-scale model of this system based on NETL's Case 10 power plant is projected to produce 7-16% more net electric power over a traditional MEA system for the same plant. These full-scale model predictions are promising, however full-scale simulations are difficult to validate using only bench-scale data. There are errors in both the measurements and models that need to be considered in addition to uncertainties introduced by up-scaling. Uncertainty quantification (UQ) is a statistical framework used to better understand these uncertainties as well as data gaps in models. By constraining models to data, distributions of model parameters are estimated and then propagated through the model to obtain distributions of key outputs such as carbon capture percent, energy penalty, stripper temperature, etc. The results from this UQ analysis aid the design of experiments (DoE) which identifies data gaps to fill to improve model accuracy. Using UQ and DoE effectively can reduce the development time of new carbon capture by years and save time and money on the path to a pilot, and ultimately full-scale, plant.

Introduction

Recent studies predict a 28% increase in world energy consumption by 2040[1]. Most of this increase comes from non-OECD countries where significant growth in fossil fuel usage is expected. Renewable energy sources are predicted to increase rapidly during this period with fossil fuel usage increasing at a slower pace. Accommodating this increase in fossil fuel usage while meeting emission targets for greenhouse gases requires CO_2 capture from fossil-fuel-fired power plants. There are many technologies in development[2] including solvents, sorbents, membranes, cryogenics, and others. A common problem with these technologies is the discrepancy between the modeling scale and the testing scale. When bench-scale data are extrapolated to full-scale models there are discrepancies in the model, parameters, and data that are seldom accounted for. This leads to an incomplete understanding of model accuracy and, ultimately, the energy penalty or carbon capture percent predicted by that model. Uncertainty quantification analyzes these discrepancies and provides a distribution for model output values that informs the user on the accuracy of their model's predictions. It can also pinpoint areas in need of additional data to reduce uncertainty in the model. This paper uses the example of the CO_2BOLs solvent[3] to demonstrate the effectiveness of these tools. The process also applies to most new carbon capture systems.

CO_2BOLs System

CO_2BOLs is a novel non-aqueous polarity-swing solvent with a predicted energy penalty less than that of traditional amine systems[4]. The Pacific Northwest National Laboratory developed CO_2BOLs as a competitive alternative to amine processes. The CO_2BOLs system mirrors a standard amine process with an absorption column and a stripper column to capture and release CO_2, respectively. The CO_2BOLs solvent benefits from polarity-swing-assisted regeneration (PSAR) to lower the energy penalty when compared to monoethanolamine (MEA) systems. The PSAR design lowers the temperature in the stripper's reboiler to reduce the required heat duty[4]. More details about the CO_2BOLs system can be found in the literature [3, 5-13].

Sub-Models

A sub-model is any model that feeds into a larger model. Sub-models can be nested as is the case with viscosity feeding into mass transfer feeding into the large-scale model. The Aspen Plus® model of the CO_2BOLs process involves several sub-models with specific parameters chosen for the analysis of the CO_2BOLs PSAR system. The analyzed models are discussed in this section.

Thermodynamics

The Non-Random Two-Liquid Electrolyte (ELECNRTL) Equation of State (EOS) predicts vapor-liquid equilibrium (VLE) behavior for this model. PTx data were provided for validation of this model. PTx data use the system temperature and component mole fractions to determine the pressure exerted by the mixture. Key parameters (highlighted in bold in Equations 1-5) were selected for UQ based on sensitivity and available data. Well-characterized parameters, such as pure-component parameters for water, kept their original values in this study. Thermodynamic consistency was maintained throughout the analysis by keeping enthalpy of formation, heat capacity, and chemical equilibrium parameters constant. In these equations γ_i is the activity coefficient, H_{ia} is the Henry's constant, P_i^{vap} is the vapor pressure, X_i is the mole fraction, and $\alpha_{i,j}$ is a fitting parameter for the binary interaction of species i and j. Equation 1 is the standard NRTL EOS for mixtures containing only molecular components. More terms are added when ions appear in the mixture.

Equation 1

$$\ln \gamma_i = \frac{\sum_j X_j G_{j,i} \tau_{j,i}}{\sum_k X_k G_{k,i}} + \sum_j \frac{G_{i,j} X_j}{\sum_k G_{k,j} X_k} \cdot \left(\tau_{i,j} - \frac{\sum_m X_m G_{m,j} \tau_{m,j}}{\sum_k X_k G_{k,j}} \right)$$

Equation 2

$$\ln H_{ia} = \boldsymbol{A_{ia}} + \frac{\boldsymbol{B_{ia}}}{T}$$

Equation 3

$$\tau_{i,j} = \boldsymbol{A_{i,j}} + \frac{\boldsymbol{B_{i,j}}}{T}$$

Equation 4

$$\ln P_i^{vap} = \boldsymbol{A_i} + \frac{\boldsymbol{B_i}}{T + \boldsymbol{C_i}}$$

Equation 5

$$G_{i,j} = e^{(-\alpha_{i,j}\tau_{i,j})}$$

Mass Transfer

Mass transfer relies on empirically calculated dimensionless numbers in addition to stream properties to calculate the mass flux from the vapor to the liquid stream. The key parameter (highlighted in bold) in this equation was used for UQ. Equation 6 is based on the Billet and Schultes equation[14]. k_G is the mass transfer coefficient and C_G is a fitting parameter.

Equation 6

$$k_G = D_G \boldsymbol{C_G} \left(\frac{a}{d_H}\right)^{0.5} Sc_G^{\ 0.333} Re_G^{0.75} \sqrt{\frac{1}{\varepsilon - h_L}}$$

Viscosity

The equations below calculate the viscosities of individual components as well as mixtures. Additional components not listed were added to Equation 10 using the standard Andrade mixing model. The key parameters (highlighted in bold) are used in UQ. In these equations μ is viscosity, T is temperature, MW is the molecular weight of a component, and α is the ratio of a component to the solvent.

Equation 7

$$\ln \mu_{BOLs} = \boldsymbol{A_{BOLs}} + \frac{\boldsymbol{B_{BOLs}}}{T} + \boldsymbol{C_{BOLs}} \cdot \ln T$$

Equation 8

$$w_{H2O} = \frac{\alpha_{H2O} \cdot MW_{H2O}}{\alpha_{H2O} \cdot MW_{H2O} + MW_{BOL}}$$

Equation 9

$$\mu_{BOLs-H2O} = \mu_{BOLs} \cdot (1 - w_{H2O}) + \mu_{H2O} \cdot w_{H2O} + \boldsymbol{D_{Binary}} \cdot w_{H2O} \cdot (1 - w_{H2O})$$

Equation 10

$$\mu_{BOLs-CO2-H2O} = \mu_{BOLs-H2O} \cdot \mathrm{e}^{\alpha_{CO2} \cdot \boldsymbol{E_{CO2}}}$$

Kinetics

Chemical kinetics equation parameters (highlighted in bold) were chosen to maintain thermodynamic consistency while still allowing UQ analysis. Each of the reaction rates corresponds to a single reaction mechanism as shown above the equation. In these equations r is the reaction rate, a is the component activity defined as the product of the mole fraction and the activity coefficient, T^{ref} is 313.15 Kelvin, and K is the chemical equilibrium constant. Other terms are taken from the standard Arrhenius equation.

$$2CO_2 + 2BOL \leftrightarrow \{BOL - CD^+\} + \{BOL - CD^-\}$$

Equation 11

$$r_1 = \boldsymbol{k_1} \exp\left(-\frac{\boldsymbol{E_1}}{R}\left(\frac{1}{T} - \frac{1}{T^{ref}}\right)\right) a_{CO_2} a_{BOL} \left(1 - \frac{a_{BOL-CD^+} a_{BOL-CD^-}}{(a_{CO_2} a_{BOL})^2 K_1}\right)$$

$$CO_2 + BOL + H_2O \leftrightarrow \{BOLH^+\} + \{HCO_3^-\}$$

Equation 12

$$r_2 = \boldsymbol{k_2} \exp\left(-\frac{\boldsymbol{E_2}}{R}\left(\frac{1}{T} - \frac{1}{T^{ref}}\right)\right) a_{CO_2} a_{BOL} a_{H_2O} \left(1 - \frac{a_{BOLH^+} a_{HCO_3^-}}{K_2}\right)$$

Uncertainty Quantification

Uncertainty quantification (UQ) is a statistical framework used to better understand uncertainties as well as data gaps in models. By constraining models to data, distributions of model parameters are estimated and then propagated through the full-scale model to obtain distributions of key outputs such as carbon capture percent, energy penalty, etc.

Standard Practice: Least-Squares Fit

For most new systems standard practice minimizes the sum of squared errors between model and data values for each sub-model to obtain fitting parameters. Additionally, expert knowledge often provides values for parameters that have insufficient data to regress. The overall result is a fixed set of parameters for the model. These parameters provide one prediction for each data point and one overall prediction for energy penalty and carbon capture percentage. This approach effectively assumes that the model is completely accurate in its point estimate of all properties and results. A sensitivity analysis is typically performed on these systems to demonstrate robustness, but these studies typically focus on model inputs rather than model parameters. This provides a good sense of how the system is predicted to react to changes in its inputs, but it does little to show how model results will change if parameter values are inaccurate.

Uncertainty Quantification

UQ progresses through each stage of the model with several steps. For simplicity the process is divided into 7 steps listed below.

1. For each sub-model, identify the equations, experimental data, inputs (x), outputs (y), and model parameters (θ) relevant for UQ analysis.

For each sub-model, the model form needs to be known with its associated parameters. The inputs and outputs for the model should match those of the data that will be used for calibration. For example, Pxy data would have temperature and feed concentrations as the input and pressure, vapor phase composition, and liquid phase composition as the outputs with vapor pressure and binary interaction parameters as the parameters. Parameter down-selection occurs during this step. Aspen Plus' property package has hundreds of default parameters for well-characterized components that can often be left at default values for UQ because no additional data were obtained. Parameters from sub-models with no supporting data would not benefit from UQ and are also left at original values. Parameters which affect thermodynamic consistency should be paired or left at original values to ensure that consistency remains during UQ. This process reduces the parameter count from several hundred to tens of parameters, which focuses UQ efforts.

2. Obtain physically reasonable ranges for inputs and model parameters. These may have to be calculated from other quantities.

Expert judgment and system knowledge determine physically reasonable ranges for the inputs and model parameters. If reasonable ranges for a property are known instead of individual parameters then the parameter values are determined using that metric instead. In the CO_2BOLs system, the latter was performed for vapor pressure to obtain parameter ranges based on property ranges.

3. Identify prior distributions for model parameters.

Prior distributions are identified only if enough information exists to justify their use. These prior distributions come from expert judgement, literature sources, or a sub-model analysis. Otherwise a uniform distribution is assumed over the ranges determined in the previous step.

4. Using an emulator (surrogate model)

If the model computes rapidly then it is used directly for the UQ calibration. If the model is more complex, such as a full-scale Aspen model, and requires seconds or minutes to compute, then an emulator should be used instead of the model. Steps for this process are outlined as follows:

4.1 Develop a "space-filling" design over inputs/parameters that will be used to train the emulator (surrogate model).

A space-filling design is first developed over the input and parameter space that covers the data and simulation ranges. This is typically done with Latin hypercube sampling[15-17] (LHS) over the viable ranges for the inputs and parameters.

4.2 Run model (Aspen) at design locations to obtain desired outputs that are comparable to experimental data. Reconsider design space if there are too many failures.

The space-filling design runs through the model and results are saved. If too many combinations result in failures then the design space is reevaluated. Parameter distributions could simply be narrowed or more stringent measures may need to be taken such as analyzing parameter interactions to jointly narrow distributions.

4.3 Development of the emulator for the model.

The emulator[18, 19] trains on the results from the actual model. A random 90% of the simulation results train the emulator while the remaining 10% validate the emulator's accuracy. R^2 and root-mean-squared values determine if the emulator is consistent with the model output. If the emulator does not predict the remaining 10% of the data, another training set consisting of 90% of the simulation results is randomly selected, and the training and validation procedures are repeated. More model runs may be necessary if the emulator fails validation too many times.

5. UQ analysis: calibrate the model to experimental data to compute posterior distributions of model parameters. Involved in this step is the development of the emulator for the model.

The model calibrates[20-22] to the experimental data as part of the uncertainty quantification process which uses Bayesian inference. The result is a set of posterior distributions for model parameters that indicate the likelihood of the parameters to be a specific value. A broad range means that the parameter is fairly insensitive and that a broad range of parameter values fit the data well. A narrow band indicates that the parameter is sensitive and only a narrow range of parameters fit the data well.

6. Get predictions with uncertainty of outputs.

Using the parameter distributions from the calibration step, the model predicts the data with uncertainty bounds. This helps determine which models suffer from poor data and can be used as the basis of a new experimental design.

7. Propagate posterior distributions to next stage if needed.

If the first-stage models are sub-models of a larger system then the posterior distributions are propagated through the large system to determine the uncertainty in overall model predictions. For the CO_2BOLs system the full-scale carbon capture model was the large system and uncertainty was predicted for the carbon capture percent.

Results

Parameter Distributions

Most parameters start with a uniform distribution as shown in Figure 1. The UQ procedure, previously outlined, determines the distribution that results in the best fit to data for each parameter. These are called posterior distributions and are shown in Figure 2.

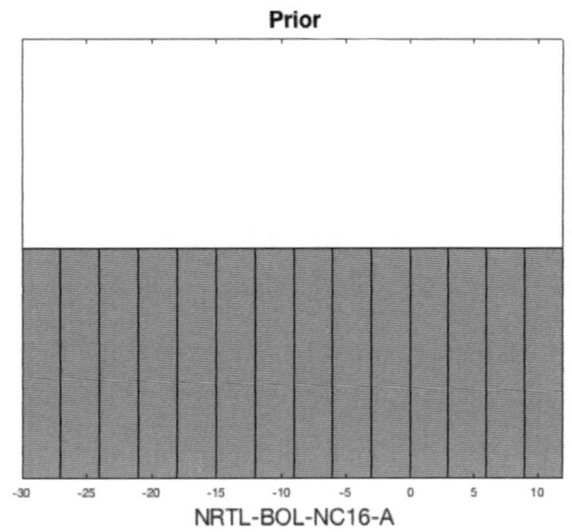

Figure 1 Uniform Prior Distribution for an NRTL Binary Interaction Parameter

Figure 2 shows univariate distributions, distributions for only one parameter without respect for any other. In Figures 2-3, Elec denotes a binary electrolyte parameter, NRTL denotes an NRTL binary interaction parameter (A and B corresponding to the same letters in Equation 3 and C corresponding to α in Equation 5), Viscosity denotes a viscosity parameter from Equations 7-10, and Vapor Pressure denotes parameters of the same letter in Equation 4. Most of these distributions are still centered on the base parameter value, but several have distributions that suggest the model will provide a better data fit by adjusting the parameter to a different value. While these parameters have fairly narrow distributions, others had fairly wide distributions indicating data were lacking, the parameter is insensitive, or the parameter is redundant.

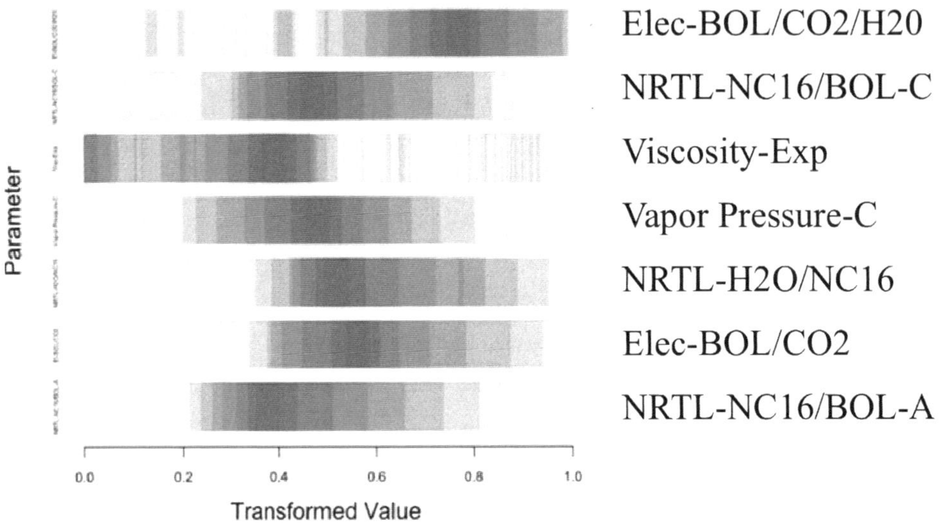

Figure 2 Posterior Distributions for 7 Key Parameters. Red indicates high probability for parameter value. 0.5 indicates the parameter's base value with 0.0 being the lowest value and 1.0 being the highest value in the original uniform distribution.

Figure 3 shows bivariate distributions, distribution densities that show how the distribution of one is affected by the value of another parameter, with univariate distributions on the diagonal. The dark regions indicate parameter combinations with high likelihood. These distributions, using bivariate distributions of several parameters for the sake of visualization, are derived from an nth-order multivariate distribution, where n is the total number of parameters in the full distribution, which is used to determine how all parameter values affect each other. This is especially useful for parameters that are part of a set of correlated parameters, such as binary interaction parameters, Henry's constant parameters, and viscosity parameters.

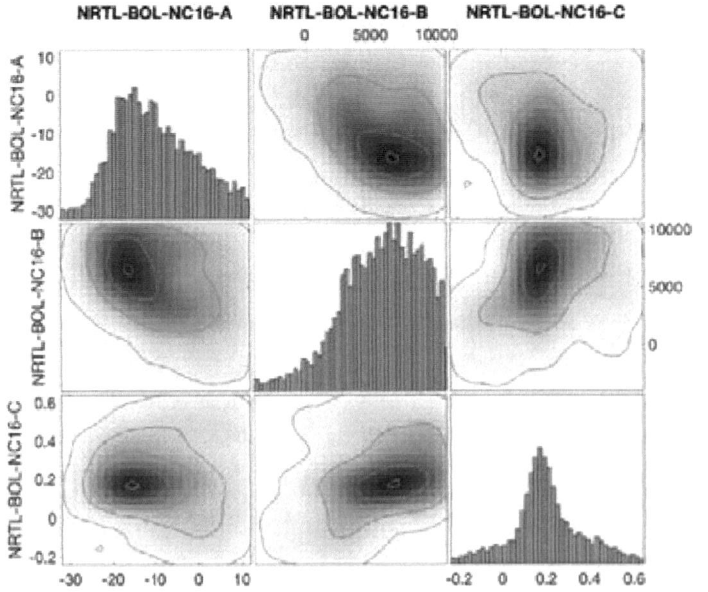

Figure 3 Bivariate Posterior Distributions for 3 Parameters

Sub-Model Results

Parameter distributions from the model calibration are used to determine model uncertainty for the individual sub-models. They are also used to determine a discrepancy function that can improve model accuracy. Figure 4 shows how including the discrepancy term can shrink the 95% confidence interval. This discrepancy function can also improve the full-scale model by modifying the sub-model equations to incorporate discrepancy. In Aspen Plus this would be the equivalent of specifying a FORTRAN subroutine to replace a property sub-model. It can also help determine a better sub-model form. Figure 5 shows model results with discrepancy for part of the thermodynamic sub-model.

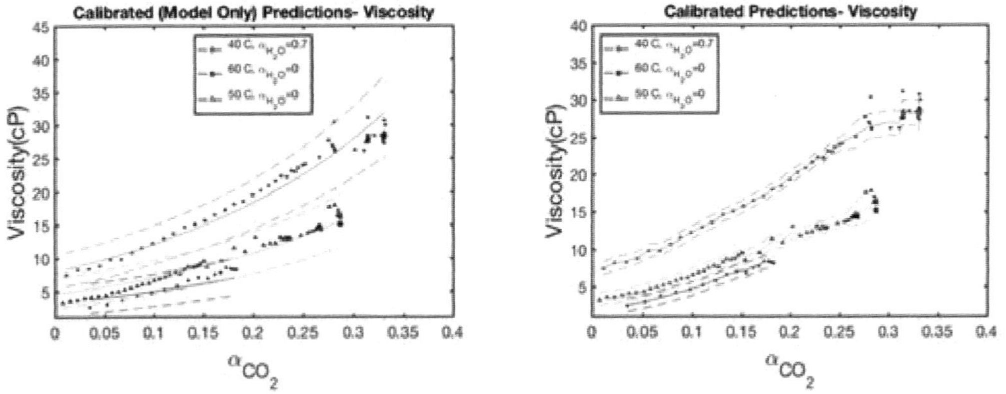

Figure 4 Calibrated Predictions for the Viscosity Sub-Model. Left: Model-Only Results (Model-Fit Uses Average Temperature) Right: Model with Discrepancy Term (Model-Fit Uses Temperature Taken from Each Data Point). Model is Solid Line and Dashed Line is 95% Confidence Interval.

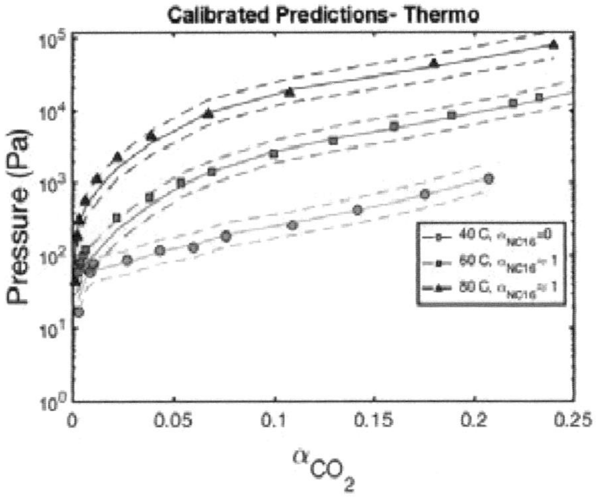

Figure 5 Predictions with Discrepancy Term (solid line), 95% uncertainty bounds for predictions (dashed line), PTx data (points) for CO2/BOL/NC16 system

Full-Scale Model Results

When analysis of the parameters and sub-models finishes, the parameter distributions are pushed through the full-scale model to determine the uncertainty in full-scale model predictions. For the CO_2BOLs system, the full-scale model predicts 90% capture of CO_2 under normal operating conditions. Figure 6 shows the resulting distribution in CO_2 capture percent when the model parameters are varied according to their posterior distributions with the same operating conditions as the base model; no discrepancy terms are included here. It is unsurprising that a novel solvent with limited data has a wide distribution in full-scale results. However, the results are still centered about 90% capture with some parameter values actually increasing the predicted percentage of CO_2 captured by the system under these operating conditions.

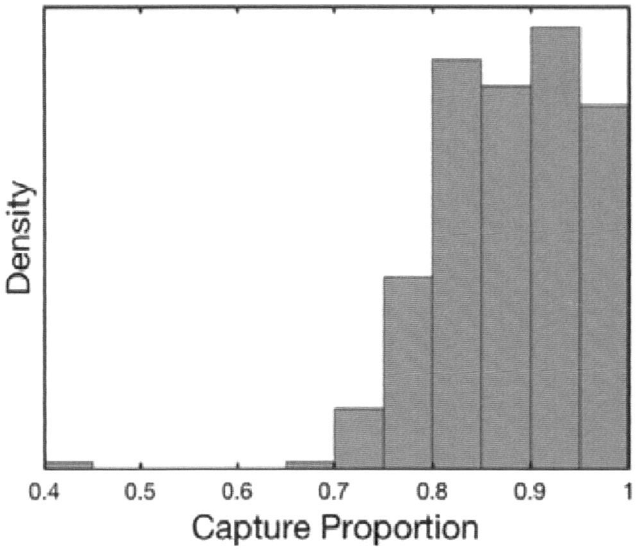

Figure 6 Carbon Capture Percent Distribution from Model Calibration Results

Conclusion

Uncertainty quantification is a powerful tool that provides valuable information about new models and their underlying data. This study has shown the steps to perform uncertainty quantification for the novel CO_2BOLs system. Novel solvent systems, like the CO_2BOLs system, often lack sufficient data to accurately model a full-scale simulation. However, model improvements require additional useful data. UQ results can help guide new data acquisition by determining regions of high uncertainty that would benefit from additional data. This can save time, effort, and money in achieving full-scale model results and can bring a project towards larger-scale experiments more quickly.

Acknowledgements

The authors acknowledge support from the Center for Non-Linear Studies at Los Alamos National Lab (LANL) and from the U. S. Dept. of Energy (DOE) through the Carbon Capture Simulation for Industry Impact (CCSI2) project. LANL is operated by LANS, LLC for the NNSA of the U.S. DOE under Contract No. DE-AC52-06NA25396.

Disclaimer

1. U. S. E. I. Administration, *2017 International Energy Outlook*, U.S. Energy Information Administration, 2017.
2. NETL, Post-Combustion CO2 Capture, https://www.netl.doe.gov/research/coal/carbon-capture/post-combustion, (accessed 8/23/2018, 2018).
3. D. J. Heldebrant, P. K. Koech, V. A. Glezakou, R. Rousseau, D. Malhotra and D. C. Cantu, *Chem Rev*, 2017, **117**, 9594-9624.
4. F. Zheng, D. J. Heldebrant, P. M. Mathias, P. Koech, M. Bhakta, C. J. Freeman, M. D. Bearden and A. Zwoster, *Energy & Fuels*, 2016, DOI: 10.1021/acs.energyfuels.5b02437.
5. D. C. Cantu, D. Malhotra, P. K. Koech, D. J. Heldebrant, F. Zheng, C. J. Freeman, R. Rousseau and V.-A. Glezakou, *Green Chemistry*, 2016, **18**, 6004-6011.
6. D. J. Heldebrant, P. K. Koech, J. E. Rainbolt and F. Zheng, *Energy Procedia*, 2011, **4**, 216-223.
7. D. J. Heldebrant, P. K. Koech, J. E. Rainbolt, F. Zheng, T. Smurthwaite, C. J. Freeman, M. Oss and I. Leito, *Chemical Engineering Journal*, 2011, **171**, 794-800.
8. D. J. Heldebrant, P. K. Koech, R. Rousseau, V.-A. Glezakou, D. Cantu, D. Malhotra, F. Zheng, G. Whyatt, C. J. Freeman and M. D. Bearden, *Energy Procedia*, 2017, **114**, 756-763.
9. P. M. Mathias, K. Afshar, F. Zheng, M. D. Bearden, C. J. Freeman, T. Andrea, P. K. Koech, I. Kutnyakov, A. Zwoster, A. R. Smith, P. G. Jessop, O. G. Nik and D. J. Heldebrant, *Energy & Environmental Science*, 2013, **6**.
10. P. M. Mathias, F. Zheng, D. J. Heldebrant, A. Zwoster, G. Whyatt, C. M. Freeman, M. D. Bearden and P. Koech, *ChemSusChem*, 2015, **8**, 3617-3625.
11. G. A. Whyatt, C. J. Freeman, A. Zwoster and D. J. Heldebrant, *Industrial & Engineering Chemistry Research*, 2016, **55**, 4720-4725.
12. G. A. Whyatt, A. Zwoster, F. Zheng, R. J. Perry, B. R. Wood, I. Spiry, C. J. Freeman and D. J. Heldebrant, *Industrial & Engineering Chemistry Research*, 2017, **56**, 4830-4836.
13. J. Zhang, I. Kutnyakov, P. K. Koech, A. Zwoster, C. Howard, F. Zheng, C. J. Freeman and D. J. Heldebrant, *Energy Procedia*, 2013, **37**, 285-291.
14. R. Billet and M. Schultes, *Trans. IChemE.*, 1999.
15. S. Leary, A. Bhaskar and A. Keane, *Journal of Applied Statistics*, **30**, 585-598.
16. M. D. McKay, R. J. Beckman and W. J. Conover, *Technometrics*, **21**, 239-245.
17. B. Tang, *Journal of the American Statistical Association*, **88**, 1392-1397.
18. J. Sacks, W. J. Welch, T. J. Mitchell and H. P. Wynn, *Statist. Sci.*, 1989, **4**, 409-423.
19. W. J. Welch, R. J. Buck, J. Sacks, H. P. Wynn, T. J. Mitchell and M. D. Morris, *Technometrics*, 1992, **34**, 15-25.
20. D. Higdon, J. Gattiker, B. Williams and M. Rightley, *Journal of the American Statistical Association*, 2008, **103**, 570-583.
21. M. C. Kennedy and A. O'Hagan, *Journal Of The Royal Statistical Society Series B-Statistical Methodology*, **63**, 425-450.
22. C. B. Storlie, W. A. Lane, E. M. Ryan, J. R. Gattiker and D. M. Higdon, *Journal of the American Statistical Association*, 2015, **110**, 68-82.

535085 Evaluating Methane Pyrolysis As a Component for CO_2 Emission Negative Renewable Energy

Abstract:

The majority of countries agreed that world wide net CO_2 emission must be zero in 2050. Based on the current emission trajectories, most climate researchers agree that CO_2 must also be actively removed from the atmosphere. Therefore, the future requires both a completely regenerative energy sector and measures to remove CO_2 from the atmosphere.

For the latter, several methods have been suggested and some of them are being evaluated in pilot scale. However, most methods have substantial obstacles, are of experimental nature and often have a substantial energy demand. As a consequence, investment and operation costs are high. Especially demanding and energy consuming is the irreversible storage of CO_2, e.g. by transferring highly compressed gas into deep earth layers.

In this study biologically removed CO_2 from the atmosphere is the starting point. Biomass is transformed into the easily storable and transportable energy carrier methane, which in turn could be pyrolysed into depositable carbon and hydrogen retaining parts of the energy. In principle, this could be achieved by a process encompassing de-centralized biogas production, followed by feeding of pre-treated methane into the existing gas grid. Methane pyrolysis can be performed using a newly developed energy efficient moving bed reactors and translation into electric energy in gas power plants. Ideally, biogas is produced continuously, whereas gas power plants are operated in periods, namely winter, where less renewable energy is available and off-heat can be utilized optimally. For electricity production, existing coal power plants might be re-constructed and the coal infrastructure might be employed for depositing the produced carbon.

The process is analysed for ecological (net CO_2 removal), energetical (net energy production) and economical (viable investment and operating costs compared to other methods) performance and the results are presented.

536584 Development of an Integrated Mass Transfer and Kinetic Model from Multi-Scale Data for CO_2 Capture Using Concentrated Piperazine

Abstract:

Process modeling and simulation with rigorous, validated models is critical for accelerating the development and deployment of CO_2 capture technologies while simultaneously reducing the risk in process scale-up [1,2]. The U.S. DOE's Carbon Capture Simulation for Industry Impact (CCSI^2) program is applying the Carbon Capture Simulation Initiative (CCSI) Toolset to accelerate the development of carbon capture technologies. CCSI developed a multiscale, simultaneous regression approach for mass transfer and kinetic parameters resulting in a highly accurate fundamental process model of CO_2 absorption into MEA solvent [3,4]. Concentrated piperazine (PZ) solution has been proposed as potential alternative solvent with improved performance due to faster kinetics, higher capacity, and a lower degradation rate compared with MEA [5]. In this work, an integrated mass transfer and kinetic model the PZ solvent will be presented.

Rigorous rate-based process models consist of several submodels with varying complexity. These models must accurately represent mass and energy balances of the phases, non-ideal thermodynamics, submodels for packing hydraulics, mass and heat transfer coefficients, and physico-chemical properties. Moreover, solvent based post-combustion CO_2 absorption modeling is complex due to the coupled effects of thermodynamics, mass transfer, and reactions. It is therefore imperative that the gas-liquid interface model accounts for complex chemical reactions.

To facilitate the absorption process, packed columns are often used for gas-liquid contacting. Several empirical correlations have been developed for the hydrodynamics and mass transfer coefficients in these packed columns; however, most of these packing models were developed using data based on simplified experiments of different systems and, in certain cases, relied on assumptions not generally valid for CO_2 capture technologies [6]. In general, the development of these models is done sequentially whereas in the real chemical solvent based CO_2 capture process, it is difficult to separate the mass transfer and chemical reaction phenomena due to fast reaction kinetics. In addition, data is often collected at different scales and different operating regimes while using the following methodologies:

- Mass transfer coefficients and reaction kinetics are estimated from wetted wall column (WWC) experiments.
    1. The physical liquid film mass transfer coefficient is measured by disregarding the gas fim resistance from absorption rate data of CO_2 into water.
    2. The gas side film mass transfer coefficient is measured using the instanteneous irreverasible reaction system, i.e., from absorption data of SO_2 into aqueous NaOH solution.
- Assuming the mass transfer coefficient models to be same, the experiments from packed columns are

then used to obtain an interfacial area model.

The sequential development of these submodels results in the uncertainty for each submodel being fixed and carried to the next model. It was shown that there is a large uncertainty associated with applying these correlations for CO_2 capture applications [7,8].

To minimize the total combined uncertainty, a simultaneous parameter regression of the mass transfer coefficients, pressure drop, interfacial area, and kinetic reaction models are performed using multi-scale experimental data from literature for pilot-scale packed columns and lab-scale wetted wall columns experiments. This simulatneous regression methodology was shown to be more aacurate for MEA solvent than the sequential regression methodology enabling resulting models to accurately predict process performance at different scales (pilot and semi-industrial) [2,4,9,10].

Concentrated piperazine with advanced process configurations was tested at the pilot scale and was found to be less energy intesive than MEA [11–13]. The integrated mass transfer, hydrodynamic, and kinetic models for concentrated piperazine solvent using the simultaneous regression is accomlished through the use of the Framework for Optimization Quantification of Uncertainty and Surrogates (FOQUS) toolset [14,15] linked with Aspen Plus. The developed model will be validated against independent experimental data [16–18].

References

[1] Putta KR, Svendsen HF, Knuutila HK. CO_2 absorption into loaded aqueous MEA solutions: Impact of different model parameter correlations and thermodynamic models on the absorption rate model predictions. Chem Eng J 2017;327:868–80. doi:10.1016/j.cej.2017.06.134.

[2] Chinen AS, Morgan JC, Omell B, Bhattacharyya D, Tong C, Miller DC. Development of a Gold-Standard Model for Solvent-Based CO2 Capture. Part 1: Hydraulic and Mass Transfer Models and Their Uncertainty Quantification. Ind Eng Chem Res 2018;submitted.

[3] Miller DC, Syamlal M, Mebane DS, Storlie C, Bhattacharyya D, Sahinidis NV, et al. Carbon Capture Simulation Initiative: A Case Study in Multiscale Modeling and New Challenges. Annu Rev Chem Biomol Eng 2014;5:301–23. doi:10.1146/annurev-chembioeng-060713-040321.

[4] Chinen AS, Morgan JC, Omell B, Bhattacharyya D, Miller DC, Wheeldon J. Predictive dynamic model of a carbon capture system: Pilot scale validation at national carbon capture center. vol. 1, 2015, p. 52–67.

[5] Rochelle G, Chen E, Freeman S, Van Wagener D, Xu Q, Voice A. Aqueous piperazine as the new standard for CO2 capture technology. Chem Eng J 2011;171:725–33. doi:10.1016/j.cej.2011.02.011.

[6] Razi N, Bolland O, Svendsen H. Review of design correlations for CO2 absorption into MEA using structured packings. Int J Greenh Gas Control 2012;9:193–219. doi:10.1016/j.ijggc.2012.03.003.

[7] Razi N, Svendsen HF, Bolland O. The Impact of Design Correlations on Rate-based Modeling of a Large Scale CO2 Capture with MEA. Energy Procedia 2013;37:1977–86. doi:10.1016/j.egypro.2013.06.078.

[8] Razi N, Svendsen HF, Bolland O. Assessment of mass transfer correlations in rate-based modeling of a large-scale CO2 capture with MEA. Int J Greenh Gas Control 2014;26:93–108. doi:10.1016/j.ijggc.2014.04.019.

[9] Morgan JC, Chinen AS, Omell B, Bhattacharyya D, Tong C, Miller DC. Thermodynamic modeling and uncertainty quantification of CO 2 -loaded aqueous MEA solutions. Chem Eng Sci 2017;168:309–24. doi:10.1016/j.ces.2017.04.049.

[10] Morgan JC, Chinen AS, Omell B, Bhattacharyya D, Tong C, Miller DC, et al. Development of a Gold-Standard Model for Solvent-Based CO2 Capture. Part 2: Steady-State Validation and Uncertainty Quantification with Pilot Plant Data. Ind Eng Chem Res 2018;submitted.

[11] Cousins A, Huang S, Cottrell A, Feron PHM, Chen E, Rochelle GT. Pilot-scale parametric evaluation of concentrated piperazine for CO_2 capture at an Australian coal-fired power station. Greenh Gases Sci Technol 2015;5:7–16. doi:10.1002/ghg.1462.

[12] Rabensteiner M, Kinger G, Koller M, Gronald G, Hochenauer C. Investigation of carbon dioxide capture with aqueous piperazine on a post combustion pilot plant–Part I: Energetic review of the process. Int J Greenh Gas Control 2015;39:79–90. doi:10.1016/j.ijggc.2015.05.003.

[13] Lin Y-J, Chen E, Rochelle GT. Pilot plant test of the advanced flash stripper for CO2 capture. Faraday Discuss 2016;192:37–58. doi:10.1039/C6FD00029K.

[14] Eslick JC, Ng B, Gao Q, Tong CH, Sahinidis NV, Miller DC. A Framework for Optimization and Quantification of Uncertainty and Sensitivity for Developing Carbon Capture Systems. Energy Procedia 2014;63:1055–63. doi:10.1016/j.egypro.2014.11.113.

[15] Miller DC, Eslick J, Boverhof J, Leek J, Tong C, Chen Y, et al. FOQUS: A Framework for Optimization and Quantification of Uncertainty 2015.

[16] Chen E, Madan T, Sachde D, Walters MS, Nielsen P, Rochelle GT. Pilot Plant Results with Piperazine. Energy Procedia 2013;37:1572–83. doi:10.1016/j.egypro.2013.06.033.

[17] Chen E, Fulk S, Sache D, Lin Y, Rochelle GT. Pilot Plant Activities with Concentrated Piperazine. Energy Procedia 2014;63:1376–91. doi:10.1016/j.egypro.2014.11.147.

[18] Chen E, Zhang Y, Lin Y, Nielsenb P, Rochelle G. Review of Recent Pilot Plant Activities with Concentrated Piperazine. Energy Procedia 2017;114:1110–27. doi:10.1016/j.egypro.2017.03.1266.

538304 Morphology of Sodium Carbonate and Its CO_2 Adsorption Performance: A First-Principles Investigation

Abstract:

Sodium carbonate (Na_2CO_3) is a promising candidate for CO_2 capture from flue gas. The morphology of sodium carbonate and co-adsorption of CO_2 and H_2O on several low-index surfaces and step edges are investigated by density functional theory (DFT) methods. We have identified six different bulk structures that are a few meV per formula unit lower in energy than the optimized experimental structure of Na_2CO_3. These structures mainly differ from one another in the orientation of the carbonate groups. This means that there are many nearly degenerate structures available to Na_2CO_3 because of rotation of the carbonate groups. The ascending order of surface energy is (101) < (001) < (010) < (111) < (011) < (100) < (110) and the crystal shape of sodium carbonate is predicted by performing a Wulff construction. We find only Na-termination surface is likely to have significant presence by taking a Boltzmann distribution of the energies of surfaces having different terminations. We also find that CO_2 and H_2O adsorb competitively on the surface O sites. The calculated adsorption energies depend significantly on the surface structure, which indicates that the CO_2 adsorption performance of Na_2CO_3 can be tuned by exposing different surface facets. Reaction barriers for bicarbonate formation on terrace, step edge, and defect sites will be discussed.

538795 Development of Nano-TiO_2 Promoted CaO Adsorbent for Enhanced CO_2 Capture at High Temperatures: Role of Crystal Level Properties

Abstract:

Calcium looping (CaL) is a promising regenerative process involves CO_2 capture from flue gas using multi-cycle carbonation and calcination at high temperature. In this work, we investigate the sorption capacity, and kinetics of nano-TiO_2 promoted CaO synthesized from the commercially available micron size calcium carbonate. The optimum wt. % of nano-TiO_2 and sorption temperature of the sorbents was explored. The morphology of the uncoated and coated $CaCO_3$ with different wt. % (1 to 10 wt. %) of nano-TiO_2 was analyzed by FESEM. The CO_2 sorption experiments were carried out in a Thermogravimetric analyzer under CO_2 atmosphere (0.02 MPa) at different temperatures (600°C, 650°C and 700°C). The crystallite size, strain and lattice parameters were estimated through Rietveld refinement of XRD data. The results demonstrate that the sorption capacity is enhanced several times for nano-TiO_2 promoted CaO than pure CaO obtained from micron and nano-size $CaCO_3$. The result shows that small amount of nano-TiO_2 provide porous structure, smaller crystallite size and high strain inside the CaO crystal during decomposition of $CaCO_3$, which improve the sorption capacity of CaO by 10 times in the first step. In general a substantial improvement is observed for subsequent cycles indicating nano-TiO_2 promoted CaO is better materials for CO_2 capture. The experimental results were fitted with theoretical models based on the shrinking core model and kinetic parameters were calculated and correlated with the crystal properties of CaO in presence of nano-TiO_2.

541778 Food-Energy-Water Nexus Systems Engineering

Abstract:

Due to population growth, economic development, urbanization, climate change, and natural resource degradation, the global demand for food, energy and clean/freshwater is rapidly increasing [1]. As demand grows the stresses and interdependences between these three resource systems, which are commonly referred to as Food-Energy-Water Nexus (FEW-N), are becoming more apparent [2,3]. Decision making can be very challenging due to multiple dimensions of the biophysical water, energy and food systems, and the players and stakeholders connected with them [4]. A holistic systems engineering approach is thus clearly needed to navigate the multi-faceted FEW-N space, identify opportunities for synergistic benefits and systematically explore interactions and trade-offs [5].

In this work, we present the foundations of a systems engineering framework and quantitative decision-making tools for the analysis and trade-off optimization of stressed interconnected resource FEW-N networks. The framework combines data analytics and mixed-integer modeling and optimization methods establishing the interdependencies and potentially competing interests amongst the food, energy and water elements in the system, along with policy, sustainability and feedback from the various stakeholders. A multi-objective optimization strategy is followed for the analysis of the trade-offs empowered by the introduction of composite FEW-N metrics as means to facilitate decision making and compare alternative process and technological options. The versatility, potential and applicability of the proposed framework is demonstrated by three nexus case studies, (i) a complex power generating system in the South-Central Texas region [6], (ii) a dairy production and processing plant [7], and (iii) a land use allocation problem on a regional farming system in China [8].

References

[1] D'Odorico, P., Davis, K. F., Rosa, L., Carr, J. A., Chiarelli, D., Dell'Angelo, J., ... & Rulli, M. C. (2018). The Global Foodâ€Energyâ€Water Nexus. Reviews of Geophysics.

[2] McCarl, B. A., Yang, Y., Schwabe, K., Engel, B. A., Mondal, A. H., Ringler, C., & Pistikopoulos, E. N. (2017). Model Use in WEF Nexus Analysis: a Review of Issues. *Current Sustainable/Renewable Energy Reports*, *4*(3), 144-152.

[3] McCarl, B. A., Yang, Y., Srinivasan, R., Pistikopoulos, E. N., & Mohtar, R. H. (2017). Data for WEF Nexus Analysis: a Review of Issues. *Current Sustainable/Renewable Energy Reports*, *4*(3), 137-143.

[4] Daher, B., Mohtar, R.H., Pistikopoulos, E.N., Portney, K.E., Kaiser, R. and Saad, W. (2018) Developing Socio-Techno-Economic-Political (STEP) Solutions for Addressing Resource Nexus Hotspots, Sustainability 2018, 10, 512; doi:10.3390/su10020512

[5] Garcia, D. J., & You, F. (2016). The water-energy-food nexus and process systems engineering: a new focus. *Computers & Chemical Engineering*, 91, 49-67.

[6] Diangelakis, N. A., & Pistikopoulos, E. N. (2017). A multi-scale energy systems engineering approach to residential combined heat and power systems. *Computers & Chemical Engineering*, 102, 128-138.

[7] Avraamidou, S., Milhorn, A., Sarwar, O., & Pistikopoulos, E. N. (2018). Towards a quantitative Food-Energy-Water Nexus metric to facilitate decision making in process systems: A case study on a dairy production plant. *28th European Symposium on Computer-Aided Process Engineering*: Accepted.

[8] Nie, Y., Avraamidou, S., Li, J., Xiao, X., & Pistikopoulos, E. N. (2018). Land use modeling and optimization based on food-energy-water nexus: a case study on crop-livestock systems. *13th International Symposium on Process Systems Engineering*: Accepted.

Renewable Carbons from Food Waste for Separation and Catalysis Technologies

This study explores the production of renewable and sustainable carbons from waste biomass resources, for adsorption and catalytic applications. Specifically, the research will explore the production of carbons derived from **food waste** and their utilization as sorbents of aromatic contaminants in **water** and catalyst supports for the production of clean **energy and fuels**.

The idea of using renewable and sustainable resources to replace the carbon derived from fossil fuels is not new. Researchers have been working on the utilization of various biomass resources, such as rice husk, coconut shells and others for years.[1–5] However, the utilization of food waste as a carbon resource has not been extensively explored. Food waste could be the ultimate resource for carbon production, since it is produced in excessive volumes and it represents one of the most significant environmental problems, which is urgently looking for a solution. The critical question is whether the activated carbons derived by food waste have the potential to be used for purification and/or catalysis purposes. Food waste is heterogeneous in nature and contains various types of minerals, which may provide advantages or disadvantages, depending on the application.

In our lab, we receive the food waste from the Dining Halls of the University of Connecticut and after pretreatment, that includes washing, grinding and sieving, we perform pyrolysis and activation to turn the waste into carbons. The produced carbons are then characterized and they are tested as sorbents and catalyst supports. Our goal is to understand how the conditions of pyrolysis and activation affect the properties of the carbons and, consequently, their effectiveness as sorbents and catalyst supports. During the pyrolysis reaction, temperature and residence time are the most important parameters that affect the yield and the properties of the produced biochar. Activation of the biochar is performed either using steam, physical activation, or using chemicals (e.g. H_3PO_4), chemical activation. During the former method of activation, temperature and residence time are, again, among the most important parameters that affect the surface area, the carbon content and the porosity of the produced carbons. During the latter method, the impregnation ratio between the chemical and the biochar is the most critical.

During this talk we will discuss the properties of the activated carbons produced by food waste and how these properties are affected by the reaction conditions and the method of activation. Surface area, porosity, carbon content, surface functional groups, morphologies are the properties of interest which will be discussed. Then, we will discuss the potential of using the activated carbons produced by food waste, as sorbents of aromatic hydrocarbons in water and catalyst supports for the hydrodeoxygenation of bio-oil.

References

(1) Tripathi, M.; Sahu, J. N.; Ganesan, P. Effect of Process Parameters on Production of Biochar from Biomass Waste through Pyrolysis: A Review. *Renew. Sustain. Energy Rev.* **2016**, *55*, 467–481.

(2) Kwapinski, W.; Byrne, C. M. P.; Kryachko, E.; Wolfram, P.; Adley, C.; Leahy, J. J.; Novotny, E. H.; Hayes, M. H. B. Biochar from Biomass and Waste. *Waste and Biomass Valorization* **2010**, *1* (2), 177–189.

(3) Ahmad, M.; Rajapaksha, A. U.; Lim, J. E.; Zhang, M.; Bolan, N.; Mohan, D.; Vithanage, M.; Lee, S. S.; Ok, Y. S. Biochar as a Sorbent for Contaminant Management in Soil and Water: A Review. *Chemosphere* **2014**, *99* (November 2013), 19–23.

(4) Tan, Z.; Lin, C. S. K.; Ji, X.; Rainey, T. J. Returning Biochar to Fields: A Review. *Appl. Soil Ecol.* **2017**, *116* (September 2016), 1–11.

(5) Azargohar, R.; Dalai, A. K. Biochar as a Precursor of Activated Carbon. *Appl. Biochem. Biotechnol.* **2006**, *131* (1–3), 762–773.

541779 A Computational Framework for Sustainable Waste Management and Simultaneous Recovery of Nutrients and Energy

Abstract:

To protect human health and the environment, communities should have sustainable infrastructures to manage waste. Critically, livestock waste may cause some air quality degradation from ammonia and methane emissions, soil quality detriment due to in-excess nutrients and acidification, and water pollution issues resulting from nutrient and pathogens runoff to water bodies, which leads to eutrophication, hypoxia, and harmful algal blooms (HABs).

In addition, the recovery of carbon from organic waste offers an alternative for replacing fossil fuels with renewable sources for energy production. Despite the significant environmental benefits by performing pollution management of these organic materials, the deployment of technologies for obtaining value-added products and renewable energy from livestock waste is not a current practice due to the high investment costs and low product market values.

This contribution describes a multi-scale approach for the development of a supply chain framework to identify and evaluate techniques and logistics for the management of solid waste generated from primary sources of nutrient pollution. This effort will inform stakeholdersâ€™ waste management options while assessing potential economic and environmental benefits. Other potential benefits include the reduction of nutrient pollution and HABs, recovery of valuable products, and adoption of nutrient-reducing technologies in different locations (e.g., community, state, regional, etc.). In addition, the proposed framework captures techno-economic and logistical issues and can accommodate diverse types of material management incentives obtained at federal and state levels.

537690 Density Functional Theory Insights into Corrosion and By-Product Formation Caused By Drinking Water Disinfection

Abstract:

Density Functional Theory insights into corrosion and by-product formation caused by drinking water disinfection

Currently, treatment plants most commonly disinfect drinking water with free chlorine to remove pathogens. Free chlorine, however, is prone to forming carcinogenic disinfection by-products from reactions with organic matter. Ill-considered application of disinfectants and complications from cleanup of their by-products, recently resulted in catastrophic corrosion and discharge of lead into the water supplies of Washington D.C. and Flint, Michigan. To avoid a recurrence of such mistakes, the strategy for drinking water disinfection in vulnerable, lead-containing supply networks was reconsidered. Specifically, we evaluated the merits and drawbacks of replacing free chlorine with chlorine dioxide in systems with lead infrastructure.

Free chlorine, the most commonly used disinfectant in drinking water treatment, is a powerful oxidant that readily reacts with viral and bacterial pathogens, but also with other water constituents, including dissolved organic matter and metallic infrastructure surfaces. Reactions with organic matter produce hazardous by-products, while reactions with infrastructure surfaces can increase the concentration of metals, including lead, in drinking water. Some water treatment facilities have substituted free chlorine with chloramines, an alternative disinfectant, to meet disinfection by-product regulations set by the Environmental Protection Agency. However, chloramines produce their own set of toxic by-products and have lower disinfection efficiency than free chlorine. Removal of disinfection by-products and the precursors to said by-products can exacerbate corrosion in the metal pipes of water distribution systems.

Unlike free chlorine, chlorine dioxide has a significantly lower propensity to form organic by-products, and it is as potent, and in some cases, more potent than free chlorine for the inactivation of bacteria and viruses. In a properly managed dose, chlorine dioxide can deliver effective disinfection of drinking water[1]. Chlorine dioxide is more expensive to use for drinking water treatment than free chlorine and it can decay in water to form chlorite and chlorate, two potentially hazardous inorganic by-products. It has been assumed that chlorine dioxide will behave similarly to free chlorine with regards to lead corrosion mechanisms because of their similar oxidation reduction potentials, but there have been few studies to confirm.

Using three different minerals, a series of experiments were performed to measure the kinetics of chlorine dioxide decay and by-product formation, considering chlorine dioxide concentration, and pH. In the presence of lead minerals, chlorine dioxide does not form chlorate as a by-product. These results are curious as copper minerals were shown to produce chlorite and chlorate through a disproportionation reaction. The lead mineral reaction was also significantly faster than the copper reaction and produced only chlorite.

To explain the disparity in by-product formation between copper and lead minerals, computational chemistry was employed. Density Functional Theory was used to calculate activation energy barriers and minimum energy structures for both lead and copper systems. These calculations provided insights into reactant, products, intermediates, and transition structures to explain experimental findings.

For experimental work, spectrophotometry was used to standardized chlorine dioxide concentrations at 359nm. Ion chromatography, by a Dionex 5000 Reagent-Free ion chromatograph, was used to measure by-products (chlorite and chlorate) and specified anions. All experiments were batch reactions, conducted in the dark under constant agitation of a stir plate. Computational studies were preformed using the Vienna Ab Simulation Package and Gaussian. Simulations were done on solvated surfaces and in the gas phase. The influencing factors on the kinetics of chlorine dioxide decay and by-product formation in the presence of lead minerals will be presented and discussed. Our data will be contrasted with the results of published results on other metals including nickel, copper, and iron.

Aging lead infrastructure will continue to be an issue in the United States. The replacement of distribution systems could take decades and at high environmental, economic, and social costs due to our poor understanding of corrosion and by-product formation mechanisms. There is a demand for alternative disinfectants that have the capability to protect society from waterborne diseases without polluting our environment with lead and toxic by-products. This work is intended to provide guidance, on the viability of chlorine dioxide as a disinfectant, to avoid costs and dangers associated with lead corrosion.

References

(1) Gagnon, G. A.; O'Leary, K. C.; Volk, C. J.; Chauret, C.; Stover, L.; Andrews, R. C. *J. Environ. Eng.* **2004**, *130* (11), 1269–1279.

(2) Edwards, M.; Triantafyllidou, S. *J. / Am. Water Work. Assoc.* **2007**, *99* (7).

(3) Dodrill, D.; Edwards, M. *Am. Water Work. Assoc.* **1995**, No. C, 74–85.

(4) Liu, C.; Von Gunten, U.; Croue, J. P. *Environ. Sci. Technol.* **2013**, *47* (15), 8365–8372.

535908 Parameter Estimation of a Model of Advanced Oxidation Processes

Abstract:

In a previous work, a mathematical model of advanced oxidation processes (AOPs) was developed.[1] It was based on the mole balance equations of species and the kinetic information offered by the reaction mechanism. The model described the treatment of ciprofloxacin (CIP) in aqueous solution using the ozonation process. Intermediate compounds of CIP degradation were neglected, and the key model parameters were the volumetric mass transfer coefficient in the liquid phase ($k_L a$) and the liquid hold up (Î²).

The present work focusses on parameter estimation of the model describing the CIP oxidation considering the reactions of intermediate compounds and total organic carbon (TOC) with the ozone molecule and hydroxyl radical. These reactions were inserted in the reaction mechanism according to a possible degradation mechanism of CIP reported in the literature.[2] A sensitivity analysis was performed in order to evaluate the influence of kinetic constants on the model predictions. The kinetic constants were varied one by one and the predicted concentration profiles of CIP and ozone in the gas phase at the column outlet were analyzed. The model parameters $k_L a$, Î² and kinetic constants were estimated from available experimental data minimizing the differences between the experimental and predicted concentrations of CIP and ozone in the gas phase.[3] In addition, reparameterization for the kinetic constants was used. The optimization problem was solved using the fmincon solver of Matlab$^{Â®}$. Values for the initial guesses of the model parameters were taken from the literature, and upper and lower bounds for the parameters were defined.[4,5] In all cases, the accuracy of the model predictions was estimated by means of the root mean square error (RMSE).

Once the parameters were estimated, simulations of the process taking into consideration the experimental conditions were performed. The model was capable to fit adequately the experimental CIP and gas phase ozone concentrations. Process simulations with the estimated parameters considering the intermediates ($k_L a$, Î² and kinetic constants) were compared with simulations using estimated parameters ($k_L a$ and Î²) neglecting the reactions of intermediates and TOC. The comparison allowed to study the role of intermediate compounds. These compounds consume ozone affecting the ozonation process. When the intermediates were neglected, the model overestimated the ozone concentrations in the gas phase for all the experimental conditions. The model was not capable to fit adequately the ozone concentrations in the gas phase even working with estimated parameters from the available experimental data. It is necessary to take into consideration the reactions of intermediate compounds of CIP degradation and TOC in order to obtain a good agreement between the model predictions and the experimental data. The mathematical model of advanced oxidation processes with the estimated parameters can be used to test different experimental conditions and design experiments that lead to a maximization of degradation.

References

1.Abreu Zamora MA, Teixeira ACSC, Le Roux GAC. Modelling and simulation of advanced oxidation process: application to the treatment of ciprofloxacin in aqueous solution by ozonation process. *Comput. Aided Chem. Eng.* 2017; 40: 553-558. In: 27th European Symposium on Computer Aided Process Engineering.

2.De Witte B, Dewulf J, Demeestere K, Van Langenhove H. Ozonation and advanced oxidation by the peroxone process of ciprofloxacin in water. *J. Hazard. Mater*. 2009; 161(2-3):701-708.

3.Bardauil Baptistucci C. Degradação do antibiótico ciprofloxacina em solução aquosa por meio de processo oxidativo avançado baseado em ozônio. *Master's Thesis*. Chemical Engineering Department, University of São Paulo, 2012.

4.Beltrán FJ, González M, Acedo B, Rivas FJ. Kinetic modelling of aqueous atrazine ozonation processes in a continuous flow bubble contactor. *J. Hazard. Mater*. 2000; 80(1-3):189-206.

5.Dodd, MC, Buffle MO, Von Gunten U. Oxidation of antibacterial molecules by aqueous ozone: Moiety-specific reaction kinetics and application to ozone-based wastewater treatment. *Environ. Sci. Technol*. 2006; 40(6):1969-1977.

ACKNOWLEDGEMENT: Antonio Carlos S. C. Teixeira would like to express his acknowledgement to FAPESP for its financial support through grant number 2013/50218-2

538185 Kinetics and CFD Model Validation for Combustion of Coal Char Using Cu-Based Chemical Looping with Oxygen Uncoupling (CLOU) Carriers

Abstract:

The chemical-looping combustion with oxygen uncoupling (CLOU) process uses gaseous or solid hydrocarbon fuels and is a promising carbon capture and storage (CCS) technology. In CLOU, the uncoupling of gaseous O2 from the oxygen carrier CuO is thermodynamically favored at temperatures above ~ 1000Â° C and pressures less than ~ 0.1 bar O2. The CLOU carrier chosen for this study was the CuO/Cu2O oxygen carrier system, while the choice of fuel was coal char obtained from the pyrolysis of Powder River Basin (PRB) coal. The overall reduction rates of the CuO oxygen carriers were investigated using a drop-tube fluidized bed reactor in the temperature range of 850Â°C â€“ 1100Â°C. The intrinsic activation energy and Arrhenius prefactor for the CLOU reaction were determined from these experiments, along with the apparent activation energy and Arrhenius prefactor for the overall CuO reduction. In addition, a computational fluid dynamics (CFD) model was developed and validated for prediction of the overall CuO reduction kinetics based on the amount of PRB coal char charged to the reactor and the known kinetics of the individual CLOU, combustion, gasification, and CuO reduction reactions occurring within the system. The PRB coal char added to the reactor consumes the O2 uncoupled from the CuO, driving the thermodynamic equilibrium of the CLOU reaction in the direction of Cu2O product formation such that this reaction is now favored at temperatures as low as 900° C.

530065 Dependence of Photocatalytic Degradation Pathway on Surface Planes of a Catalyst: A Case of Diuron Degradation on ZnO

Abstract:

Heterogeneous photocatalytic reaction has been used as a technique for elimination of toxic contaminants in water. The process relies on oxidation of the adsorbed pollutant by highly active radicals, mainly hydroxyl radical, which are produced when a catalyst is activated by light having energy higher than the bandgap of the catalyst. The attack of the hydroxyl radical on an organic compound leads to its degradation toward total mineralization, but it usually involves several intermediates. Interestingly, the intermediates from the degradation of a compound reported in various papers are usually inconsistent. This discrepancy should be addressed so that appropriate catalyst can be designed, especially for the degradation of toxic compounds because some intermediates may be more toxic than their parent compounds.

In this work, the dependence of the photocatalytic degradation pathway on surface planes of the catalyst is demonstrated. Diuron [3-(3, 4-dichlorophenyl)-1, 1-dimethyl urea], which is one of the most highly persistent and highly toxic herbicides, was used as a model compound. Zinc oxide (ZnO) was chosen as the catalyst because it can be synthesized such that the dominating surface on a particle can be easily controlled. A hexagonal crystal of wurtzite ZnO is consisted of two main sets of surfaces; (i) zinc-terminated (0001) and oxygen-terminated (000-1) polar surfaces locating at the top and bottom planes of the crystal, and (ii) mixed-terminated (10-10) nonpolar surface as the side planes. Herein, two types of ZnO, i.e., conventional hexagonal particles with polar surfaces as the dominating planes, and nanorods with the mixed-terminated surfaces, were studied.

The conventional particles were prepared by the sol-gel technique at room temperature, after which the calcination at 500°C was performed. On the other hand, the nanorods were synthesized by the hydrothermal technique at 140°C. Zinc acetate was used as the precursor for both syntheses. Various characterization techniques, including X-ray diffraction (XRD), X-ray photoelectron spectroscopy (XPS), UV-Visible spectroscopy, selected area electron diffraction, and nitrogen adsorption/desorption, were performed to confirm that both catalysts are nonporous single-crystalline ZnO in wurtzite phase with similar surface compositions. Their optical bandgaps are the same at 3.1 eV. The major differences between these catalysts are their morphologies, sizes and specific surface areas. The conventional ZnO powders are micron-sized particles with (0001) or (000-1) plane, while the nanorods exposes mostly the (10-10) surface. Because of the much smaller particle size of the nanorods, their specific surface area is much higher than that of the conventional particles, i.e., 17 m^2/g versus 1.4 m^2/g.

The adsorption experiments were conducted in batch in the dark, using diuron aqueous solution with concentrations in the range of 0-25 mg/l. The Freundlich model was found to represent the experimental data well. The fitted parameter relating to adsorption intensity shows unambiguously that diuron adsorb more strongly on the conventional ZnO than on the ZnO nanorods. The diuron-surface interaction is suggested to be electrostatic. Interestingly, the greater amount of diuron was adsorbed on the conventional ZnO particles than on the nanorods, although the former has one-order of magnitude lower surface area than the latter.

Molecular calculations using density functional theory were applied to support the experimental findings. The adsorption configuration of diuron was found to be surface dependent. On the zinc-terminated surface, diuron adsorbs in planar configuration because both ends of diuron, which are negatively charged, are attracted to the positively charged surface zinc atoms at the same time. In case of the oxygen-terminated surface, diuron tilts its molecule almost side-way so that the electron-rich amide oxygen moves away from the negatively charged surface oxygen.

On the mixed-terminated surface of ZnO, molecule of diuron is bent so that the most negatively charged part of the molecule, i.e., amide oxygen and nitrogen attaching to the aromatic ring, could approach positively charged zinc atoms on the surface. At the same time, the aromatic end of the molecule is lifted from the surface because of the repulsion between chlorine atoms attaching to it and surface oxygen atoms. If zinc atom on the surface is defined as the adsorption site, it can be seen that only half of the surface atoms of the mixed-terminated surface are adsorption sites, unlike the polar surfaces where all atoms are able to be the adsorption sites. Furthermore, the adsorbate molecule that already adsorbed on the surface will present steric effect toward adsorption on nearby sites. This steric effect will become more severe toward the adsorption capacity on the mixed-terminated surface, where not all atoms are the adsorption sites. That explains why the adsorption capacity on the conventional ZnO particles is much higher than that on the nanorods.

The photocatalytic degradation experiments were conducted in a microreactor to eliminate mass transfer resistance for the transport of diuron from bulk liquid to the surface of the catalyst. The outlet stream from the reactor was collected and analyzed by liquid chromatography equipped with tandem mass spectroscopy (LC-MS/MS). For the degradation on the ZnO nanorods, the first set of intermediates detected using the shortest residence time (1 min), which are assumed to be the direct intermediates from the degradation of diuron, are formed by hydroxylation and demethylation at the methyl group. These reactions take place at the aliphatic moiety of diuron, which is the part of the molecule adsorbed on the mixed-terminated surface of the nanorods. On the other hand, on the conventional ZnO particles, the intermediates detected are the results from diuron being attacked by hydroxyl radicals on both aliphatic and aromatic sides simultaneously. As previously discussed, diuron adsorbs in a planar configuration on the zinc-terminated surface, while it turns perpendicularly to the oxygen-terminated surface. Both of these give rise to equal probabilities of attack by the hydroxyl radicals on either parts of diuron. The structure of the subsequent intermediates should therefore be consequently affected. Hence, we propose that the difference in pathways from using different catalysts is caused by the adsorption configurations of diuron onto the different surfaces.

Finally, the toxicity of the generated intermediates was investigated. Both the phytotoxicity and cytogenotoxicity were evaluated in comparison with diuron. The results show that the initial intermediates formed from diuron degradation on the ZnO nanorods are less toxic than those generated on the conventional ZnO particles. Nevertheless, with prolonged residence time, e.g., 15 min, the toxicity of the degradation product was dramatically reduced to a level similar to that of deionized water.

522623 The Formation of Peroxymonocarbonate (HCO_4^-) and Its Impact on the Degradation of Organic Contaminants during Hydrogen Peroxide (H_2O_2) *in Situ* chemical Oxidation (ISCO)

Abstract:

Hydrogen peroxide (H_2O_2) reacts with iron and manganese adsorbed on surfaces and on minerals to produce hydroxyl radical and other reactive oxygen species during *in situ* chemical oxidation of contaminated groundwater. Under the conditions used for remediation, high concentrations of hydrogen peroxide (e.g. up to ~10 M) are encountered. Using ^{13}C NMR we have shown that under these conditions, an alternative pathway for H_2O_2 loss occurs in which H_2O_2 reacts with HCO_3^- to produce HCO_4^-, at a HCO_3^- concentration of 10~1000 mM, HCO_4^- was formed and quickly reacted with phenol to produce di- and tri-hydroxyl phenols. The rate of transformation of phenol was determined by the concentration of HCO_3^-. HCO_4^- also reacted with substituted phenols in an electrophilic behavior. Results of this study indicate that HCO_4^- is a potentially important but previously unrecognized oxidative species generated during H_2O_2 ISCO and that it may be possible to adjust treatment conditions to enhance its formation as a means of accelerating rates of contaminant removal.

539306 Monitoring and Classification System for Water Recycling

Abstract:

Minimizing water consumption and disposal has huge benefits in terms of safety and cost savings. However, to maximize the amount of water that is recycled at a site generating multiple complex waste streams, an intelligent water monitoring system is needed to ensure that water from various sources is processed appropriately. Mainstream Engineering has developed a chemometric water monitoring system (CWMS) to fully characterize water prior to recycle or disposal. This technology relies on statistical methods that relate data to properties of interest. Specifically, we used a combination of several rapid, reagentless sensors to characterize the water according to several classes.

We demonstrated the CWMS for a variety of wastewater sources including laundry, kitchen, and latrine. We then validated our models with additional samples from other water sources such as wastewater treatment facilities to demonstrate the robustness of the system. After specifying all of the components, we designed a prototype system. Principal component analysis was used to classify water and draw distinctions between surface non-potable water, greywater and blackwater. We were also able to eliminate the need for dedicated COD, BOD_5, turbidity and chlorine sensors by processing spectroscopic data with chemometrics.

522532 Water Recovery and Reuse in Soluble Coffee Production Using a Dynamic Membrane Process

Abstract:

Through a grant from the U. S. Environmental Protection Agency, Pollution Prevention program, we have partnered with NestlÃ© USA to explore the use of novel membrane technology to improve the manufacturing efficiency of soluble coffee manufacturing operations. Our case study at their Freehold, NJ factory, evaluates water recovery methods for a spray dried and freeze dried coffee manufacturing plant. The particular focus of this paper is on a dynamic / shear-enhanced vibratory field membrane method to recover water for plant operations, thus reducing operating costs and environmental footprint. This research helps address a major issue in the food manufacturing sector â€“ the efficient use of water and waste reduction.

The production of spray dried and freeze dried coffee involves many highly complex processes that rely on large amounts of water. These operations include process water for cooling, water to be heated to produce steam, water for process equipment operations, water for intermediate production steps, and water for cleaning and sterilization. When utility generation is included, such as operation of cooling towers, water consumption can be significant.

Many of the problems associated with effective recovery, involve the separation of dissolved, colloidal, and particulate matter, to obtain water of suitable quality for reuse. Suspended and colloidal matter is one of the biggest causes of fouling and performance degradation in operating these water recovery systems. This causes additional expense, and in many cases industry abandons this approach, since it will not produce sufficient quantity and quality of water for extended periods of time. One potential solution that is described in this paper, is a vibratory-field membrane process for food/biomass waste processing.

Our study describes the benefits and limitations of this technology, and how laboratory investigations help guide industrial use and scale-up considerations. Various membrane types were evaluated to help inform decision makers based on the reuse goals. The study evaluated the overall savings and reduction in life cycle emissions for the recovery of water from a waste stream, for use in cooling tower operation. The assessment included savings, both economically and environmentally, in the reduction in fresh water intake, waste pretreatment, and final disposal. These are compared to the increased cost and operational factors of a new recovery system. The case study found that the greatest cost and environmental impact reductions were associated with the reduction in wastewater discharge. Long-term economics were more highly influenced based on equipment size. This case study is applicable to the food and beverage manufacturing sector.

525339 How Active Is Too Active?Â a Catalyst Selection Study

Abstract:

When selecting a catalyst formulation as part of a new reactor build, the process can be designed around the catalyst such that nearly any commercially-viable catalyst can be used. For these systems, higher activity catalysts generally give rise to higher product rates or reduced capital investment. However, most of industryâ€™s active processes are not newly built, but have been in service for many years (if not decades). For these systems, higher activity catalysts can also enable higher product rates, *but typically only to the extent constrained by the original process design*. Indeed, many if not most of todayâ€™s active systems are now multiply constrained, such that it is not always true that increasingly higher activity catalysts will yield better performance.

As one example, many fixed-bed, vapor-solid reactor systems are constrained by heat transfer. For these systems, a circulating coolant absorbs the heat of reaction thereby controlling the temperature rise and making possible higher product rates. In this context, there is generally a maximum achievable product rate corresponding to the maximum heat removal capability of the system. This maximum product rate is itself a function of the catalyst (in particular, the catalyst activity and selectivity).

In choosing from next-generation, high activity catalysts for a system constrained by heat transfer, the impact of varying activity and selectivity must be balanced, taking into account also the free variables for optimization and their bounds. Here, we utilize detailed reactor simulations to investigate the catalyst selection problem for one such industrial system. Our steady-state simulations incorporate mass and energy balances and reaction kinetics while predicting temperatures and concentrations that vary in two spatial dimensions. We begin by demonstrating how such simulations, when informed by laboratory experiments, can predict the heat transfer-constrained operation as observed in the plant. Then, we utilize the tuned reactor model to explore how process optimization accounting for a catalyst change (e.g., changes to the base reaction rates, activation energies, and partial reaction orders) leads to varying and sometimes counterintuitive changes in reactor performance. To finish, we develop simplified analytical models from first principles and compare their predictions to those generated using the full reactor model. These simplified models can be utilized as a coarse guide for catalyst selection in the case that more detailed and highly tuned reactor models are not available.

530554 Glycol Loss Minimization for a Natural Gas Dehydration Plant Under Upset Conditions

Abstract:

Low-temperature separation with Mono Ethylene Glycol (MEG) injection process is a common dehydration technique for natural gas processing, which needs to maintain a hydrocarbon dew point. However, dehydration systems based on glycol frequently result in a significant loss of glycol during plant upset conditions such as rapid flow fluctuation, refrigeration system failure and inadequate control of stripper column [1]. These losses occur due to carryover with gas, hydrocarbon and vaporization from stripper column overhead. For this reason, it is important to develop process operation methodologies for reducing glycol losses and thereby reducing emissions.

In this research, steady-state and dynamic models are developed using the Aspen Plus simulation framework. A parametric analysis of inlet flow fluctuation, temperature fluctuation of the low-temperature separator (LTS), stripper column overhead temperature fluctuation and reboiler temperature fluctuation were conducted under steady-state conditions. Simulation results indicated that that most of the glycol is lost via vaporization through the stripper column overhead and some glycols are carried over with gas and hydrocarbon from LTS. Dynamic simulations were utilized to develop a new control strategy for optimum column operation along with an improved stripper model. It is shown that the simulation results can be used to study plant upset conditions, minimize glycol loss, improve product gas specifications and reduce plant operating costs. In particular, it is shown that the improved design has the potential to reduce glycol losses by over 70%, lower emissions and provide plant stability.

References

1. Haque, M. E.; Ethylene Glycol Regeneration Plan: A Systematic Approach to Troubleshoot the Common Problems. *Journal of Chemical Engineering, IEB, 27, No. 1, 21-26, 2012.*

Hydrodeoxygenation of Karanja Oil for the Production of Green Diesel: Process Design with Heat Integration and Economic Analysis

Swarnalatha Mailaram and Sunil Kumar Maity,
IIT Hyderabad, Hyderabad, India

Introduction

Major concern on environmental effects and depletion of fossil fuels led to search for the alternative resources for energy. Biomass is the best possible alternate source in terms of its availability in nature and it also has the potential to transform to all forms of energy including transportation liquid fuel which is the leading energy consuming sector with ~27% share globally [1]. Much research and technological development has been done so far in the production of biofuels ex: bioethanol, biobutanol, biodiesel etc. Biomass derived oxygenated biofuels suffer with low mileage because of the presence of oxygen and these biofuels are not compatible with existing infrastructure. Transesterification is one of the technology producing bio diesel which is limited to blending with petro-diesel due to the presence of oxygen which lowers the fuel mileage. Therefore, there is a need of hydrocarbon fuels similar to petro-diesel, which is compatible with existing combustion engine. Inedible vegetable oil consisting of long hydrocarbon chain with low oxygen content and lesser complexity of chemical structure compared to lignocellulosic biomass is the promising biomass to produce transportation fuels. Karanja oil production potential is more in south east Asia with 78.6wt% C_{18}, 11.3wt% C_{16}, 3.7wt % C_{20}, 6.4wt% C_{22} fatty acid composition [2]. Hence, Karanja oil is chosen as the feedstock.

Hydrodeoxygenation process

Hydrodeoxygenation (HDO) is the route which removes oxygen content in the triglyceride in the form of water to form hydrocarbon diesel range fuel called as green diesel. Green diesel has higher energy density (44 MJ/kg), high cetane number (70-90) and can directly be used in existing combustion engines [3][4]. UOP has recently developed a two-step ecofining process aiming to produce green diesel by HDO [5]. Studies on various oils by HDO process has been performed by researchers. HDO is generally carried out at high hydrogen pressure (25–150 bar) over supported metal catalyst in the temperature range of 523 K to 773 K [6]. Generally, HDO of Karanja oil can be envisaged through two different routes: (i) direct HDO and (ii) two-step HDO (hydrolysis of vegetable oil followed by HDO of fatty acid). In the Direct HDO, hydrogenation of unsaturated triglyceride in Karanja oil gets saturated triglyceride and undergo HDO to form diesel range hydrocarbons (C_{18}-C_{22}) termed as green diesel, water, traces of CO_2 with propane as a co-product. In the second route, Karanja oil is first hydrolyzed with water to form fatty acids and glycerol (co-product). These fatty acids then undergo HDO forming green diesel, water and traces of CO_2 [3][4][7].

Process design with heat integration

In the present work, process design for both the routes were performed using Aspen Plus software for various capacities of plant ranging from 0.01 to 0.15MMT of Karanja oil /annum. Schematic process for direct HDO process is shown in Fig. 1. In direct HDO, karanja oil and hydrogen enter the HDO reactor operating at 623K, 100bar. Excess hydrogen is removed in a separator following reactor. The product stream containing green diesel, water, CO_2, propane is sent to downstream processing. Water, green diesel and gases (propane and CO_2) are separated and gases are further sent to absorber and stripping unit in which monoethanolamine (MEA) is used as a solvent to separate CO_2 from propane and propane is the revenue generating co-product. As shown in Fig. 2., In two step HDO, karanja oil is hydrolysed with water in hydrolysis reactor at 533K and 56bar to form fatty acids and glycerol. Glycerol and water is separated and sent to the glycerol refining section to obtain 99.99wt% glycerol which is the co-product in this process. Fatty acids formed are sent to the HDO reactor to form where green diesel, water and traces of CO_2 are formed. Green diesel is separated from gases and water. Heat integration for both the processes has been done to reduce the utility requirement and optimize the heat transfer within the plant. Pinch analysis has been chosen to obtain the minimum energy consumption in the plant and design heat exchange network for both the processes. Resulting grand composite curve (GCC) by pinch methodology, was useful to obtain the minimum hot and cold utility requirement and heat recovery within the system. Based on the possibility of heat recovery in the process and pinch analysis criteria, heat exchange network has been designed. This helped in redesigning the processes with the heat exchange network system.

Economic analysis

The economic evaluation has been carried out for different plant capacities ranging from 0.01 to 0.15MMT/annum for both direct HDO and two-step HDO process. All possible cost contributing factors were considered to obtain capital and operating expenses for both the processes. Capital investment for both the processes is practically identical. Impact of plant capacity on both processes is huge. The production cost was reduced with increasing plant capacity. The reduction in the production cost was, however, negligibly small with increasing plant capacity beyond 0.12 MMT/annum. Hence, 0.12 MMT/annum has been considered as the optimum plant capacity. The production cost of green diesel was estimated as 0.84 $/kg and 0.798$/kg for direct and two-step HDO respectively for the plant capacity of 0.12 MMT Karanja oil per annum. Also, contribution of various costs to production cost has been studied. Feedstock alone is contributing around 73-77% to the production cost which is very high. Contribution of co-product credits has significantly reduced the production cost of green diesel in two-step process. Utility consumption reduction with the help of heat recovery in the process has shared 1.8-4.18% which was ~35% contribution without heat recovery. Price of co-products has also a significant impact on the production cost. Production cost was decreased by 0.5% and 1.6% for 10% increase in the price of co-products for direct and two-step HDO respectively. The discounted cash flow analysis was further carried out to evaluate the economic performance and determine the minimum selling price of green diesel. The price of green diesel excluding taxes was found to be 1.186 $/kg and 1.21 $/kg for an acceptable rate of return of 8.5% for direct and two-step HDO respectively. The green diesel can be considered as economically competitive with the retail price of the petro-diesel.

Summary

Conceptual processes of direct and twostep HDO is designed using Karanja oil as feedstock. Maximum energy recovery using pinch methodology has been done. Heat exchange network has been designed for both the processes. Heat recovery within the system with the use of pinch analysis has reduced the utility consumption in the process. Thereby, reducing the production cost of green diesel significantly. Economic analysis for modified processes with heat exchange network was evaluated. Capital, operating expenses, production cost of green diesel is estimated. Feedstock cost plays key role in reducing the cost of the diesel. If reduced, drastic decline in the cost can be observed. Discounted cash flow analysis was used to find the minimum selling price of green diesel.

References

1. Annual Energy Outlook. U.S. Energy Information Administration. Available at: http://www.eia.gov/forecasts/aeo/index.cfm: 2015.
2. P. Kumar, S.R. Yenumala, S.K. Maity, D. Shee, Kinetics of hydrodeoxygenation of stearic acid using supported nickel catalysts: Effects of supports, Appl. Catal. A Gen. 471 (2014) 28–38.
3. S.R. Yenumala, S.K. Maity, D. Shee, Reaction mechanism and kinetic modeling for the hydrodeoxygenation of triglycerides over alumina supported nickel catalyst, React. Kinet. Mech. Catal. 120 (2017) 109–128.
4. Kalnes TN, Marker T, Shonnard DR, Koers KP. Green diesel production by hydrorefining renewable feedstocks. Biofuels Technol 2008:7–11.
5. UOP LLC, A Honeywell Company. Honeywell Green Jet Fuel: Sustainable fuels for cleaner skies. 2011. Available at: https://www.uop.com/?document=honeywell-green-jet- fuel-brochure&download=1
6. Glisic SB, Pajnik JM, Orlovic´ AM. Process and techno-economic analysis of green diesel production from waste vegetable oil and the comparison with ester type biodiesel production. Appl Energy 2016;170:176–85.
7. S.R. Yenumala, S.K. Maity, D. Shee, Hydrodeoxygenation of karanja oil over supported nickel catalysts : influence of support and nickel loading, Catal. Sci. Technol. 6 (2016) 3156–3165.

Figures

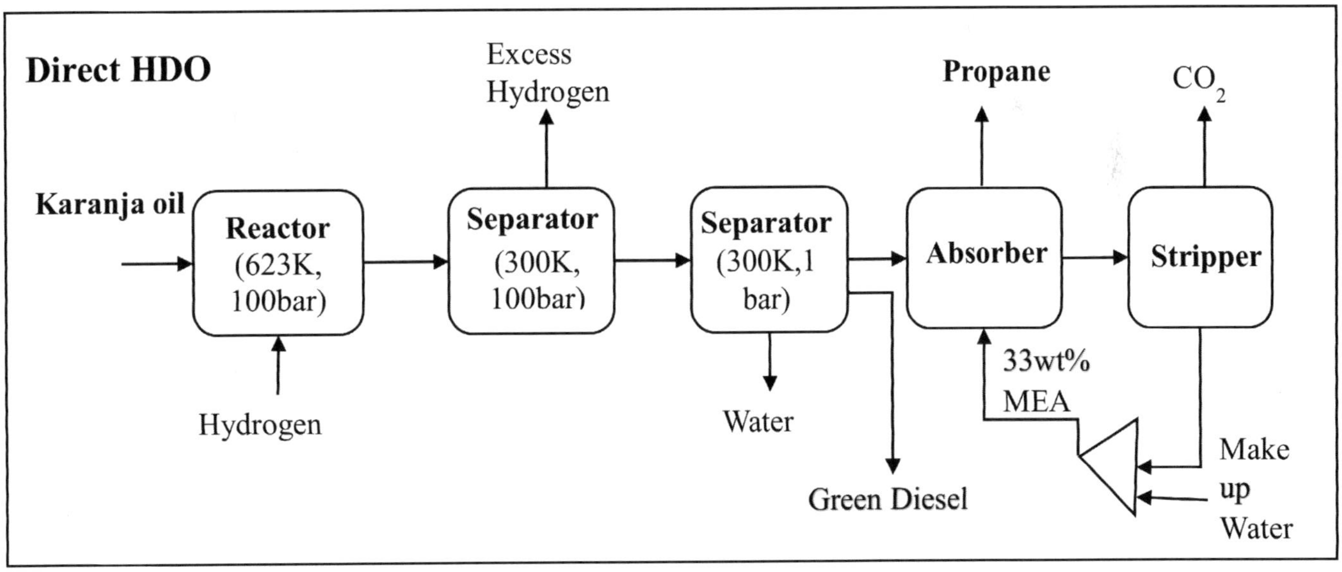

Fig. 1. Schematic process flowsheet of Direct HDO process.

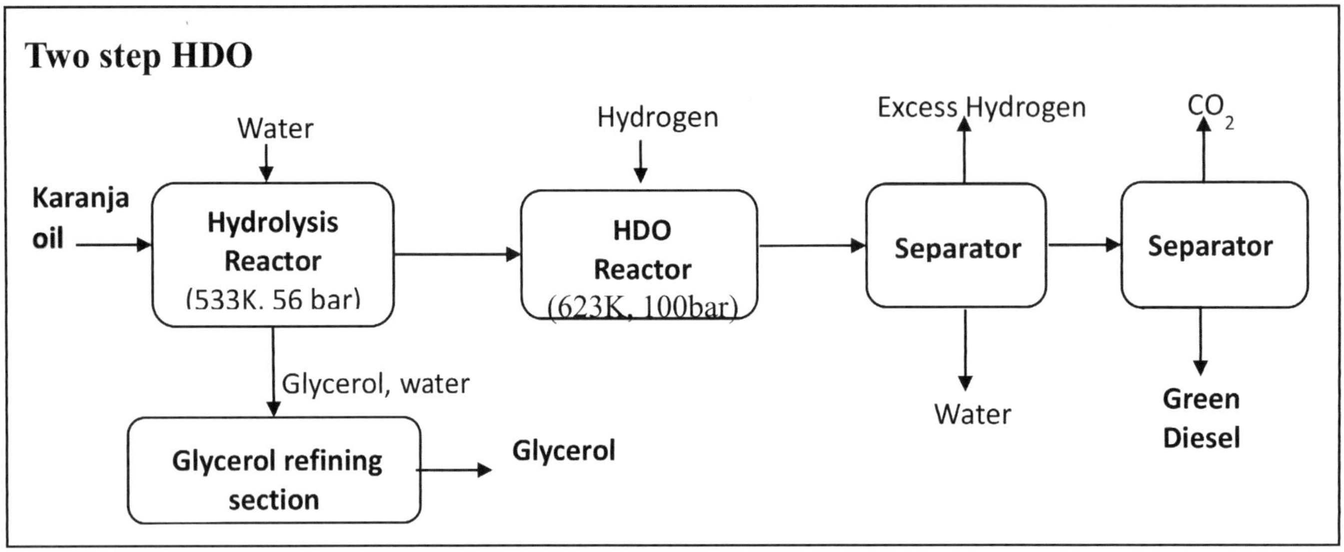

Fig. 2. Schematic process flowsheet of Twostep HDO process.

531870 An Integrated Process for Desulfurization (HDS-Extractive-Oxidative)

Abstract:

An integrated process for Desulfurization (HDS-Extractive-Oxidative)

Mokhtarian, M.R. Dehghani, F. Banisharif

Chemical Engineering Department, Iran University of Science and Technology, Narmak, Tehran, Iran

Abstract:

Available sulfur compounds in the fuels such as thiophens are considered as one of the main source of air pollution. They converted to SOx and released to the atmosphere as a harmful chemical for human health. Hence, removal fuel desulfurization is mandatory from environmental point of view as well as public health issues. Nowadays, in many countries, stringent regulations are being enacted to reduce the sulfur content to lower than 10 ppm for both diesel and gas oil. During recent years many attentions have been paid to extractive-oxidative desulfurization (EDS-ODS) while economic evaluations show that they are not economically feasible compared to the hydro desulfurization (HDS) process. On the other hand, HDS process is considered as an energy insensitive process and needs to be operated at High pressure and temperatures. To present an economic process, in this work an integrated process of HDS and EDS-ODS were modeled. The HDS section was modeled based on operational data from Tehran refinery and extractive-catalytic oxidative desulfurization (ECODS) part were modeled based on our novel pilot experimental data. Gas oil as the feedstock with total sulfur content of 22600 ppmw was fed to the process. Optimal conditions were obtained using the response surface methodology and box-behnken design (BBD), based on experimental design results, hydro desulfurization from the 22600 ppmw to 1930 ppmw level was done under the following conditions: hydrodesulfurization temperature: 311 â„ƒ, hydrodesulfurization pressure: 24 bar and hydrogen purity of 35 wt%. Then gas oil with the total sulfur content of 1930 ppmw was entered into the oxidative-extractive desulfurization unit. The effects of different operating parameters such as temperature, oxidant to sulfur molar ratio and the type of solvent in the extraction stage were investigated. The best performance was obtained at oxidant to sulfur molar ratio of 4:1 and acid to oxidant volumetric ratio of 1:1. It must be mentioned that hydrogen peroxide (30%) and acetic acid (95%) were used as oxidant. Finally, a liquid-liquid extraction process was utilized for deep desulfurization step. Acetonitrile used as solvent at 1:1 ratio towards gas oil. It was found that the proposed process has lower costs and also is inherently safer than that of conventional HDS, ODS and EDS.

Keywords: Optimization, Hydrodesulfurization, Extractive- Catalytic oxidative desulfurization, Ultra-low Sulfur Fuel

540773 2018 Outlook for Energy: A View to 2040

Abstract:

The Outlook for Energy is ExxonMobil's global view of energy demand and supply through 2040. It also highlights the dual challenge of providing affordable energy to support prosperity while reducing environmental impacts – including the risks of climate change. We use the Outlook to help inform our long-term business strategies and investment plans.
The Outlook anticipates significant changes through 2040 across the world to boost living standards, reshape the use of energy, broaden access to abundant energy supplies, and accelerate decarbonization of the world's energy system to address the risks of climate change.

- Global energy needs will rise about 25 percent over the period to 2040
- Electricity demand will nearly double in non-OECD countries
- Worldwide electricity from solar and wind will increase about 400 percent
- Natural gas will expand its role, led by growth for electricity generation
- Growth in oil demand will be driven by commercial transportation and chemical needs
- Efficiency gains and growing use of less-carbon intensive energy sources will contribute to a nearly 45 percent decline in the carbon intensity of global GDP

Energy is fundamental to modern life, and as the world's population approaches 9 billion people in 2040, we are challenged to improve living standards everywhere. Understanding the factors that drive the world's energy needs – and likely choices to meet those needs – is the mission of the Outlook. By sharing the Outlook with the public, we hope to broaden that understanding. Energy matters to everyone, and we all play a role in shaping its future.

541543 Energy Decarbonisation Scenarios

Abstract:

With the energy sector being associated with 2/3 of the annual greenhouse gas emissions, decarbonizing the energy is a key step to reduce their level. The National Determined Contributions (NDC's) announced so far by the participating countries do not meet the objectives of the Paris-climate agreement, as far as reaching a peak in emissions in the near future or decreasing to a level that is compatible with the stated long-term targets.

Using different scenarios for energy demand and supply, options for energy sector decarbonisation are evaluated. Meeting the 2-degree target requires a major transformation in the power sector, efficiency improvements – specially in the buildings and in the industry, and a revolution in transport towards fuel and modal shifts. With a significant increase in renewable energy, system integration and optimization will also be key in reducing the cost of the energy transition. Meeting more ambitious climate goals ("well below 2 degrees") will push further the technological innovation and implementation requirements throughout the entire energy value chain. A scenario aiming at zero net-emissions by 2060 is presented, along with a comparison to the 2-degree case.

542625 Fundamental Research Needs to Advance Energy Technologies

Abstract:

The U. S. Department of Energy (DOE), Office of Science, Office of Basic Energy Sciences (BES) supports fundamental research in chemical and materials sciences to provide the foundations for new energy technologies and to support DOE missions in energy, environment, and national security. This presentation will discuss opportunities for fundamental research to impact DOE's energy mission "to catalyze the timely, material, and efficient transformation of the nation's energy system and secure U.S. leadership in energy technologies" with a focus on the way we generate, store and use energy nationally. I will provide an overview of BES strategic planning over the past decade that identified priority research directions for advancing energy applications, highlight key scientific advances in these areas, and discuss some future opportunities for modern science, particularly chemical sciences, to accelerate the transformation of the U. S. energy portfolio.

547479 Energy, Environment, and Engagement Beyond the Lab

Abstract:

Engagement of researchers with the public, policy makers and the press is important to increasing the influence of science and engineering on societal choices about the environment, energy, climate, etc., on a finite planet. This talk will recount experiences, impacts, and lessons learned in the process of stepping out from the laboratory into public-facing academic leadership roles.

547442 Sustainability in Manufacturing beyond ROI

Abstract:

Mass production of goods requires enormous amounts of natural resources, such as energy precursors and water. Therefore, the contribution from manufacturing sustainability initiatives to overall resource conservation is sufficiently greater than those from, for example, changes in household behavior. Most industrial manufacturers are driven by financial metrics and the same rules apply in assessing viability of resource conservation opportunities compared to other investment options. Without beneficial returns on investment (ROI), sustainability projects are often left unpursued. A growing number of scientific studies have shown positive impacts of corporate social responsibility, and particularly corporate environmental performance (CEP), on financial performance. Unfortunately, CEP-driven financial variables are yet to be introduced and widely adopted by manufacturing industries for integration into ROI assessment of sustainability projects. The growing notion that â€œit pays to be greenâ€ has been infiltrating the scientific community, but under increasing natural resource scarcity and climate change stresses, more work needs to be done to establish benchmarked CEP-inclusive ROI analysis and expand the amount of vital sustainability initiatives in manufacturing.

542591 Electrochemical Conversion of Ammonia and Nitrogen for Sustainable Food-Energy-Water

Abstract:

Electrochemical Conversion of Ammonia and Nitrogen for Sustainable Food-Energy-Water

Gerardine G. Botte

Center for Electrochemical Engineering Research

Department of Chemical and Biomolecular Engineering

Russ College of Engineering and Technology

Ohio University, Athens

OH 45701

The production of reactive nitrogen fertilizer via the Haber-Bosch (HB) process, combined with industrialization and intensification of agriculture, has created an abundant global food supply to support population growth, but has doubled the amount of reactive nitrogen compared to pre-industrial levels.[1] With those perspectives, ammonia can be seen as an important chemical of use or as an environmental challenge. Ammonia emissions into air (ambient ammonia) and water represent an environmental challenge. Ambient ammonia not only contributes to inorganic PM2.5 (particulate matter with an aerodynamic diameter of less than 2.5 Î¼ m) directly but also plays an important role in secondary organic aerosol formation by interacting with gaseous phase organic acids and forming condensable salts. In addition, ammonia emissions in water are associated with environmental problems such as algae bloom. Various industries and other operations are considered ammonia emitters, but the ones related to sustainable food production such as livestock management and fertilizer application are among the major contributors accounting for 60-75% of the ammonia emissions.[2-3]

On the other hand, ammonia produced commercially via the HB process is usually known as one of the greatest inventions of the 20th Century. The HB process made it possible for the first time to produce synthetic fertilizers and produce sufficient food for the Earthâ€™s growing population. However, the HB process requires high temperatures and pressures and as a result it can only be done in large plants which consume large amounts of power and generate large quantities of CO_2 from the fuels used to power the process, such as natural gas. It is estimated that the HB process accounts for nearly 1% of the entire global power consumption. Due to the sheer scale of the HB process, distribution from the point of production to the point of use becomes an additional carbon and energy burden. Alternative methods of manufacturing ammonia that enable small scale, distributed generation at the point of use using renewable energy and sustainable feedstocks could have a great impact on global CO_2 emissions.

A closed loop approach that includes producing fertilizer at the point of use, applying nitrogen fertilizer as close to the time that crops needs it, and recovering and recycling nitrogen fertilizer can contribute to address the challenges related to ammonia emissions, while decreasing water and energy consumption. Electrochemical conversion of nitrogen and ammonia presents an excellent platform to achieve this closed loop approach. Dr. Botte and members of her research group at the Center for Electrochemical Engineering Research (CEER) are working on the electrochemical conversion[4]/synthesis of ammonia[5] in alkaline media. In this talk, recent advances in the platform process for ammonia and nitrogen removal/recycling and ammonia synthesis will be discussed as well as their application for sustainable food, energy, and water.

References:

1. Grand Challenges - Manage the Nitrogen Cycle [WWW Document], n.d. URL http://www.engineeringchallenges.org/9132.aspx (accessed 5.28.18).

2. G. E. Mansell, "An Improved Ammonia Inventory for the WRAP Domain (Final Report)", ENVIRON International Corporation, Report No. 415.899.0700, March 7 (2005).

3. FAO. Global estimates of gaseous emissions of NH_3, NO, and N_2O from agricultural land, International Fertilizer, Industry Association. Rome. 2001.

4. Botte, G. G., Cooper, M. and Vitse, F., (2009) Electro-catalysts for the oxidation of ammonia in alkaline Media, U.S. Patent No. 7,485,211.

5. B. L. Sheets, G. G. Botte, Electrochemical nitrogen reduction to ammonia under mild conditions enabled by a polymer gel electrolyte, Chemical Communications 54 (2018), 4250-4253.

542589 Encapsulation and Nanoparticle Formation for "Non-Standard" Applications

Abstract:

Encapsulation and nanoparticle formation for "non-standard" applications

Prof. Robert K. Prud'homme[1], Prof. Rodney D. Priestley[1], Leslie Shor[2] ,Douglas Scott[1], Jie Feng[1]

[1] Princeton University, Department of Chemical and Biological Engineering, Princeton University, Princeton NJ 08544

[2] University of Connecticut, Department of Chemical and Biomolecular Engineering, Storrs, CT 06269-3222

Presenting author: Robert K. Prud'homme, prudhomm@princeton.edu

Traditionally, encapsulation and nanoparticles have been associated with high value applications such as drug delivery in medicine. The two main reasons have been the high cost of nanoparticle assembly processes for complex nanoparticles that encapsulate and release actives in a controlled way, and the fact that encapsulation technologies for larger scale macro-particles do not translate to nano-scale particles. Recent advances in low-cost nanoparticle assembly present the opportunity to open up new applications. First, we present an overview of encapsulation and release technologies. Then we provide a review of the rapid-precipitation technologies we have developed that rely on the control of the fluid dynamics, diffusion-limited aggregation, and steric stabilization of nanoparticles. The processes are scaleable, low cost, and produce nanoparticles from 50-400 nm. Several novel applications will be presented: (1) nanoparticle delivery of agricultural actives s to plant roots by protist active transport, (2) triggered indicator inks for sterilization processes, (3) "green chemistry" encapsulation of personal care compounds using purely aqueous processing.

528954 Life Cycle Analysis of Carbon Capture Retrofit Using the Petra Nova Model

Abstract:

The Petra Nova Project near Houston, TX captures 90 percent of a slipstream of flue gas from W.A. Parish Unit 8 equivalent to 240 MWe [1]. Petra Nova utilizes an amine based capture process (Kansai Mitsubishi Carbon Dioxide Recovery Process [KM CDR Process]) and began commercial operation on January 11, 2017 [2, 3]. This installation marks the world's largest implementation of post-combustion capture at a power plant. One of the unique attributes about the Petra Nova Project is that the steam and power required to run the capture process are not provided by Unit 8, but rather a separate 75 MW cogen natural gas plant. That 75 MW facility previously operated as a peaker unit, but was converted to cogeneration service specifically for the Petra Nova Project [4]. Approximately 60% of the output from the cogen plant will be used to operate the capture equipment and the remainder will be sold onto the grid [5]. Because of this configuration, the total power output from Unit 8 was not affected by addition of carbon capture. This configuration or similar approaches avoids the use of steam and/or power from the base coal-fired power plant and is anticipated to be used for future deployments of carbon capture of both first-generation solvents and other advanced cogen concepts like the Fuel Cell Energy Combined Electric Power and Carbon-dioxide Separation (CEPACS).

Cogen systems provide unique challenges in evaluating both the cost and environmental performance due to multiple fuel streams and accounting of trade-offs between the different services provided by the system. As technology concepts, such as carbon capture, move from design to demonstration, understanding how technology developers, investors, and the public perceive the environmental strengths and weaknesses of disparate technology choices is critical to making fair and accurate comparisons. Understanding and communicating the environmental performance of technology options can identify R&D improvement opportunities during the design and demonstration phase as well as improve the acceptance of new technology for commercialization.

This analysis compares the life cycle environmental impacts for an amine-based capture retrofit with an external co-gen source providing the utility inputs for capture compared to a more traditional retrofit in which the plant/unit output is de-rated with the utility inputs for operating the capture system coming directly from the plant. One of the major differences between the designs is that in the case of the external cogen unit, the combustion emissions associated with operating the capture unit are not part of the flue gas stream that is directed to the capture unit. Another unique consideration in this analysis is the inclusion of the upstream natural gas life cycle in the boundaries of the cogen retrofit. This analysis compares the life cycle environmental impacts for an amine-based capture retrofit with an external co-gen source providing the utility inputs for capture compared to a more traditional retrofit in which the plant/unit output is de-rated with the utility inputs for operating the capture system coming directly from the plant. To our knowledge this comparative analysis has not been performed to-date and offers an opportunity to fully understand the environmental impacts of the Petra Nova Project and offer insight into how this type of project compares to more conventional carbon capture technologies.

References

1. ***PETRA NOVA PARISH HOLDINGS W.A. Parish Post-Combustion CO2 Capture and Sequestration Project***. 2017; Available from: https://www.netl.doe.gov/research/coal/project-information/fe0003311-ppp.
2. ***Capturing Carbon and Seizing Innovation: Petra Nova Is POWER's Plant of the Year***. 2017; Available from: http://www.powermag.com/capturing-carbon-and-seizing-innovation-petra-nova-is-powers-plant-of-the-year/.
3. DOE Office of Fossil Energy. ***Petra Nova, World's Largest Post-Combustion Carbon-Capture Project, Begins Commercial Operation***. 2017; Available from: https://energy.gov/fe/articles/petra-nova-world-s-largest-post-combustion-carbon-capture-project-begins-commercial.
4. S&P Global Platts. ***NRG to return Petra Nova peaker plant in Texas to operations in mid-July***. 2016; Available from: https://www.platts.com/latest-news/electric-power/houston/nrg-to-return-petra-nova-peaker-plant-in-texas-21746588.
5. Wang, U. ***NRG's $1B Bet To Show How Carbon Capture Could Be Feasible For Coal Power Plants***. 2014; Available from: https://www.forbes.com/sites/uciliawang/2014/07/15/nrgs-1b-bet-to-show-how-carbon-capture-could-be-feasible-for-coal-power-plants/2/#7307d66d321b.

Attribution

KeyLogic Systems, Inc.'s contributions to this work were funded by the National Energy Technology Laboratory under the Mission Execution and Strategic Analysis contract (DE-FE0025912) for support services.

530389 A Model for Packed and Fluidized Bed Absorbers with Micro-Encapsulated CO_2 Sorbents

Abstract:

Micro-encapsulated CO_2 sorbents (MECS) are a carbon capture technology that possesses the benefits of both liquid solvents and solid sorbents. They consist of a CO_2-absorbing liquid or slurry, encased in spherical, gas-permeable, polymer shells. By greatly increasing the surface area of the liquid, microencapsulation enables the use of slow-reacting, viscous, and phase-changing solvents to reap the benefits of good thermodynamic properties, low water content, and/or low volatility. However, in terms of process design, MECS have more in common with solid sorbents. They behave as a light powder, with particle diameter adjustable from ~50-600 μm. Several candidate solvents have been successfully encapsulated and tested for longevity, CO_2 capacity, and CO_2 absorption rates. One of the most promising candidates is Na_2CO_3 solution, due to its stability, low cost, and CO_2 capacity. The rate of CO_2 absorption in this sorbent is well-characterized and known to be relatively slow. However, the CO_2 absorption rate can be enhanced by catalysts or promoters, such as the Zn complex of 1,4,7,10-tetraazacyclododecane (â€œcyclenâ€), a biomimetic catalyst. Encapsulated Na_2CO_3 solution with cyclen is the capsule design that is implemented in all the experiments and models presented here. However, these results and conclusions can be readily adapted to other solvents (e.g. ionic liquids) that have been successfully encapsulated in the lab.

The goal of this work is to estimate the size of a large absorber filled with these Na_2CO_3 + cyclen capsules that have been developed at the bench scale. In order to accomplish this task, a model is first developed for the CO_2 diffusion and reaction in an individual capsule. This model is fitted to match in-house experimental CO_2 absorption data for small batches of capsules. This individual capsule model is then validated against experimental data collected on a larger batch of capsules fluidized with simulated coal flue gas. Matlab models are then developed for two absorber bed designs: 1) a multi-stage counter-flow fluidized bed, and 2) an annular packed bed with radial flow. These two bed models are then optimized for minimum reactor size at both the pilot (1 MW) and power plant (500 MW) scales. Based on that optimization, a modular packed bed absorber is proposed for a 500 MW coal plant that has similar dimensions to state-of-the-art MEA absorbers under the same conditions. This 500 MW absorber design has the advantage of being highly modular, so the layout and dimensions of the system are flexible. Furthermore, these large vessels are solely filled with beads of Na_2CO_3 and water: cheap, abundant materials that demonstrate great potential for large-scale manufacturing.

Finally, a parametric study is performed in order to guide future capsule design based on absorber size. The results of that study demonstrate that the absorber size can be dramatically reduced by a combination of the following capsule improvements: 1) making the capsules smaller to increase total surface area, 2) making the capsule shell thinner to maximize solvent volume fraction, and/or 3) using a different solvent to raise the mass transfer coefficient. By making modest improvements to these three design parameters, the absorber bed diameter could be reduced from 21.5m to 17.5m, making this absorber narrower than a conventional MEA tower.

535308 Flue Gas to Food Security: Radical Reductions in Greenhouse Water Usage through Direct Utilization of CO_2

Abstract:

Agriculture is mankind's single largest usage of water, comprising 70% of all water usage. Optimizing water usage in agriculture is therefore crucial to ensuring global water security. Only a tiny fraction of water taken up by a plant's roots is ever incorporated into plant matter; the vast majority is lost through transpiration. There are essentially two types of transpiration; useful transpiration, a necessary mechanism for the transport of nutrient transport. The other type, useless transpiration, involves unnecessary evaporation from leaf and other surfaces. Useless transpiration is estimated to exceed useful transpiration in most agricultural conditions and can therefore be considered mankind's largest avenue of water wastage.

In this research, a greenhouse is quantitatively modelled as a bio-reactor and it is shown that the high air-flow rates required to supply adequate CO_2 necessarily impose a high rate of evaporation, as air leaves the greenhouse in the same warm and humid conditions as the air inside the greenhouse. These evaporative losses make up a clear majority of mankind's total water usage and can be mitigated by enrichment of CO_2 fed to the greenhouse, which reduces the airflow required to supply sufficient CO_2, reducing evaporative losses commensurately.

It is shown that using a typical gas-fired power station's flue gas composition significantly reduces the required airflow and therefore evaporative losses by as much as 95% in many cases. Savings of this magnitude will not be entirely achievable because some amount of transpiration is necessary but these results nevertheless indicate that useless transpiration can be largely eliminated, drastically reducing water requirements for agriculture. Energy balance analysis also reveals that large amounts of energy are consumed by the heating and humidification of the air streams and that CO_2 enrichment therefore also promises to reduce heating requirements in cold-climate greenhouses.

535982 Redesigning the Regulated Power Plant: Optimizing Energy Allocation to Electricity Generation, Carbon Capture, and Water Treatment Processes at Coal-Fired Power Plants

Abstract:

Coal fired power plants (CFPPs) are the largest industrial emitters of air toxics, aqueous contaminants, and greenhouse gasses in the US. Compliance with promulgated and potential regulations to address these emissions will require the installation of new carbon and water emissions control technologies. These separation processes require energy, either in the form of heat or electricity, and may significantly reduce the generation efficiency and revenue of CFPPs. A systematic reevaluation of the use of all thermal and electricity sources at CFPPs will aid power plant designers in maximizing the efficiency and cost-effectiveness of emissions control retrofits. To assist in this reevaluation, we first build mass- and energy-balance models of the turbines at CFPPs. We then seek to maximizing revenue by varying the allocation of steam among three sinks: the turbine for electricity generation, the solvent regeneration for carbon capture, and the wastewater treatment unit. We demonstrate this approach by evaluating trade-offs for the National Energy Technology Laboratoryâ€™s 550 MW pulverized coal combustion power plant model. We find that the revenue maximizing balance of electricity or steam allocation to emissions control processes is a function of the efficiency of conversion, the efficiency of the process, and the value of the emission control (e.g. the price on carbon). As a result, the optimal allocation of energy sources is likely to vary significantly between individual plants.

522330 Cost Analysis Model for Air Capture Sorbents

Abstract:

Air capture is the technology of capturing carbon dioxide (CO_2) from ambient air to mitigate climate change impacts. For air capture, a sorbent is used to bind carbon dioxide either chemically or physically. Being in its early development stages, air capture sorbent costs are difficult to estimate. The sorbent is the most significant part of an air capture system and can contribute as much as half of the total weight. If we allocate a fraction of the revenue of CO_2 collection to sorbent cost, then we can estimate the value of a kilogram of sorbent by setting the net present value of the allocated CO_2 revenue minus the sorbent cost to zero. Using this assumption, we developed a model to estimate the maximum allowed budget for an air capture sorbent with known characteristics. For every sorbent with a known capacity ([kg CO_2/kg sorbent]), cycle duration ([min]) and degradation rate (), we can calculate the maximum allowed budget for the sorbent ([$/kg sorbent]) in a market with a known price of carbon dioxide ($_{CO2}$ [$/ ton CO_2]) and a given discount rate ().

To date, the authors have not been able to find a comprehensive cost analysis for any of air capture sorbents. There are lab-scale sorbents with exceptional capacity and kinetic characteristics, but these features do not guarantee an affordable large-scale sorbent and a cost-effective air capture process as a result of that. When an air capture sorbent has a high degradation rate (resulting in a short lifetime), the model shows that the sorbent's economy cannot be significantly improved even with substantial improvements in capacity and cycle duration.

Although approximate, our model provides a baseline estimation for commercial cost limits of air capture sorbents produced and tested on lab-scales. The analysis points to the importance of the lifetime of the sorbent. Additionally, the model provides a useful insight in comparison between air capture and flue gas scrubbing and even predicts a budget for new post-combustion sorbent, since the analysis readily carries over to other separation problems.

517308 A New 12-Step Climate Change Mechanism, based on Formation of Carbon Dioxide Hydrates Around Deep Ocean Submarine Volcanoes and Hydrothermal Vents

Abstract:

We are living in perilous times, according to a significant anxious portion of the world's population. We are being told that our planet is in a path that leads to uncontrolled warming from the sun, because we have been burning fossil fuels at dangerously high rates for too long now. The critical connection established by a vocal segment of the scientific community is the presence of ever-increasing levels of carbon dioxide gas from various energy/materials-producing operations, or just from plainly existing in this world. Another significant fraction of the world's population does not believe that human existence is in any danger at all despite what the anthropogenic or man-made climate change (CC) proponents have been preaching for more than a decade now. To round out the camps, some people just do not care; but one can infer from their actions the lack of decisiveness or the tendency to retreat into cozier and/or simpler lifestyles. This work is meant to augment the current debate on the CC issue, based on scientific concepts and methods used by the author in the field of chemical engineering. Ideas that are presented here are not meant to be the last word on the CC debate. Instead they are formulated to provide a middle and hopeful ground on this contentious subject.

In this presentation, the author will unveil critical mechanistic connections between the formation of underwater waves from catastrophic thermodynamic phase changes involving carbon dioxide hydrates deep in the ocean and prevailing climate changes, including the more disruptive climatic events (such as the El Nino phenomenon). To someone with enough knowledge of energetic systems, significant heat absorptions/releases and volume changes per fluid mass resulting from the formation of condensed phases seems to be quite obvious. But this concept has been overlooked and/or disregarded. It seemed that mass and energy effects of natural forces that resulted in these massive underwater waves have been averaged out in the entire ocean ecosystem or not involved in the measurements. And yet it will be shown that while their origins on the ocean floor are local, their climatic effects can be global in scale.

It should be expected that time-dependent (dynamic) measured climatic variables (such as global carbon dioxide ocean uptakes, sea surface temperatures, atmospheric temperatures, sea levels, etc.) from human-caused (anthropogenic) carbon dioxide emissions will have a well-behaved smooth-curve dynamic pattern. This is because the atmosphere where emissions are mostly released is a relatively good fluid dispersion system for anthropogenic carbon dioxide that is also uniformly absorbing of the sun's energy. Whereas, it will be shown here that effects of natural intense localized temporal energy releases from the oceans will have a distinct dynamic signature, and should show as a series of peaks in various time-dependent plots. If the dynamic signature coming from anthropogenic carbon dioxide emissions is dominant, then the overall result of its dynamic signature in the global system should be dominant as well. As it can be seen from the implications of the Fifth Intergovernmental Panel for Climate Change (IPCC) report, this is not the case. Time-dependent peaks of climatic experimental data are still significant instead of being completely overwhelmed by the anthropogenic dynamic signature. Various climatic data will be shown here to be mechanistically linked to dynamic peaks associated with catastrophic energy absorptions/releases, with cycles currently revealed in mean seasonally adjusted global measurement data that are recurring at 1-7 peaks per decade. These climate change cycles have not been adequately explained by current concepts, even though they have usually been recently occurring as a yearly cycle.

In conclusion, with better understanding of the conceptual underpinnings of global carbon dioxide releases, it seems to be feasible to formulate engineering solutions to manage these above-mentioned catastrophic climate cycles from the deepest sections of the oceans. The possible attainment of a manageable global equilibrium of climatic behavior is the goal that the author aspires to communicate to the public. It should also be noted that the 12-step mechanism outlined in this work to explain climate change-related behavior has been used by the author to explain some other ocean mysteries, such as El Nino Southern Oscillation (ENSO) events, the volcanic torus, large deep ocean solitary waves or solitons, and other interesting revelations.

538432 Marine Exoskeleton-Based Biosorption of Heavy Metals: Performance and Cost Analysis

Abstract:

Heavy metals are extensively found within the natural environment contributing significantly to potable water pollution. Heavy metals are found in potable water because of industrial and agricultural activities, fossil-fuels combustion, volcanic activity, or mineral erosion. Although specific heavy metals are essential for human, animal, and plant metabolic needs, only small amounts are needed and exceeding these amounts can be toxic. As a result, the development of new water remediation technologies, especially for heavy metal removal from water, has been pronounced recently. Among the methods for metal ion removal, adsorption is one of the most reliable, robust, and hence prevalent. The technical feasibility and cost effectiveness of this process is highly related to the materials used. Crustacean shells, a waste from the seafood industry, have been identified as a viable material for the biosorption of lead, cadmium, chromium and zinc. This biomaterial is a low-cost alternative for water remediation. Its performance has been evaluated in batch experiments for heavy metals under pH, temperature, initial concentration, and time. Results demonstrate a removal efficiency of ~ 99.5% and for lead, 97.0% for zinc, 99.3% for chromium and 6.8% for cadmium. The results were compared with commercially available products and. In addition, a preliminary cost analysis of the material was performed obtaining an estimated of US$0.03/m^3 of treated water with a minimum operation cost associated to the use of this material.

532888 RO Denitrification and Desalting of Impaired Brackish Water in Remote Communities

Abstract:

Groundwater nitrate contamination poses a threat to public health for community water systems especially in agricultural areas and in rural communities that cannot afford and/or capable of operating and maintaining the necessary small water treatment systems. In recent years, RO has been promoted as a technical solution to providing such communities with safe drinking water. However, the challenge to RO application in distributed remote communities is that the operation must be flexible with respect to water demand and quality. It is also critical to establish a cost-effective approach to assess the adequacy of deployed RO water treatment process for nitrate removal (real-time nitrate monitoring might still be costly for a number of distributed small treatment systems) as well as to provide an economical solution to handle the nitrate-laden residual stream locally in a robust and environmentally manner. In order to address the above challenges, the technical feasibility of RO water treatment in remote communities was evaluated based systematic analysis of process simulation for pilot communitiesâ€™ water treatment system based on site-specific information (e.g., water consumption, water quality). RO system design was evaluated for operation with concentrate recycle, over a wide range of external variables (e.g., water use demand, feed nitrate, and salinity levels), for target nitrate removal at water recovery effective for reducing system foot print and suitable for deployment in remote communities. The correlation between nitrate and salt passage was established via computer simulation as well in a series of laboratory experiments using a pilot-scale brackish water RO unit (developed by UCLA) integrated with the portable online nitrate monitoring system. In addition, the feasibility of utilizing existing septic tanks in remote communities for handling nitrate containing RO concentrate was evaluated with respect to technical and environmental viewpoints given. Results demonstrated that system operation at the optimal concentrate recycle ratio enabled recovery up to ~90%, while producing permeate product water at nitrate levels well below the regulatory established MCL for nitrate. Through establishing a correlation between nitrate passage and salt passage, small water system owners/operators, and regulatory agencies can assess the performance of the distributed small water treatment systems in an economical manner in order to gain confidence in the reliability of treatment system to handle nitrate removal. The current study suggests that local RO treatment of community groundwater and handling RO residual stream in a septic tank would have a minimal environmental impact while providing a significant beneficial use of safe drinking water.

516532 Optoelectrokinetic Trapping of *Escherichia coli* in Water

Abstract:

At least 30% of the worldâ€™s population drink water from sources contaminated with feces and as a result, high loads of *Escherichia coli*. It is also one of the risk factors responsible for the death of 361,000 children yearly, due to preventable water sanitation-related diarrheal diseases.

This procedure traps *E. coli* and other bio-particles by the simultaneous application of laser power and Alternating Current(AC) frequency through a microfluidic chip. Its success with microorganisms is based on critical frequency, laser intensity, voltage and bio-particle size. The trapping ability of this technique will ensure continual access to particle-free water for the populace.

Preliminary experiments have shown that increasing concentrations of *E. coli* in the microfluidic chamber varies directly with the electrical resistance of the entire system and indicates that the trapping AC frequency of *E. coli* is within the range of 15kHz - 40KHz.This on-going research can be successfully used to isolate, monitor and quantify *E. coli* in streams and other drinking water sources.

528273 Biocomposite Material from Fique and Iron Nanoparticles As Adsorbent for Mercury Removal from Aqueous Solutions: Kinetic and Equilibrium Studies

Abstract:

The wide range of contaminants in surface and groundwater has become a critical issue worldwide, due to population growth, rapid industrialization and development of long-term drought. Persistent pollutants in wastewater include heavy metals, inorganic and organic molecules, and many other complex compounds. As a result, the number of available water resources in nature to sustain life has been decreasing over time, becoming one of the biggest problems nowadays, because of the diversity of pollutants and the rapid depletion of available sources. Therefore, there is the need for controlling the harmful effects of pollutants and to improve the conditions of human life, for example according to WHO, mercury is one of ten products or groups of chemicals that pose special public health problem due to its high contamination potential and its widespread use in not regulated industries [1].

Mainly, mercury has become an important contamination vector in areas were mining of noble metals is active (especially in illegal extraction operations), one of the major source of mercury contamination is the gold extraction process [2]. This issue is of special interest in Colombia, where illegal mineral extraction uses extensive quantities of mercury, affecting atmospheric and water quality [3]. This problem is particularly alarming in the small-scale gold mining process (usually artisanal performed), where one kilogram of gold obtained released into the environment three to five kilograms of mercury [4]. Increasing the statistics of Colombians health problems related to neurological symptoms, decreased motor skills, impaired hearing and vision, among others [2]. Although gold mining is an activity with significant importance in Colombian economy (producing 30 tons per year), environmental control is poorly structured, especially in the areas where handmade extraction is practiced; creating at the same time environmental liabilities, turning this practice into an unsustainable process [4] [5].

Mercury separation by adsorption, has been an alternative that takes advantage of leverage the physical interactions between the solid surface and adsorbate. Different materials can be used as adsorbents, depending on the application, price, mechanical and chemical resistance, abundance, easiness to use, easy regeneration, adsorption capacity and surface area. In this works, a new economic and efficient alternative to remove mercury (II) from aqueous effluents has been investigated. A biocomposite material using iron nanoparticles anchored onto Fique fiber (lignocellulosic biomass), were developed as adsorbent.

In the adsorbent synthesis raw fique fiber was pretreated during 60 minutes under sonication with the aim to increases the crystalline cellulose and clean fiber, then the pretreated fiber was functionalized by cationization treatment under acid and basic conditions. In order to obtain iron nanoparticles anchored into the fique fiber, an impregnation method was implemented, the functionalized fiber was submerged in chloride ferric changing the time on one, two and three days. After, the impregnated fiber was soaked in a reducing solution of 1M $NaBH_4$ (Merck MilliporeÂ®) per 10 minutes, holding the molar ratio 2:1 (BH_4Ë‰ / Fe^{+3}), finally the adsorbent was dried at room temperature.This material was characterized by Atomic Absorption Spectroscopy (AAS), X-Ray Diffraction (XRD), Fourier-Transform Infrared Spectroscopy (FTIR) and Scanning electron microscope (SEM).

Preliminary experiments were performed to determine the amount of adsorbent material to use in the adsorption (400, 800, 1000 and 1500 mg). The adsorption experiments were arranged in a jacketed glass batch reactor, with continuous agitation (240rpm), using 800 mg of adsorbent in contact with 150 mL of mercury solutions. Each run lasted 30 minutes. All experiments were performed at 20Â°C and 551mmHg. Every 5 minutes an aliquot was taken and analyzed for mercury (II) using vapor atomic absorption spectrometry (CVAAS). A Box-Behnken experimental design evaluated the effect of pH (4-10), initial concentration of mercury ions [Hg^{2+}] (1-14 ppm), and load of iron in the adsorbent material (10.9-14.9 Fe wt.%), in the removal of mercury (II).

The equilibrium was reached after 15 minutes, with a maximum value of 92.43% mercury removal of mercury (II) ions from aqueous solution, under following conditions: pH: 4, mercury initial concentration:10 ppm and 10.9%wt load of iron in the material. The adsorption process was found to be pH dependent, due to pH has no apparent effect on adsorption in the range of pH 4 to 7. Nevertheless, for pH greater than 8, the adsorption capacity increases as the pH value also increase. The thermodynamic study confirmed that the adsorption was spontaneous and exothermic in nature.

In addition, eight models were adjusted to understand how the adsorption process was carried out, these models included: adsorption diffusion models, adsorption reaction models, chemisorptions kinetics model and adsorption isotherms. It was evidenced the chemisorption mechanism predominates respect to physisorption, Langmuir, Freundlich and Tempkin isotherms were applied to describe equilibrium data. However, they failed to properly describe most of the experimental isotherms obtained. It was found that the kinetics of mercury adsorption onto the surface of the biocomposite is described by the pseudo-second order model supporting chemisorption (chemical reaction) as rate controlling mechanism which indicates that the adsorption process is irreversible. Thus, biocomposite was found to be superior adsorbent compared to other adsorbents for mercury (II) ions from aqueous solutions.

536617 Synthesis of Graphene from Mango Biomass Via PECVD

Abstract:

Introduction:

In the year 2004, a wonder material only one atom thick was discovered. The materials is an allotrope of carbon known as Graphene. The carbon atoms are arranged in hexagonal shapes with sp^2 hybridization state. Graphene has found numerous applications in catalysis, Li-ion batteries, water purification, gas separation, super capacitors, quantum computing, building materials, composites and several others. Graphene has been synthesized by several methods such as mechanical exfoliation, hydrothermal treatment and chemical vapor deposition.

In traditional CVD process, the reaction time is in the order hours, if not days. Also, the use of non-conventional sources of carbon like biomass, plastic waste, etc takes even more time with growth temperature of > 900 â„ƒ. Recently, with the advancements in plasma technology, the traditional CVD process has been modified to Plasma Enhanced Chemical Vapor Deposition. A carbon source is usually required to form graphene, traditionally methane and copper is used as a substrate.

Materials and Methods:

Mango peels are a waste in the food industry but rich in carbon. Biomass derived from mango peel has been used as the carbon source in this study. The biomass is placed on a copper substrate in a Quartz tube used as a the PECVD reactor. The plasma gas used was 10% Hydrogen and 90% Argon. The PECVD reactor operates at a temperature of 750 â„ƒ. The plasma chamber was under vacuum and a power of 300 W was used. A 15 minute annealing time was used for all synthesis.

Results:

The synthesis was carried out for 5-60 minutes growth period to monitor the effect on quality and elucidate a possible mechanism fo the synthesis. Monolayer graphene was obtained at synthesis time of 60 minutes with least defects. At lower synthesis time either the defects were too high or multilayer graphene was obtained. Plasma treatment was essential for synthesis of graphene. With only thermal annealing, graphene was not obtained.

537873 Hydrothermal Carbonization of Wet Biomass Materials for the Production of Hydrochar

Abstract:

Organic waste is a major component of landfills, which is the largest source of methane emission in the USA. From harvesting to consumption, excluding the non-edible parts, 30-40% of the food is wasted, equaling more than 35 million tons per year or approximately 240 pounds of food per year per person. Finding ways to utilize this wet organic waste into useful products will be desirable from both economic and environmental sustainability points of view.

Villanova University and Gray Brothers-Somax Environmental Company are working together in developing a comprehensive biorefinery system involving multiple conversion process technologies, specifically anaerobic digestion and thermochemical processes, for utilizing municipal and agricultural solid waste into various valuable bio-based products. One conversion process technology being studied is the hydrothermal carbonization (HTC) process. HTC is a thermochemical process that converts wet biomass materials to a coal-like product. Under sub-critical conditions, water in the HTC reactor stays in liquid form and acts as a reaction medium to promote the breakdown and cleavage of chemical bonds in the solid biomass. The absence of liquid-to-vapor phase change of the water makes the process significantly less energy intensive compared to a process that involves water vaporization or drying. The temperature range for HTC is from 180-240°C producing solid hydrochar as the main product.

The focus of this research was to evaluate the effects of feedstock and HTC process parameters on the yields and physiochemical properties of hydrochar products. HTC reaction temperature and retention time were the major parameters that were varied. Retention times varied from thirty minutes to two hours and the temperature range was focused between 180C and 240°C. Nine different food groups and other biomasses were used as feedstocks for the HTC process, which include: vegetable proteins, mixed vegetables, carbohydrates, dairy, fruits (and fruit waste), animal proteins, brown grease, anaerobic digestor seed, and a mixed food waste.

Various analytical and characterization testing will be conducted on the liquid, solid, and gas products of the HTC process. The characterization techniques will include CHNOS elemental analysis, approximate analysis (TGA), specific surface area, porosity, heating value, and microstructure (SEM analysis) of the hydrochar products. The liquid and gas analysis will be performed by using gas chromatography techniques. Results from this study will be analyzed to determine the potential optimum process conditions for the HTC process for different organic material feedstocks.

*Contact: Dr. Justinus Satrio (justinus.satrio@villanova.edu)

534236 A Feasibility Study on Biofuel Production Using Anaerobic Digestion and Thermochemical Catalysis

Abstract:

The use of biomass as a potential feedstock for the production of liquid hydrocarbon fuels has been under investigation in the last few decades. This paper discusses a preliminary design and a feasibility study of producing liquid hydrocarbon fuels from biomass through a combined biochemical and thermochemical route. The process involves anaerobic digestion (AD) of the biodegradable portion of the biomass to produce methane rich gas. The methane rich biogas stream is purified by removing contaminants and upgraded to liquid hydrocarbon fuel in a gas to liquid facility (GTL) via thermochemical conversion route. The biogas conversion involves two major steps tri-reforming step to produce syngas (a mixture of CO and H_2), and Fischer-Tropsch Synthesis (FTS) step to convert the syngas to a spectrum of hydrocarbons. Separation and upgrading of the produced hydrocarbon mixture allows production of synthetic transportation fuels. AD is ranked as one of the best waste management options as it allows for energy recovery, nutrient recovery, and reduction in greenhouse gases emission.

A detailed process modeling of the process was carried out using Aspen Plus process design software package. Data for the process was based on literature on AD combined with laboratory results on the biogas to liquid conversion process. The composition of the final liquid hydrocarbon from the Aspen Plus model has been compared to the composition of commercial diesel fuel, and results have shown good agreement. As a result, the most current commercial diesel prices were used to evaluate the potential revenue from selling the product in the open market.

The total capital investment to construct the plant with a capacity of handling 100,000 ton per year of wet biomass is \$11.8 million with a potential of producing 3 million gallons of diesel. The base case feedstock is corn stover. The annual operating cost to run the plant is estimated to be \$8.05 million. An annual revenue from selling the diesel product is estimated to be \$10.8 million taking into account a green energy incentive of \$1.00/gallon of diesel sold. The net present worth at the end of the plant life is \$7.26 million with a discounted cash flow of return of 27.1% (if \$3/gallon incentive is provided the process would have NPW of \$28 million, DCFROR of 54.5%, and discounted payback period of 1.5 years). The breakeven cost of diesel is determined to be \$2.96/gallon assuming no tipping fees are charged for handling the waste.

Sensitivity analyses results concluded that the profitability of the process is most sensitive to variation in diesel selling price. Based on these results, it can be concluded that the process is profitable only if incentives are provided for renewable fuels due to the current low prices of fossil fuels.

530438 Identifying Co-Products from Guar and Guayule Processing Residues

Abstract:

Guayule is a woody shrub native to the southwestern US that produces rubber in its stems. In the production of guayule hard rubber for tires, two residue streams are produced: a complex resin liquid and a ground, dry woody bagasse. Guar is a bean plant that grows well in semi-arid regions and produces a polysaccharide in the endosperm of its seeds. This polysaccharide is processed into gum products for hydraulic fracking and food applications. Residues from guar gum production include a meal that is used as an animal feed and the non-seed parts of the plants. Identifying higher-value products and applications for these residues the goal of the Logistics thrust of the Sustainable Bioeconomy for Arid Regions (SBAR) USDA-funded Biofuels and Bioproducts Challenge Area Coordinated Agricultural Project. Early work on this project has focused on characterization of the residue materials and the potential to use those materials to produce renewable fuels and chemicals.

538699 Continuous Calcination of Pyrolysis Oil Derived Renewable Coke in a Rotary Tube Furnace

Abstract:

The advancement of innovative pyrolysis methods has enabled flexible processing of pyrolysis oil (bio-oil), which could accelerate integration into a traditional petroleum refinery infrastructure. Both catalytic fast pyrolysis (CFP) and tail-gas reactive pyrolysis (TGRP) enable ease of distillation, leaving a solid residuum that is suitable for conversion into a renewable calcined coke (bio-coke), with potential to replace petroleum coke (petcoke) used in a wide range of industrial applications. For example aluminum smelting employs calcined petroleum coke to manufacture anodes for large-scale conversion of alumina into aluminum in the Hallâ€"HÃ©roult process. The industry faces ongoing coke quality issues associated with petcoke , such as increasing impurity content (sulfur and heavy metals), decreasing bulk density, and increasing concentrations of amorphous textures. The environmental impacts of the smelting process could also be mitigated from a reduction in CO_2 and SO_2 production via use of bio-coke. Production of calcined coke from bio-oil distillation residues could allow for production of coke with sufficiently anisotropic texture and low metal impurities. This talk will cover advances in renewable calcined coke technology. We employed a rotary tube furnace to mimic the processing and most phenomena that occur with renewable coke calcination in industrial rotary kilns. Bio-oil distillate bottoms were fed continuously into the tube furnace at fixed mass flow rates. Tube rotation speed and tube inclination angle are adjusted, such that both residence time and product morphology can be accurately controlled. Properties of the calcined renewable coke are compared with previously obtained results from batch processing.

529829 Experimental Investigation and Process Design for the Conversion of Carbon Black Waste into Valuable Resources

Abstract:

Carbon black waste generated in oil refinery may contain a high concentration of heavy metals, for instance, vanadium. It is therefore of great interest to recover the vanadium as value-added products instead of just disposing in landfill. In this study, three different types of vanadium compounds have been developed from a set of leaching and post-processing steps, including monoclinic bismuth vanadate, vanadium-based metalâ€"organic frameworks (V-BDC and V-NDC), and alkaline earth vanadate nanomaterial. A variety of applications of these compounds are investigated in this study. Both V-BDC and V-NDC showed high catalytic activities for benzyl alcohol oxidation towards benzaldehyde. Alkaline earth vanadate nanomaterials are potential semiconductors for semiconducting glasses. To evaluate the profitability and feasibility of implementing these technologies on an industrial-scale process, a conceptual process design was developed for the scale-up production and cost estimates to determine the profitability of the developed process. Sensitivity analysis was then conducted based on perturbations to several economic parameters and the results can be used to estimate the variability of the profit margins to uncertainties in these economic parameters.

533128 Thermodynamics and Kinetics for KOH Leaching of Potassium Alunite from Copper tailings

Abstract:

Thermodynamics and kinetics for KOH leaching of potassium alunite from copper tailings

Mengjie Luo, Youfa Jiang, Ping Li, Xingfu Song, Jianguo Yu**

National Engineering Research Center for Integrated Utilization of Salt Lake Resources, East China University of Science and Technology, Meilong Road 130, Shanghai, China, Postcode 200237.

mengjie_luo@ecust.edu.cn, xfsong@ecust.edu.cn, jgyu@ecust.edu.cn

Abstract:

Five to fourteen billion tons of mine tailings per year worldwide will be discharged from the contemporary mining industry. Tailings as waste is not only the loss of valuable resources, but also brings serious environmental impacts and ecological damages. In Fu Jian province of China, about 18000 ton/day copper tailings are discharged from the production line of the extraction of Cu from copper mine. Through mineral flotation, we can get the refined copper tailings with 72.67% alunite, 19.30% kaolinite, and 6.86% quartz. Alunite is considered a potential resource for both the production of alumina and potash fertilizer. In this work, a novel gradient leaching process is proposed to recover potassium and aluminum resoureces from copper tailings. In the proposed gradient leaching process, alunite from tailings is selectively extracted, while impurities of kaolinite and quartz, are still remained in the leaching residues. Here, we shall investigate reaction thermodynamics, reaction kinetics and dissolution mechanism of alunite from copper tailings in KOH solution.

It is deduced that there exist three reactions during tailings leaching with potassium hydroxide, as shown in Eq(1)-(3). Using the HSC chemistry software, $\Delta_r G^\circ$ and $\Delta_r H^\circ$ for three reactions with KOH in molten state and aqueous state are calculated at 303 K-493 K, as shown in Figure 1. It is found that under the same reaction conditions, the thermodynamic tendency of reactions with KOH is potassium alunite > kaolinite > quartz.

$KAl_3(OH)_6(SO_4)_2 + 6KOH = 2K_2SO_4 + 3KAlO_2 + 6H_2O$ (1)

$Al_2O_3 \cdot 2SiO_2 \cdot 2H_2O + 6KOH = 2K_2SiO_3 + 2KAlO_2 + 5H_2O$ (2)

$SiO_2 + 2KOH = K_2SiO_3 + H_2O$ (3)

Thermodynamically, the reaction of potassium alunite in KOH solution might become preferential and dominant in the leaching process of tailings. Therefore, the gradient leaching of potassium alunite from tailings is proposed, where potassium alunite is dissolved firstly, and then following kaolinite and quartz dissolution. Leaching experiments are carried out to test the thermodynamical prediction, as shown in Figure 2. Using 50% KOH to leach tailings at 323 K, potassium alunite is dissolved into solution with above 90% recovery, while kaolinite and quartz as impurities are still kept in residue. It is recommended to control the silicon ions concentration as small as possible in the solution in order to maintain the high potassium leaching ratio and to reduce the expensive KOH consumption. Therefore, the leaching temperature should be maintained below 373 K.

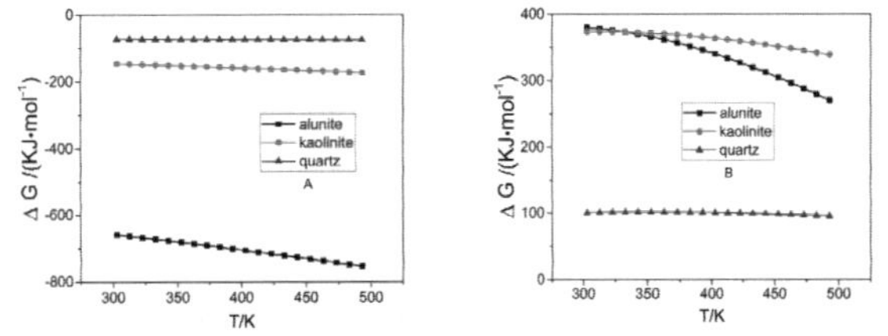

Figure 1. $\Delta_r G^\circ$ for reactions (1)-(3) with KOH in molten state (A) and aqueous state (B) calculated by HSC chemistry software

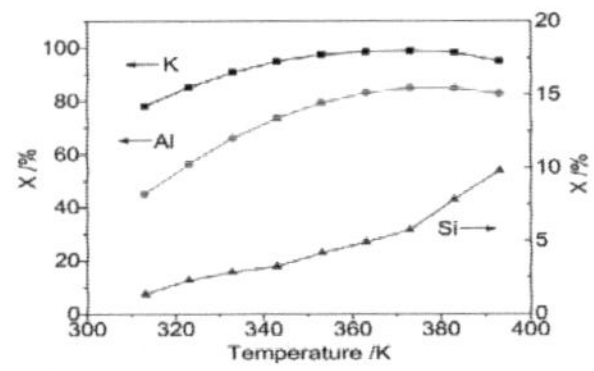

Figure 2. Experimental results of K, Al, Si leaching ratio at different leaching temperature

Thermodynamic analysis and experimental results prove that it is feasible to extract potassium alunite from copper tailings into the alkaline solution at 323 K-363 K and to keep kaolinite and quartz in the leaching residues avoiding the traditional desilication step, being a green and closed-loop gradient leaching process. The dissolution mechanism of potassium alunite from tailings is that individual aluminum atoms at the surface react with hydroxide ions to form activated complexes and sulfate tetrahedral at the surface depart intact, as shown in Figure 3. According to the proposed dissolution mechanism, the shrinking core model with the surface chemical reaction control is used to predict the leaching kinetics of alunite from tailings at 323 K-363 K with an acceptable accuracy, where the apparent activation energy is 44.48 kJ•mol^{-1} and the order of reaction with respect to KOH is 1.76, as shown in Eq(4).

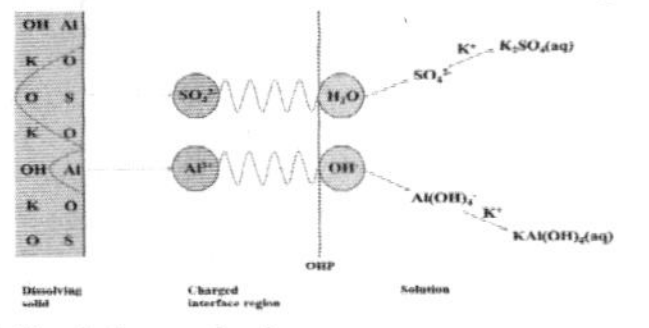

Figure 3. The proposed dissolution mechanism

Reaction model: $1-(1-X)^{1/3} = 2122.16\, C_{KOH}^{1.76} e^{-44480/(8.314T)} t$ (4)

Keywords: copper tailings, gradient leaching, thermodynamics, kinetics, dissolution mechanism

MANUFACTURE OF SUSTAINABLE LIGHT-WEIGHT CONCRETE BLOCKS BY UTILIZING INDUSTRIAL SOLID WASTE

Arunima Shukla[1] *and Ashok N. Bhaskarwar*[2]
Indian Institute of Technology Delhi, New Delhi, India

Abstract

Light weight or aerated concrete has been accepted worldwide as an eco-friendly building material. The literature survey reveals that the waste from the construction-materials manufacturing process can be utilized and the process of utilization established as a secondary industry in the area of solid waste recycling. We propose here an eco-friendly and less energy intensive solution to utilize the industrial solid waste to manufacture aerated concrete than the currently commercialized autoclaved aerated concrete (AAC) manufacture. We have used Colloidal Gas Aphrons (CGAs), which are surfactant stabilized spherical microbubbles, as an aerating medium to manufacture light weight concrete. CGAs are a new route of aeration which not only allow industrial waste incorporation, but also provide a solution to the pursuit of a zero-waste manufacturing process for the aerated-concrete products.

Keywords: Supplementary cementitious material; Solid-waste utilization; CGAs-based aerated concrete.

Introduction

Precast fabricated light weight concrete blocks have gained a lot of importance in recent times owing to their light-weight, thermal insulation, heat resistance, and sound absorption-properties [1]. Properties of aerated concrete depend mainly on the porosity and microstructure, which in turn are functions of the type of binder and aerating medium used. The technology which is currently being used commercially is the use of aluminium-powder derived hydrogen as an aerating medium to manufacuture the autoclaved aerated concrete blocks (AAC). These blocks are kept in pre-curing room for several hours for initial setting. After reaching the semi-solid stage, these blocks are cut into smaller sized blocks for ease of transportation and utilization selling using high precision cutting wires technology [2]. These smaller aerated blocks are then autoclaved at about 190 ^{0}C under 12 bar pressure for about 14 hours [3]. Autoclaving results in generation of thermal stresses and expansion which may lead to development of cracks in the AAC blocks. These damaged AAC blocks are removed and sold away by the manufacturers at throw-away prices. The rate of solid-wastes generation because of the handling in automated aerated-concrete industries and subsequent transportation is estimated to be approximately 4 to 10 % of the total production rate. Some solutions were proposed to use this solid waste after crushing, such as using crushed autoclaved aerated concrete as filters for waste water treatment [4], but these solutions cannot be used at a large scale because of their energy intensive nature. The whole AAC manufacturing process is energy intensive which makes it costly as well as less eco-friendly. Aiming to develop an environment-friendly, energy intensive process which can cut down the costs of aeration and autoclaving and result in a zero-waste industry, we recently proposed the use of colloidal gas aphrons (CGAs) as an aerating medium to manufacture light-weight concrete blocks [5, 6]. CGAs are spherical microbubbles, typically in the range of 10-100 µm, which last long and have water-like flow properties [7, 9]. CGAs were invented by Sebba (1987), and are composed of an inner gas core, surrounded by hydrogen-bonded viscous water shell stabilized by an inner and outer surface consisting of adsorbed surfactant monolayers, a surfactant bilayer, and an outer electrical double layer as shown in figure 1. Using the

[2]Professor, Department of Chemical Engineering, IIT, Hauz Khas, New Delhi-110016, INDIA. Author to whom all correspondence should be addressed; e-mail: ashoknbhaskarwar@yahoo.co.in
[1]Research Scholar, e-mail: arunimas3@gmail.com
#

conventional route of making CGAs as proposed by Sebba (1987), CGAs have an air hold-up of typically 0.65-0.70 [8]. By changing some features of the CGAs production process, we were able to achieve an air hold-up in the range of 0.75-0.80, and used it here to make the light-weight aerated concrete blocks.

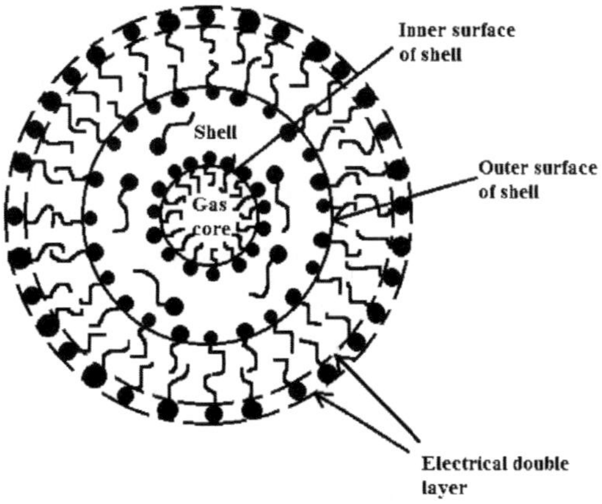

Figure 1. Conceptual schematic diagram of CGAs [10].

This paper mainly focuses on the use of CGAs to manufacture light-weight aerated concrete also utilizing industrial wastes such as fly ash and crushed autoclaved aerated concrete (CAAC) to make integrated aerated concrete blocks. We are able to utilize CAAC of 25 % w/w of total solids in the CGAs-based aerated concrete to make the re-integrated blocks.

Experimental work

Experimental Set-up

CGAs generator consists of an open cylindrical vessel with inverted-L shaped baffles inside it. A horizontal disc is attached to a shaft which is rotated with a electric motor which provides high rpm of up to 8000. Outlet for CGAs is from the bottom of CGAs generator, which is further connected to a centrifugal pump for withdrawing the CGAs from the generator as well as for recirculating CGAs in the CGAs generator.

Raw Materials Used

Sodium lauryl sulfate (SLS, Fisher Scientific) and tap water were used to generate CGAs. Ordinary Portland cement (43 grade), fly ash, and autoclaved aerated concrete waste (Biltech Ltd.) were used in making the re-integrated aerated concrete blocks.

Equipment Used

CGAs generator set-up was used to make the CGAs, which were then mixed with the solid raw materials in a planetary mixer (Ferrotek equipments). The resulted slurry was then poured into the moulds of 5 cm^3 (Aimil Ltd.) capacity, and cured in a steam-curing chamber (Ferrotek equipments) after the initial setting. The samples were dried in a hot air oven (Macro Scientific Works Pvt. Ltd.) at 110±10°C. Scanning Electron Microscopy (Zeiss EVO 50) was used to study the surface morphology of the cured aerated concrete samples. An optical microscope (Motic make) was used to take the photomicrographs of CGAs.

Experimental Procedure

A fresh batch of CGAs was prepared every time to make the aerated concrete. Tap water and surfactant were taken in the CGAs generator, which were then mixed at a low rpm to make the aqueous solution. This aqueous solution was stirred at 5500 rpm to make CGAs. CGAs were then mixed with cement, fly ash, and crushed autoclaved aerated concrete waste (CAAC) to make the aerated concrete slurry of the required composition. Prepared samples were then dried in air for 24 hours, and after the initial setting, these samples were de-molded. The de-molded samples were steam cured in a steam-curing chamber at 65°C for 4 to 5 hours. After curing, the samples were dried in an oven at 110±10°C for 24 hours to measure the actual weight for calculating their oven-dried density of the samples.

Results and Discussion

CGAs were formed in the CGAs generator and appeared like a milky-white dispersion to the naked eye. The dispersion was then observed under an optical microscope. The photomicrographs of CGAs were taken at different magnifications as shown in figure 2.

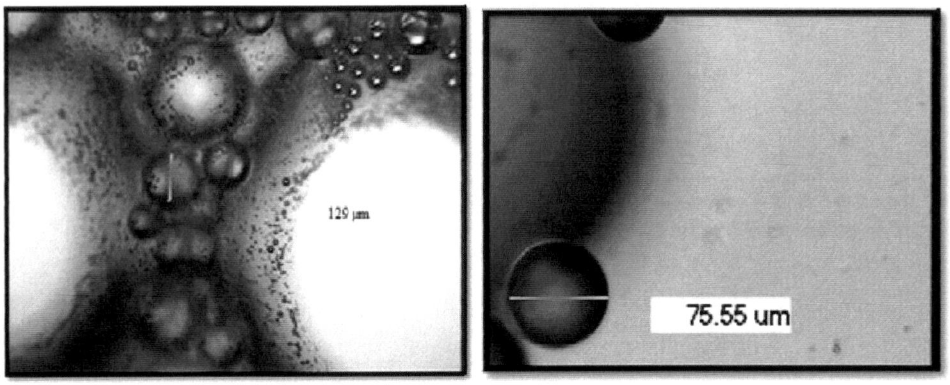

Figure 2. Photomicrographs of CGAs made using anionic surfactant, sodium lauryl sulfate (SLS), at different magnifications (100 X and 400 X).

The air hold-up achieved in CGAs was in the range of 0.75-0.80, and was measured by weighing 100 mL of the dispersion in a measuring cylinder. The weight of 100 mL of the dispersion was 20 g. Considering that air has a negligible weight, this whole weight of 20 g was of liquid. Since the density of solution was about 1 g/cc, the volume of the liquid in this dispersion was 20 cm^3. Therefore, the rest of the volume i.e. 80 mL was of air, which confirmed that the CGAs air hold-up was 0.80. The CGAs having typical aphron size range of 20 to 60 μm have a total specific surface area of 2400 cm^2/cm^3 to 800 cm^2/cm^3 for air hold up of 0.80. The sizes and the size range of CGAs were measured using the photomicrographs of CGAs and using Image J software to calculate the sizes and areas of the bubbles.

Chemical and physical analyses of cement and fly ash were done using Indian Standard code IS:269 and IS: 1727, respectively. These analyses confirmed that the raw materials matched the standards for further experimentation. CGAs were mixed with cement and fly ash to prepare the mortar. The prepared slurry was then poured sequentially into three moulds of 5 cm^3 each. Based on the sequence of pouring, three samples with different densities were observed, which were termed as 'cuts' as shown in figure 3. First-poured sample was named as first cut, second-poured sample as second cut, and third-poured sample as third cut. Densities of the samples, of course decreased with an increase in the weight percentage of CGAs in the slurry used. First poured sample was having the lowest density for every fixed composition. This is because the CGAs create a variation in air hold-up with height by creaming, which makes the top layer of slurry the lightest and the bottom layer the heaviest for each composition. All the experiments were repeated at least three times, and reproducibility is indicated by the error bars as shown in the figure 3.

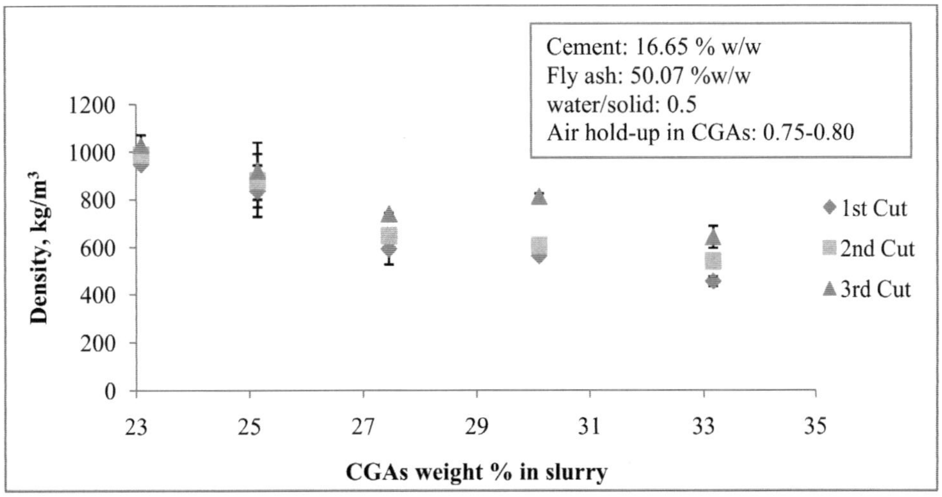

Figure 3. Densities of aerated concrete samples vs. CGAs weight % in slurry.

Compressive strength test was done for all the samples at a fixed load rate, which showed typically a decrease in the strength with an increase in the CGAs weight percentage in slurry, as expected (figure 4). With the addition of CGAs a slightly unusual trend was observe around 27 % w/w, as there was a decrease in the strength upto this and then a surprising increase with the addition of more CGAs to the slurry. This variation might be because of a narrower size range of CGAs in the slurry with the addition of more CGAs, as narrower air-void distribution is responsible for a higher strength in concrete [11]. The third-cut samples had the highest compressive strength, and the first-cut the lowest strength. A wide range of strengths was observed depending on the densities of the samples.

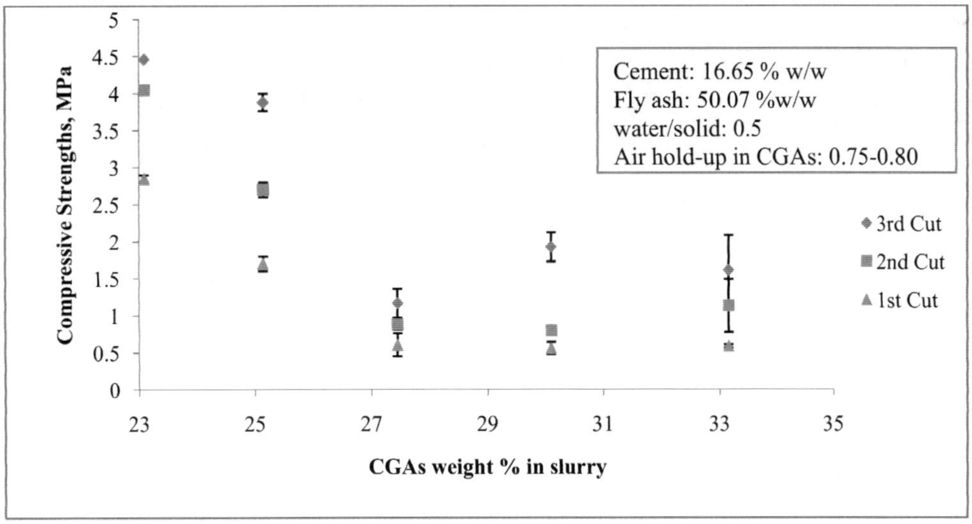

Figure 4. Variation in compressive strength with CGAs weight % in slurry.

All the data points were then plotted together to see the linear relationship between compressive strength and density (figure 5). It shows a reasonably good fit with a correlation coefficient of 0.75.

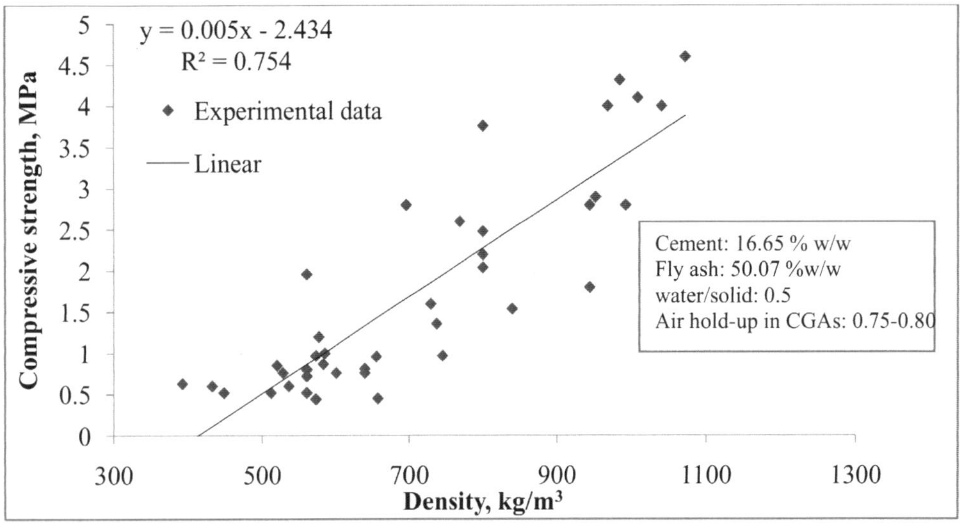

Figure 5. Compressive strength vs. density for all samples together.

Surface morphology of prepared concrete samples was observed using a scanning electron microscope (SEM) as shown in figure 6. SEM revealed the generation of hydration product in the aerated concrete. Calcium silicate hydrate was present which is responsible for the high strength of concrete in the concrete samples. Micro-pores and macro-pores were also observed in the analysis on the concrete which also signifies the porosity because of tiny micro-bubbles.

Figure. 6 SEM micrographs of aerated concrete sample at 5000X (L) and 10000X (R).

Prepared aerated concrete samples were then put in water and observed for the next 15 days. The aerated concrete blocks kept floating on the surface of water. This behavior of samples proved that the pores were not interconnected, which would make these blocks durable.

Autoclaved aerated concrete solid waste

Properties of re-integrated blocks made by utilizing CAAC waste were also measured. With varying percentages of CAAC, fly ash, and cement composition also varied which is shown in table 1. CAAC waste was used without crushing, which had a density of 610 kg/m^3.

Table 1. Various compositions with CAAC for target density (TD) of 800 kg/m^3, where CAAC density is 610 kg/m^3.

S. No.	W/S	CAAC (% of total solid mass)	Cement (weight %)	Fly ash (weight %)
1	0.5	5	21.10	42.19
2	0.5	10	19.99	39.97
3	0.5	15	18.88	37.75
4	0.5	20	17.77	35.53
5	0.5	25	16.66	33.31

The CAAC was utilized upto 25% w/w with respect to the total solids for a target density of 800 kg/m^3. All the experiments were repeated in triplicates, and standard deviations were plotted to represent the variations in the experimental data. Achieved oven dried densities were more or less in the range of target density as shown in figure 7.

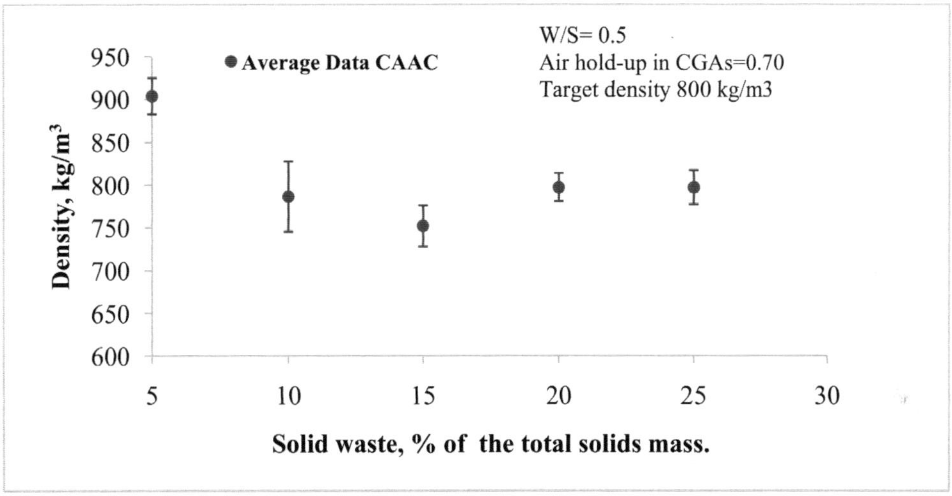

Figure 7. Densities of aerated concrete samples vs. solid waste percentages of autoclaved aerated concrete (AAC) for 800 kg/m^3 target density.

The samples were then steam-cured and their compressive strengths measured at a constant load rate. The compressive strength decreased with an increase in the solids waste percentage in the slurry as shown in figure 8. An experimental observation might explain this trend. At 25 % w/w of CAAC, almost the whole mould was filled with the solids-waste and there was very little space left for the binding and strength-imparting CGAs-based aerated cement-fly ash slurry. Hence, the decrease in the strength of the reintegrated aerated concrete samples while increasing the percentage utilization of CAAC.

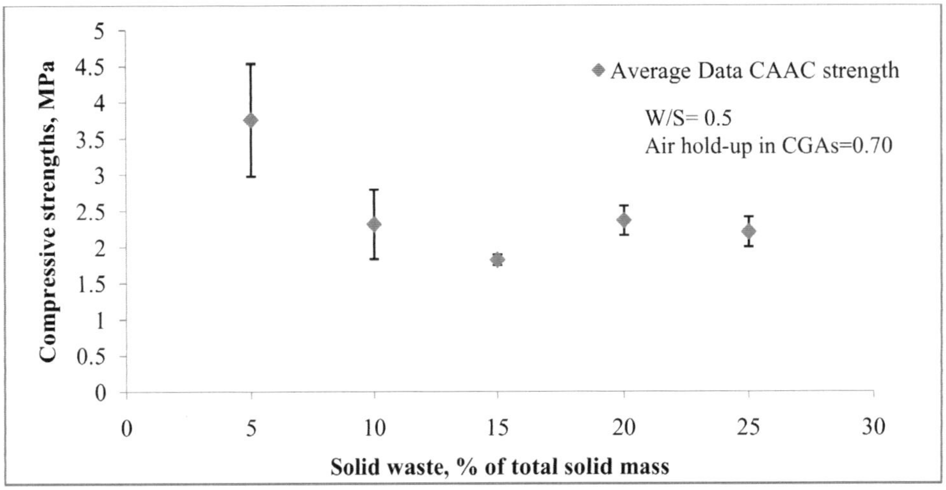

Figure 8. Compressive strength of reintegrated samples vs. solids-waste percentage of autoclaved aerated concrete (AAC) for 800 kg/m^3 target density.

There is a good correlation between the compressive strength of individual reintegrated concrete sample and the density, with a correlation coefficient of 0.84 as shown in figure 9.

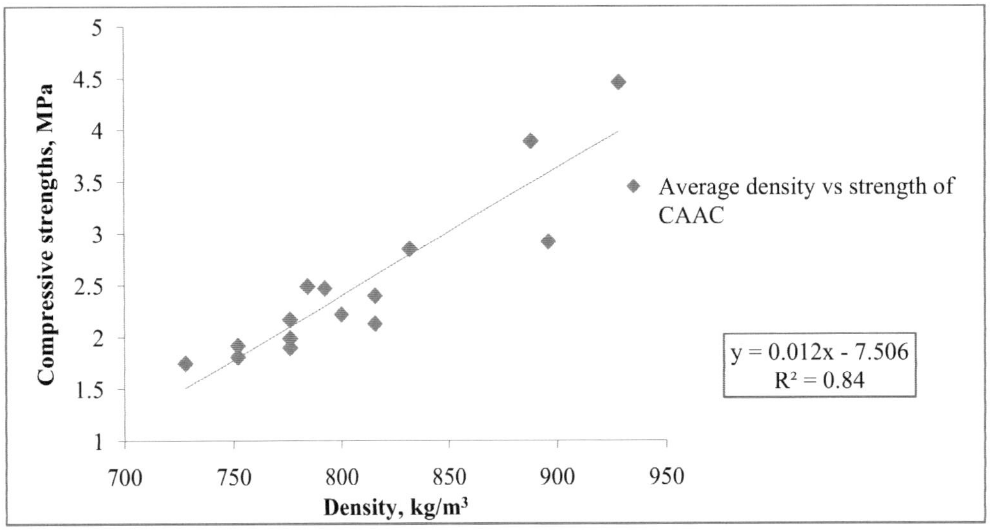

Figure 9. Compressive strength vs. density of re-integrated blocks of aerated concrete.

SEM analysis was done for the samples to examine the interface between CAAC waste particles and the CGAs-based aerated slurry as shown in figure 10. Hydration products and pores were observed, but no broken interface or de-bonding was observed.

Figure 10. Scanning electron micrographs of bonding between CAAC and CGAs based aerated slurry, at 1200X (left) and at 3000X (right), respectively.

Properties of the CGAs-based aerated concrete samples were compared with the Indian Standard (IS:2185-part 4), as shown in table 2. The properties of reintegrated samples matched well with the Indian Standard IS:2185 (2008).

Table 2. Comparison of the CGAs-based aerated concrete blocks with IS:2185 (part-4) 2008.

IS:2185 (part-4)	**CGAs-based aerated concrete samples with fly ash**	**CGAs-based aerated concrete samples with CAAC**
Strength, MPa (individual min.)	**Strength, MPa (individual min)**	**Strength, MPa (individual min)**
2.0 at min. reported density of 800 kg/m^3	2.04 at 800 kg/m^3	2.2 MPa at 800 kg/m^3
	0.63 at min. achieved density of 392 kg/m^3	1.75 at min. achieved density of 728 kg/m^3
22 at max. reported density of 1800 kg/m^3	4.6 at max. achieved density of 1072 kg/m^3	4.46 at max. achieved density of 928 kg/m^3

Conclusions

Summing up all the experimental observations and measurements, we may conclude that the CGAs are a promising medium to manufacture aerated concrete without the need for aluminium powder and autoclaving, and might make the aerated-concrete manufacturing process highly cost effective, sustainable, and environment-friendly. It is also possible to utilize fly ash and CAAC in aerated concrete to make re-integrated blocks. Fly ash is a well established raw material to manufacture concrete, but CAAC is a completely new waste/raw material for this product. CGAs are made, using an anionic surfactant, with an air hold-up in the range of 0.75-0.80. The size range of 20-60 µm, and the specific surface areas of 2400-800 cm^2/cm^3 were observed for air hold up of 0.80 in CGAs. CGAs-based aerated concrete was manufactured by utilizing fly ash for the densities range of 392-1072 kg/m^3 and compressive-strengths range of 0.63-4.6 MPa. Non-connecting pores of the reintegrated aerated concrete samples make the concrete blocks more durable. Moreover, 25 % w/w of CAAC with respect to the total solids was incorporated in the aerated slurry, and re-integrated blocks were made successfully. CGAs-based aerated concrete, by utilizing CAAC, was made with densities in the range of 728-928 kg/m^3 and compressive strengths of 1.75-4.46 MPa. The properties of the manufactured CGAs-based aerated concrete with utilizing the industrial solid waste, was found to be confirming to the IS:2185(part 4)- 2008.

Acknowledgments

The authors would like to thank Foundation for Innovation and Technology Transfer (FITT) at IIT-Delhi for their support in patenting the work, DST-Technology Systems Development Programme (DST-TSDP) for their research grant (Sanction no. DST/TSG/WM/2015262-G), and Biltech building elements limited, Palwal, for supplying the CAAC solid wastes for carrying out the experiments. We are also thankful to Mr. Satish Chand (Biltech Ltd., Delhi, INDIA) and Dr. Narendra Kumar Tiwary (NCCBM, Delhi, INDIA) for bringing the problem of utilization of solid wastes autoclaved aerated concrete to our attention.

References

[1] Geetha, S. and S. Madhavan (2015), "Light Weight Composite for Structural Wall Panels," Mater. Today Proc., vol. 2, no. 4–5, pp. 2928–2937.

[2] Chavan, P. G. and S. B. Patil (2013), "Substantial use of solid waste," Int. J. Adv. Technol. Civ. Eng., vol. 2, no. 1, pp. 61–63.

[3] Bisceglie, F., Gigante, E. and M. Bergonzoni (2014), "Utilization of waste Autoclaved Aerated Concrete as lighting material in the structure of a green roof," Constr. Build. Mater., vol. 69, pp. 351–361.

[4] T. Bao et al. (2016), "Advanced wastewater treatment with autoclaved aerated concrete particles in biological aerated filters," J. Water Process Eng., vol. 9, pp. 188–194.

[5] Bhaskarwar, A. N. and N. K. Tiwary (2013), "Air entrainment in cement mortars by colloidal gas aphrons (cga) for manufacture of aerated concrete," IN 2013 IN 295156.

[6] Bhaskarwar, A. N. and A. Shukla (2014), "Low density aerated concrete and a process of preparing the same," IN 2014DE01783 A 20160422.

[7] Jauregi, P., Mitchell, G. R. and J. Varley (2000), "Colloidal gas aphrons (CGA): Dispersion and structural features," AIChE J., vol. 46, no. 1, pp. 24–36.

[8] Bhatia, D., Goel, G., Bhimania, S. K. and A. N. Bhaskarwar (2005), "Characterization and drainage kinetics of colloidal gas aphrons," AIChE J., vol. 51, no. 11, pp. 3048–3058.

[9] Shukla, A. , Tiwary, N. K. and A. N. Bhaskarwar (Jan. 2015), "Air entrainment in cement mortars by Micro-Bubble Suspensions (MBSs) for manufacture of aerated concrete and low-density aerated concrete and a process of preparing the same.," Found. Innov. Technol. Transf. forum, IIT Delhi, vol. 21, no. ISSN-0972-2548, p. 20.

[10] Sebba, F. (1987), Foams and Biliquid Foams-Aphrons. Wiley, Chichester, Z. England.

[11] Liu, M. Y. J., Alengaram, U. J., Jumaat, M. Z. and K. H. Mo (2014), "Evaluation of thermal conductivity, mechanical and transport properties of lightweight aggregate foamed geopolymer concrete," Energy Build., vol. 72, pp. 238–245.

533434 Aerated Concrete Blocks Production Using Secondary Steelmaking Slag

Abstract:

Aerated concrete blocks production using secondary steelmaking slag

Introduction:

Steel market plays a pivotal role in the diversification of the oil-based economies to grow its non-oil based commodities and create an industrial sector that competes internationally. One of the main commercial processes for steelmaking is the Electric Arc Furnace (EAF) steelmaking. This process is commonly preceded by an ironmaking process where iron ore pellets are reduced to metal iron and then sent as direct reduced iron to the EAF. In the EAF, also called primary steelmaking step, the reduced iron is melted and combined with flux (mainly dolomite) for impurities removal forming a separate phase of liquid EAF slag. Finally, the liquid iron goes through an alloying process in the Ladle Refining Furnace (LRF), also called secondary steelmaking step, where the composition of the steel undergoes final adjustment and burnt lime is added to remove impurities forming a separate phase of liquid LRF slag. Both liquid slags (EAF and LRF) are air-cooled after being poured out of the furnaces forming solid slags.

In the steelmaking process described there are several solid wastes that is formed at each step of the iron/steelmaking process. DRI dust (containing > 50% Fe) is generated from the reduction step, EAF slag is generated from the primary steelmaking step, and LRF slag (~50% CaO) is generated from the secondary steelmaking step. While there are research directions focused on the use of these solid wastes, many steelmaking industries accumulate these solid-wastes in landfills resulting in hefty landfill costs. This project aims at reducing the environmental and economical impact of steelmaking plant by utilizing the LRF slag in the production of Aerated Autoclaved Concrete blocks (AAC).

AAC are highly insulating construction materials, with low thermal conductivity, very light weight, and exceptional fire resistant making them excellent materials for buildings insulation. Moreover, unlike conventional concrete, AAC are produced from a chemical process that has zero CO2 emissions. AAC is conventionally made from lime powder, Aluminum (Al) powder, silica sand, cement, gypsum, and water. The chemical process of producing AAC generates hydrogen gas by the reaction of the Al powder with calcium hydroxide expanding the slurry to almost double its size giving the AAC its lightweight property. In addition, the high-pressure autoclave curing process increases its mechanical strength significantly by reacting CaO with SiO_2 and H_2O forming crystalline hydrates called tobermorite. Due to the high composition of CaO and SiO_2 in the LRF slag, we propose in this project to produce AAC using LRF slag to replace the CaO and SiO_2.

Outline of method:

In an attempt to study the viability of partially replacing lime and silica with LRF slag, AAC blocks samples were produced by varying the mass composition of lime with the LRF slag while keeping the overall mass composition for both of these materials constant. The produced samples then were tested for their mechanical strength and crystal structure. Special attention was given to the amount of Al powder added and to the water-to-dry components (liquid-to-solid) ratio because of their direct effect on the mechanical strength of the AAC. Detailed experimental steps are not included for brevity.

Results brief overview:

The prepared AAC specimensÕ compressive strength properties and microstructural features are analyzed and evaluated in comparison to conventional AAC. The compressive strength was tested using MTS alliance RF / 150 Material testing system. Our results show that the AAC produced from pure slag (sample #4) meets the range of the compressive strength of a standard AAC block; hence the slag is a promising candidate for the production of AAC. Table (2) shows the compressive strength of AAC blocks with varied LRF slag compositions.

Table SEQ Table * ARABIC 1 Compressive strength of AAC blocks with varying LRF slag composition

Samples	*COMP. STRENGTH-(KN/Mpa)*
Standard	3 – 4.5
1	3.15
2	2.14
3	2.77
4	3.6
5	2.01
6	2.47

The crystal structure of the produced AAC samples was captured using Scanning Electron Microscope imaging technique and was compared to the crystal structure, tobermorite, of a standard AAC. Sample 4 (100%LRF slag) showed a needle-like tobermorite structure as compared to a lath-like structure for the conventional AAC. Nevertheless, the results indicate that inclusion of LRF slag enhances crystal growth (tobermorite) due to increase of Ca/Si ratio which increases compressive strength.

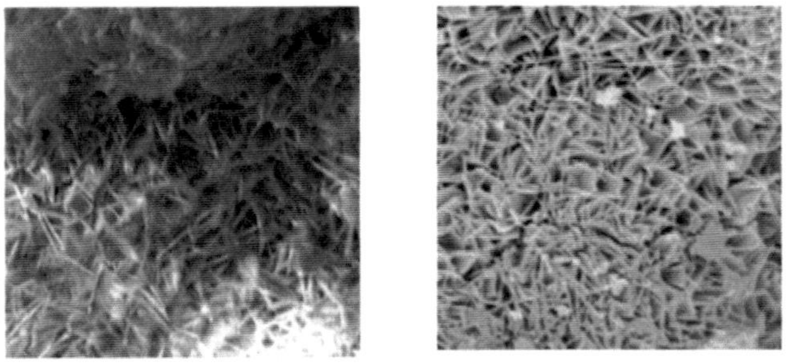

Figure SEQ Figure * ARABIC 1 Crystal structure of conventional AAC (left) and AAC made from LRF slag (right)

Acknowledgement

This work was conducted in collaboration with Emirates Steel Company, HiTech concrete company, and Al Jazeera Factory for Construction Material.

References:

Y. Chen, J. Chang, Y. Lai and M. Chou, "A comprehensive study on the production of autoclaved aerated concrete: Effects of silica-lime-cement composition and autoclaving conditions", *Construction and Building Materials*, vol. 153, pp. 622-629, 2017.

F. H. Ibraheem, R. Muhsin Fathi Shaweas and S. Mustafa Shakir Mahmood, "Production of thermostone in Koya city", *First International Symposium on Urban Development: Koya as a Case Study*, pp. 175-180.

AAC | AAC Block Technical Specification- Eco Green. [Online]. Available: http://www.ecogreenproducts.in/technical_specification.php. [Accessed: 25-Feb-2018].

532984 Integration of Algal Wastewater Treatment with Hydrothermal Liquefaction to Increase Process Net Energy Recovery

Abstract:

Current wastewater treatment processes are energy intensive and expensive. Places with limited financial resources sometimes cannot follow all the guidelines for safe wastewater treatment and discharge contaminated water into rivers or oceans. This affects people who depend on surface water. The water from rivers also leeches underground and can contaminate groundwater aquifers. By integrating low-cost biological wastewater treatment with biofuel production, the net energy intensity of wastewater treatment can be reduced and provide a more cost-effective solution to water contamination problems. In this research, algae was used to treat wastewater through uptake of carbon, nitrogen and phosphorus in the municipal wastewater for growth. Hydrothermal liquefaction of the algae grown on wastewater yields bio-crude oil, char, and nutrient-rich aqueous phase. Here, HTL was performed under different operating conditions (temperature, solids loading, reaction time) to understand the effects of these parameters on product yield. The energy content and CHNS elemental composition of the products were measured to construct mass and energy balances. The energy recovered in the HTL products, along with the lower-cost algal treatment method, increased the net energy recovery (NER) ratio, making the overall process more feasible.

534627 Supply Chains Analysis for Sustainable Production of Aviation and Marine Biofuels: A Comparative Study of Feedstocks, Technologies, Regions, and GHG Emissions

Abstract:

This study presents a techno-economic and greenhouse gas (GHG) emissions analysis of energy and material self-sufficient biorefineries for production of aviation and marine biofuels. The scope of this analysis focuses on the screening method to reduce a large number of initial possibilities and includes three main stages: *i*) biomass supply and logistics for different types of feedstocks (*i.e.* sugar bearing crops, oil bearing plants and agricultural and forest residues); *ii*) geographical sourcing of feedstocks (incl. Brazil and China for aviation biofuels, and Nordic countries for marine biofuels); and *iii*) biomass pretreatment and conversion technologies with intermediates upgrading (*i.e.* ethanol to jet fuel (ETJ), direct fermentation to jet fuel via farnesene (DFJ), fast pyrolysis (FP), gasification and Fischer Tropsch (GFT), and hydrothermal liquefaction (HTL)). Thus, more than 400 configurations of potential supply chains for aviation and marine biofuel production are compatred from a techno-economic and environmental perspective.

One of best results for aviation biofuel production are obtained when Brazilian sugarcane bagasse is directly upgraded to jet via FP with a minimum selling price (MSP) of around 2200 US$/ton biojet. The MSP could further be reduced if lignin is considered to be sold for higher value applications. Furthermore, the lowest production costs, overall the analysis, were obtained when eucalyptus, pine and macauba were used as feedstocks together with hydrothermal liquefaction or fast pyrolysis (with hydro-processing in the case of Macauba oil). These scenarios resulted in minimum selling price of around 620 to 870 US$/ton biojet. All of these scenarios resulted in a GHG emissions reduction of around 50-70% with respect to the fossil based jet fuel.

In the case of the marine biofuel production, the total production costs were estimated to be around 3.2-11.2 times higher than the costs of the fossil based marine gas oil. Furthermore, the GHG emissions were estimated in the range of -36kg/GJ to 96kg/GJ (HFO: 89kg/GJ). This large is also due to the application of different models for biogenic CO2 accounting.

The methodology here presented builds up on the combination of early-stage analysis [1], techno-economic evaluation [2], parametric analysis for combined techno-economic and environmental assessment [3-4], for integration of the three pillars of sustainability during the early-design stages of sustainable biorefineries [5-7].

References

[1] Moncada J., Posada JA, Andrea R. (2015). Early sustainability assessment for potential configurations of integrated biorefineries. Screening of bio-based derivatives from platform chemicals. Biofuels, Bioproducts and Biorefining 9:722â€“748.

[2] Santos CI; Cornelio da Silva C; Mussatto MI; Osseweijer P; van der Wielen L; Posada JA. (in press). Integrated 1st and 2nd Generation Sugarcane Bio-refinery for Jet Fuel Production in Brazil: Techno-economic and Greenhouse Gas Emissions Assessment. Renewable Energy.

[3] Posada JA, Brentner LB, Ramirez A, Patel MK. (2016). Conceptual design of sustainable integrated microalgae biorefineries: parametric analysis of energy use, greenhouse gas emissions and techno-economics. Algal Research 17:113-131.

[4] Almeida-BenalcÃ¡zar E; Gevers-Deynoot B; Noorman H; Osseweijer P; Posada JA. (2017). Production of bulk chemicals from lignocellulosic biomass via thermochemical conversion and syngas fermentation: a comparative techno-economic and environmental assessment of different site-specific supply chain configurations. Biofuels, Bioproducts and Biorefining 11:861â€“886.

[5] Palmeros Parada M, Osseweijer P, Posada JA. (2017). Sustainable biorefineries, an analysis of practices for incorporating sustainability in biorefinery design. Industrial Crops and Products 106:105-123.

[6] Palmeros Parada M, Asveld L., Osseweijer P, Posada JA. (2018). Setting the design space of biorefineries through sustainability values, a practical approach. Biofuels, Bioproducts and Biorefining 12:29-44.

[7] Pashaei-Kamali F., Borges JA, Osseweijer P, Posada JA. (in press). Towards social sustainability: screening potential social and governance issues for biojet fuel supply chains in Brazil. Renewable & Sustainable Energy Reviews.

538059 Hydrochar Obtained from Hydrothermal Carbonization from Lipid Extracted Algae and Its Use As Solid Fuel

Abstract:

Microalgae have tremendous potential as a feedstock for production of liquid biofuels such as biodiesel and renewable diesel along with value-added co-products. Both biodiesel and renewable diesel production pathways involve a lipid extraction step that results in significant solid residues (60-80%) known as â€˜lipid extracted algal reside (LEAR). Suitable utilization of LEAR will be important to the overall cost economics of algal bio-fuels production. In the present study, we evaluate LEAR as a bioenergy feedstock- we specifically investigate the pelletization and combustion properties of algal hydrochar obtained from the hydrothermal carbonization (HTC) process. HTC of *Scenedesmus dimorphous* (whole biomass and LEAR) was performed using a 2-L Parr reactor at 175 ^{o}C for 30 min holding time to obtain the hydrochar. Combustion of hydrochar and untreated biomass samples was conducted in a controlled environment using a thermogravimetric analyzer (TGA) at 25-900 ^{o}C (heating rate of 40-100 ^{o}C/min) and was compared with a conventional coal. The solids (hydrochar and untreated algae) were pelletized using a laboratory hand press pelletizer at 120 ^{o}C and 20 MPa pressure and characterized for energy densification and pellet properties. TGA studies showed that combustion of the all types of algae samples occurred in three stages. Pelletization provided higher energy densification for hydrochar obtained from the LEAR feedstocks. Pellets from hydrochar materials were found to be hydrophobic and stronger than pellets from the raw feedstocks. This presentation also details the combustion kinetics of different algal samples obtained in the study.

539803 Benthic Polyculture Biomass from Wastewater Algal Turf Scrubbers As a Feedstock for Bioeconomy

Abstract:

Algal turf scrubbers (ATS) are floway systems designed to treat wastewater, via microalgal and cyanobacterial biofilm cultivation. ATS are a low cost and low energy-intensive method of removing nutrients, mainly from agricultural wastewater, utilizing algal photosynthesis. The resulting a resilient and adaptable benthic polyculture biomass can be harvested and further processed to get bioenergy, biofuels, and bioproducts. Carbohydrate, protein, and fatty acid contents of these polyculture biomass have been studied as well as biochemical and thermochemical processes to recover them and convert them into useful end products. A 2013 study has estimated an ethanol yield of 2.24 $L.m^{-2}.year^{-1}$ based on the ~24% carbohydrate content of biomass harvested from ATS floways on the Great Wicomico river, Chesapeake Bay, VA[1]. This study reviews literature and our experimental data on culturing, harvesting, and processing of benthic polyculture biomass from ATS and evaluates its feasibility as a feedstock for bioeconomy.

Reference:

1. Adey, W. H.; Laughinghouse, H. D.; Miller, J. B.; Hayek, L. A. C.; Thompson, J. G.; Bertman, S.; Hampel, K.; Puvanendran, S.; Buschmann, A., Algal turf scrubber (ATS) floways on the Great Wicomico River, Chesapeake Bay: productivity, algal community structure, substrate and chemistry1. *Journal of Phycology* **2013,** *49* (3), 489-501.

524108 Investigation of Biochar Liquefaction to Produce Chemicals and Fuels

Abstract:

Due to the threat of global warming and its effects, emphasis on renewable fuels have increased in the recent times. To replace fossil fuels, which has been a major contributor to the global warming, production of bio-fuel from biomass has been explored by various researchers. Biomass consists of 30-35 wt% cellulose, 15-35 wt% hemicellulose and 20-35 wt% lignin. Biochar is a carbon rich hydrocarbon produced from biomass and biomass model compounds by hydrothermal carbonization. Biochar produced by hydrothermal carbonization requires low temperature, has higher carbon recovery, low ash content and has more surface oxygen containing groups when compared to other methods. In our research, we have studied the liquefaction of biochar in a hydrogen donating solvent (tetralin) by hydrocracking at 400 C where the thermal breakdown of hydrocarbon bonds form radicals, which are stabilized by the hydrogen donated by the tetralin. It was observed that the biochar produced from different biomass model compounds behave differently during dissolution and liquefaction. It was observed that the liquefaction of biochar depended on (a) biomass model compound (b) presence of biocrude and (c) the hydrothermal carbonization temperature. The products mainly consisted of pre-asphaltenes (tetrahydrofuran soluble), asphaltenes (toluene soluble) and oil (hexane soluble) and were characterized by gas chromatography-mass spectroscopy analysis.

522831 Investigation of Microwave Assisted Transesterification Reactor of Waste Cooking Oil

Abstract:

In this work, the esterification and trans-esterification reactions of waste cooking oil (WCO) with high free fatty acids (FFA) (â‰¥ 1% by weight) were investigated. This investigation used a two-step batch process with kinetic based reaction mechanisms and examined potential benefits of combining the conventional Fatty Acid Methyl Ester (FAME) trans-esterification method with microwave technology. Optimization of an acid catalyzed FAME process to minimize FFA content in the feed stock found that a feed volume ratio of 0.3 liters of methanol per liter of WCO, 2% grams of sulfuric acid (H_2SO_4) per gram of WCO for a reaction time of 200 minutes at a reaction temperature of 60°C produced a biodiesel yield of 24%. By comparison, optimization of a base catalyzed FAME process found that a feed volume ratio of 0.3 liters methanol per liter WCO oil, 1 gram of potassium hydroxide (KOH) per gram of WCO for a reaction time of 60 minutes at a reaction temperature of 60°C produced a biodiesel yield of 93-98%. When microwave irradiation was included, the reaction time for the acid catalyzed system was 80 minutes (compared to 200 minutes) and 10 minutes for the base-catalyzed system (compared to 60 minutes). These preliminary results suggest microwave technology merits further investigation for industrial application to biodiesel production.

542620 The Impact of Shale Gas and Oil on the Chemical Industry

Abstract:

Lower alkanes and aromatics are very useful as both fuels and chemical raw materials. Occasionally supply disruptions constrained availability but also encouraged investigation of alternative feedstocks and chemistries. Recent new technologies for the production of unconventional gas and oil sources from tight shale formations significantly increased availability of both methane and condensate, radically altered process economics, resulted in the construction of much new intermediates capacity, and multiplied the ways traditional and alternative chemistries might be exploited. This presentation will outline historically how natural gas components became useful as chemical feedstocks, how government regulations occasionally led to perturbations in availability and price, how supply crises led to the development of alternative chemistries, catalysts, and ultimately the exploitation of shale resources. More specifically, will be discussed how excess coproducts, especially condensate, from increasing shale gas and oil production for energy are having significant unexpected impacts on the chemical process industries and opportunities for new process chemistry research.

541186 Sustainable Energy and Chemicals: Past, Present, and Future

Abstract:

The last 40 years has seen the U.S. energy situation vacillate from gasoline rationing and concerns over Peak Oil, to projected energy and feedstock prices that were noncompetitive in global markets for chemicals production, to the economic revolution of frack gas and light tight oil. The importance of low cost energy and feedstocks in driving economic opportunity will be discussed relative to this history, along with the importance of sustainability in decision making to gain stakeholder buy-in and acceptance. Strategies for managing career paths and competencies despite changing targets and high uncertainty will be addressed.

542618 Disruptions: What the Future May Hold

Abstract:

Disruptive forces are all around us. They change the way we communicate and work. They change the way we develop goods and services. They also change the types of products that are manufactured. Examples such as the way in which we produce and consume energy or the feedstocks we use to produce product have seen many changes over the course of the last few decades, and these disruptions are expected to continue. The chemical engineering discipline and the expectations for a chemical engineerâ€™s skills has also changed and adapted as a result of these disruptive forces. A discussion of the broad themes associated with these forces and the implications that they may have to chemical engineers in the future will be discussed.

542621 Geopolitical Factors Influencing the Evolution of the Chemical Industry

Abstract:

The basic chemicals and plastics industry is built on a simple principle: find a cheap feedstock and convert it into something more valuable. However, the right feedstock depends on place and time. Certain policies and actions of various governments have influenced the evolution and profitability of the chemical industry during the past 50 years, sometimes in surprising ways. For example, high oil prices and US government funding encouraged the development and commercialization of coal gasification technology during the late 1970s and early 1980s. This enabled China to develop a competitive basic chemical industry using coal as a feedstock during the last 25 years. Similarly, an abrupt increase in oil prices during the first decade of the 21st century encouraged the present shale gas revolution, causing a major shift in the economy of producing basic chemicals and polymers in the US. If history is our guide, then what governments choose to fund and regulate may be as important as market forces in determining the shape of our industry and profession in the future.

542622 Agility & Resilience: How to Maintain Career Competitiveness in the Changing Chemical Industry

Abstract:

The world around is changing. The shift to protectionism, global instability and price volatility of raw materials and feedstocks has the potential to wreak havoc on the chemical industry. With the revolution of shale gas in North America and the sudden drop in feedstock costs, the chemical sector in the US has reasserted its historic dominance. Learn how the role of a chemical engineer can contribute to industry growth and resilience, while adapting to allow for personal agility throughout the career life cycle.

537749 Application of Shewanella Biofilms for the Dissimilatory Reduction of Iodate in Groundwater

Abstract:

Plutonium production at the U.S. Department of Energy (DOE) Hanford Site has resulted in the generation of Iodine isotopes. The release of radioactive ^{129}I to the subsurface due to leaking tanks and direct disposal has resulted in several plumes in the groundwater. Speciation of Radioiodine has shown that iodate comprises a major percent of iodine. Iodate, which is a +5 oxidation state of iodine is found to be the dominant species due to its thermodynamic stability. One way to reduce iodate to iodide (-1 oxidation state) is by the action of biofilms formed by dissimilatory metal-reducing bacteria such as *Shewanella oneidensis* MR-1, which can gain energy for growth by coupling the oxidation of various electron donors to the reduction of iodate. The polyionic nature, physical and chemical heterogeneity of biofilm matrices, combined with the reduced susceptibility of bacterial biofilms to toxicity by inorganic and organic pollutants, make them particularly suitable for bioremediation applications. This work develops a protocol for the proliferation of *S. oneidensis* MR-1 biofilms and investigates the potential of these biofilms to reduce iodate. Preliminary results indicate that the presence of optimal amounts of lactate (electron donor) enhances MR1 biofilm formation and seems to interact with the speciation of iodate. These studies could lead to novel approaches and techniques for the bioremediation of radioiodine at contaminated sites.

536986 Cost Effective Removal of Chemically Toxic Dyes Using Micro-Organism from Activated Sludge: Understanding Sorption Mechanism, Kinetics and Associated Thermodynamics

Abstract:

A systematic investigation was carried out for demonstrating the simple and cost effective removal of toxic dyes using activated sludge from municipal wastewater. The sorption process was found to follow Freundlich isotherm and pseudo 2nd order rate kinetics. Difference in sorption efficiency for activated and deactivated sludge in aerobic and anaerobic conditions revealed the involvement of different micro-organism in sorption process with the indication of biodegradation. The sorption process was found to be spontaneous but entropy driven for RBBR and EBT while enthalpy driven for CR. The process was established to be a combination of adsorption as well as biodegradation.

With a push toward more inexpensive and eco-friendly options, biosorption and biodegradation are worthwhile methods of effluent treatment. Biosorption describes the uptake of inorganic or organic molecules through attachment to the cell membrane due to a number of chemical or physical processes on the surface of the microorganism. Biodegradation is a less commonly used method of removal as many synthetic dyes are difficult to degrade naturally or without producing any toxic byproducts as this process occurs when living organisms alter the molecular structure with the possibility of using it as a source of nutrients. Currently, treatment of dye contaminated water has been studied using organisms ranging from aquatic macrophytes, to eucalyptus bark, and even including potato peels and watermelon rinds as a desire to find affordable adsorbents is becoming more essential. One study used the fungus *P. ostreatus* supplemented with an alcohol to generate oxidizing laccases that could remove the color from solution through break down the dye Remazol Brilliant Blue R in solution.

Sewage sludge is a less commonly studied adsorbent for the removal of dyes yet it has several properties that make it conducive for the use of textile effluent treatment. However, coagulants widely used in the wastewater treatment are investigated for dyes adsorption. Wastewater sludge is s a combination of several viruses, bacteria, protozoa, and other small microorganisms that form aggregates in a suspended aqueous solution. The surfaces of these aggregates contain several functional groups that provide binding sites for adsorption of the dyes in solution. The functional groups of the chemical dyes interact with the active sites on the surface of the cell surfaces via different chemical or physical attraction methods including ion exchange, complexation, chelation, and micro-precipitation. Because wastewater sludge is commonly produced as a result of city sewage treatment it is easily acquired and particularly inexpensive.

This research proposes the use of locally obtained municipal wastewater sludge for the bio-removal of three synthetic textile dyes: Congo Red (CR), Eriocrhome Black T (EBT), and Remazol Brilliant Blue R (RBBR). Both anaerobic and aerobic samples of wastewater sludge were acquired and maintained for the duration of this research allowing for dual comparisons of adsorption between conditions for each of the dyes. The adsorption of the dyes on to the sludge samples were then analyzed to determine the kinetics and isotherm model fit for the data.

A systematic study was carried out to understand the effect of activation and deactivation of sludge on the sorption properties of the dyes. The % of sorption was found to decrease on deactivation which indicated that probably biodegradation also played significant role in the present sorption process. If the % of sorption is considered as combination of adsorption on the surface + the biodegradation, then on deactivation the % of sorption should decrease as there will be no contribution from biodegradation (considering surface area and surface charge remained unchanged on deactivation). Therefore, it is required to know how surface properties modified on deactivation to evident the significant contribution from biodegradation.

The difference in sorption behavior of the sludge obtained from aerobic and anaerobic conditions indicated that, probably the micro-organism involved in sorption process from both the sludge are appreciably different. An attempt was also made to understand that on deactivation, how the particle size of the sludge gets modified. For all the cases the size of the particles was found to be concentrated on ~ 0.12 Î¼m. This study also gave us the indication that on deactivation, there was no significant change in the surface area of the sludge.

The adsorption of compounds onto the sludge which a combination of aggregated floc and suspended particles is one of the dominant removal pathway for dyes. Biodegradation is transforming substrate into new compounds through biochemical oxidation reactions inside the microorganisms with the presence and absence of dissolved oxygen. The disappearance of the dyes was monitored to evaluate their biodegradability within 240 hours of culture. RBBR and EBT have showed strong resistant to be biologically degraded by the microorganisms where slight decreasing took place during the first 100 h of incubation with 12.9 and 37.4 removal efficiency, respectively. Furthermore, the concentration of RBBR and EBT in the supernatant was approximately a straight line till the end of the test. CR have followed similar removal trend with other dyes, however, after 100 h of culture with the microorganism, its concentration kept declining over time which indicates that it has a higher biodegradation affinity to the microorganisms.

The negative value of the Gibbâ€™s free energy for sorption primarily suggested that all the sorption processes are spontaneous. The enthalpy values associated with the sorption processes revealed that, the sorption of CR and EBT on the sludge is exothermic while that of RBBR is strongly endothermic. The release of the solvent molecules during the chemical interaction of the active coordinating sites of the sludge and the dye might be responsible for the enhancement in the entropy. This entropy driven process is also an indication of the formation of inner sphere complex between coordinating sites and the dye molecules. On the other hand, CR sorption was found to be largely driven by the enthalpy change.

The sorption of chemically toxic dyes on the micro-organism was demonstrated as an efficient and cost effective way of their removal. Analysis of different isotherm models revealed the predominance of Fruedlich isotherm with chemisorption. The sorption capacity for CR and EBT was found to be superior to that of other sorbent reported in the literature, while for RBBR it is comparable. The process was found to be energetically favorable following pseudo 2nd order kinetics for all the dyes on activated sludge in aerobic and anaerobic conditions. The sorption of RBBR was found to be endothermic whereas for others they were exothermic. The sorption efficiency of sludge in aerobic condition was found to be more than that in anaerobic condition. The reduction in sorption on deactivating the sludge indicated that biodegradation is an additional channel for dye sorption or deactivation leads to the change in surface morphologies resulting the reduction. The size of the particle remained unchanged as the surface functional groups revealing biodegradation to be active during the sorption process.

"Phytoremediation of Cr-Contaminated Soil: Use of Chelators in Cr Phytoextraction"

Sandra Cutts[1], Dr. Karolina Mukhtar[2], Dr. Fanny Coutelot[3], Dr. John Seaman[3], Dr. Robin Brigmon[4], and Dr. Robert W. Peters[1]

1. Department of Civil, Construction, & Environmental Engineering, University of Alabama at Birmingham, Birmingham, AL
2. Department of Biology, UAB, Birmingham, AL
3. Savannah River Ecology Laboratory, Aiken, SC
4. Savannah River National Laboratory, Aiken, SC

Abstract

In the age of industrialization and developmental progress, technological advancements have contributed to the challenge of environmental contamination of organic and inorganic pollutants. Moreover, contamination of heavy metals is considered one of the most critical threats to the soil, the environment, human health, and water resources. There are many conventional remedial methods used for environmental contaminants, though costly and often not efficient. Consequently, phytoremediation is an environmentally friendly and cost-effective way to clean up the environment of contaminants using plants. Nonetheless, the concept of using plants to remediate the environment of contaminants is not a new technology as Hartman (1975) indicated that phytoremediation was used about 300 years ago for wastewater treatment. Therefore, research has been conducted on many inorganic metals, e.g., iron (Fe), lead (Pb), zinc (Zn), and copper (Cu), using plant species but only sparse research has been performed on the heavy metal chromium (Cr^{+6}). Moreover, hexavalent chromium is not an essential element for plants therefore, specific uptake mechanisms for Cr^{+6} have not been developed. Consequently, this research investigates the use of phytoextraction to remove Cr^{+6} from contaminated soil and the study is being conducted in three scenarios – a controlled greenhouse environment, a rooftop field setting and on a contaminated site in central Alabama. Thus, research results indicate that the use of plant species *brassica kaber* (wild mustard), *thlaspi caerulescens* (pennycress), and *cyperus esculentus L.* (yellow nutsedge)) alone statistically improve contaminant removal of Cr^{+6}. Moreover, this study enhanced the uptake process with the use of amendments, such as, ethylene diamine disuccinic acid (EDDS), citric acid ($C_6H_8O_7$) and ethylene diamine tetraacetic acid (EDTA), which have indicated that plant species statistically removed greater contamination than that conducted with the use of no amendments.

Reference Cited:

Hartman, W.J., Jr., 1975. "An Evaluation of Land Treatment of Municipal Wastewater and Physical Siting of Facility Installations", U.S. Department of Army, Washington, DC.

527190 Enhancement of Oily Sludge Biodegradation in Historic Refinery Wastewater Lagoons

Abstract:

The presence of untreated historic oily sludge in refinery wastewater ponds represents a long-term liability at older petroleum refineries. In many cases, this waste will be classified as a listed hazardous waste if it is removed from the pond as part of pond maintenance, process alteration or pond unit closure. The oil component of this sludge is typically difficult to biodegrade because it consists of long-chain or polyaromatic hydrocarbons. Most efforts to remediate this type of waste center on removal, dewatering, and solidification or stabilization for final disposal. However, this approach is equipment intensive, costly, and typically disturbs normal site operations. For a refinery that is still in operation, in situ treatment approaches that rely on biodegradation of the waste material can offer a more economical solution for this liability.

The purpose of this research effort is to evaluate different approaches to bioremediate or treat sludge-containing wastewaters impounded in lagoons from historic refinery wastewater treatment operations. The primary approach for enhancing the desired biodegradation treatment is by dispersion of the oily material to increase oil / water interfacial area and subsequent addition of microbial consortia optimized for crude oil or heavy hydrocarbon biodegradation. Laboratory experiments are conducted for each of these enhancement steps. Modeling of the dispersion and biodegradation process is also conducted for comparison with the laboratory results. The goal is to attain nearly complete oily sludge biodegradation in an aqueous environment within a period of a couple of years. This result would be an acceptable treatment goal for an operating refinery where the wastewater lagoon is still part of current operations.

The laboratory results completed for dispersing oil material indicate increased amount of material dispersed into the water column primarily with various types and energy level of mixing, and secondarily with total amount of oil applied and the type of vendor-supplied dispersant. The results of the laboratory biodegradation tests will also be presented. The modeling results obtained to date indicate significant differences in the time required for 95 percent of material to be biodegraded, depending on the particle size of the dispersed oily material. Extension of the model to incorporate greater variability in conditions is ongoing. These results indicate that the critical elements in this process are control of the particle size obtained in dispersion, and the energy used for mixing.

515931 Reactive Fe Mineral Coatings in Redox Transition Zones

Abstract:

In this research, we are studying reactive Fe mineral coatings in redox transition zones from a site with subsurface contamination. An anaerobic 60 foot core was collected from the Chambers Works Site in Deepwater, NJ. Working in a glovebox, two inch subsamples were used to evaluate redox transition zones using bulk composition, redox potential, pH, VOCs in the headspace, as well as abundant bacteria. Four redox transition zones were identified and further studied with a suite of complementary analyses to characterize the surface chemistry. Analyses have included sequential extraction (phases and speciation), x-ray diffraction (mineralogy), x-ray fluorescence (composition), field-emission scanning electron microscopy (surface morphology down to the nm scale) with energy dispersive analysis (surface composition), and bench-scale experiments. Evidence from multiple lines suggest a clear trend in bulk minerals and reactive iron (II)/(III) mineral coatings throughout transition zones. Identified iron (II) mineral coatings include pyrite, mackinawite, siderite, and pyrrhotite. While ferrihydrite, goethite, and lepidocrocite were the dominant iron (III), greigite and magnetite were observed as iron (II)/(III) minerals. In the shallowest transition zone running through an aquitard layer, spherical and sub-spherical packed framboidal pyrite and greigite were abundantly observed. Flaky mackinawite, individual grains of cubic pyrite, and pyrrhotite were dominant iron sulfides in a deeper aquitard transition zone. Siderite and magnetite were rich in two deeper transition zones found in aquifer layers that may be less reactive to targeted contaminants than aquitard zones. Overall, the most significant reactive mineral coatings were observed in redox transition zones located in aquitards where there is a strong indicator of Fe^{2+}/Fe^{3+} cycling. Iron speciation, mineral coatings, and mineral morphology observed in redox transition zones contribute to contaminant attenuation and degradation pathways. Therefore, ongoing studies are focused on abiotic degradation pathways for halogenated solvents present in the system.

535713 Enhanced Biosorption of Heavy Metals through Base Treated Coconut Husk

Abstract:

Heavy metals from industrial and domestic are considered as one of the major source of pollution. Beside these heavy metals, wastewater also contains metallic ions and other pollutants which create a huge problem in ecology and become health risk for human beings. In this work, removal of heavy metals like Copper and Zinc from aqueous solution using base treated coconut husk (BTCH) are studied at different pH. BTCH was characterized using FTIR, BET, XRD and SEM. To check the efficiency of BTCH, various parameter like contact time, pH of aqueous solution, initial concentration of metal ions and biosorbent dose over sorption kinetics of Cu and Zn using BTCH were investigated. The equilibrium condition was meet at average of 30 minutes for pH range of 3.0 to 5.5. It kinetic study showed that it followed pseudo second order rate reaction and reaction rate was calculated at 25°C. Surface status and reaction kinetic shows that base treated coconut husk has enough potential to separate Copper and Zinc ions from aqueous solution.

538047 Analysis of a Circular Economy: From Food Waste to Foods

Abstract:

The US-EPA estimated that 38 million tons of food waste went to landfill in 2014 and in 2010 66.5 million tons of food didnâ€™t make it to consumers. The food waste streams total to around a billion pounds of food waste, where the majority go to landfill. Publicly owned waste water treatment plants in U.S. process approximately 14 billion kilograms of dry solids per year.[1] Turning waste streams into valuable products is the goal of the circular economy.[2]

Hydrothermal carbonization (HTC) process is proposed to be a solution to the large streams of food waste and biosolids from waste water treatment plants that creates a highly valuable, multiuse product. HTC is a thermochemical process that converts wet biomass materials to a coal-like product. Under sub-critical conditions, water in the HTC reactor stays in liquid form and acts as a reaction medium to promote the breakdown and cleavage of chemical bonds in the solid biomass. The absence of liquid-to-vapor phase change of the water makes the process significantly less energy intensive compared to a process that involves water vaporization or drying. The temperature range for HTC is from 180-240°C producing solid hydrochar as the main product.

The solid product of hydrothermal carbonization, hydrochar, has multiple applications ranging from a combustion fuel source, a catalyst for anaerobic digestion that increases the purity of methane production, a soil amendment, and the base for other carbon product such as activated carbon.

In this study we will explore the circular economy of utilizing organic wastes into foods. The steps involved in the circular economy system include: 1) collecting organic wastes (specifically food wastes and bio-solids), 2) converting the organic waste streams to hydrochar via HTC process, 3) growing food produces by indoor gardening, and 4) using hydrochar from step 2) as fuels to power the indoor gardens.

Utilizing organic waste streams in cities such as Chicago or Philadelphia, where there are a multitude of abandoned warehouses, provides an opportunity to produce locally grown produces that would typically come from across the country from California or Mexico. Not only will this circular economy reduce environmental impacts, but it will increase positive societal impacts by benefitting the local economy, providing educational opportunities of farming that would not typically be available in cities, and serving fresher produces to the local community.

A comparative life cycle assessment of our proposed process will be applied to compare conventional farming to our circular economy proposal. We hope to prove a reduction of environmental impacts, such as global warming potential, fossil fuel depletion, natural resource depletion. The life cycle assessment results will aid in a full system analysis which evaluate the proposed circular economy idea from societal, technical, environmental, economic, and political perspectives.

- https://www.epa.gov/sustainable-management-food/sustainable-management-food-basics
- https://www.sciencedirect.com/science/article/pii/S0301479717303808

*Contact: Dr. Justinus Satrio (justinus.satrio@villanova.edu)

538097 Re-Wiring the Domestic Food Trade for Reducing Irrigation Impacts in the United States

Abstract:

The food-energy-water(FEW) nexus concept recognizes the interconnected nature of the three systems and advocates managing them together to avoid sub-optimal outcomes and unintended consequences. The FEW nexus challenges differ for developed agro-economies such as the United States(U.S.) from agriculture centric developing nations. For U.S., irrigation provides a crucial link to study interconnections between food-water-energy resources. Additionally, food trade is an integral part of food system as it bridges the growing disconnect between production and consumption centers. Depending on the origin of food production, trade can potentially reduce environmental impact of food consumption. However, environmental impacts of agriculture are not directly considered when forming trade connections and may carry risk associated with suboptimal FEW system management (e.g. sourcing food from water-depleted regions). Structural optimization of food trade provides an opportunity to assess whether FEW resources can be managed together with respect to land-water-energy constraints. This work focuses on environmental impacts associated with FEW systems, namely- water required for irrigation, greenhouse gas(GHG) emissions from commodity transport, and pumping irrigation water for food production.

We leverage publicly available datasets from agriculture census and transportation surveys to create a network of regional food trade. The study combines data on state-wide food production, irrigation statistics, and life cycle inventory databases to convert food transfer network into embedded irrigation water, and GHG emission flows. Specifically, we focus on rice production and trade in the U.S. Rice is one of the staple food crops and produced entirely through irrigation, making it an ideal crop for the irrigation impact analysis. Majority of rice production is concentrated in four regions-Arkansas Grand Prairie, Mississippi Delta, Gulf Coast, and Sacramento Valley. The model optimizes rice trade network by keeping the demand constant and evaluate a series of constraints pertaining to reduction in energy and water usage to understand whether production can be shifted realistically to optimize all three systems. We limit the potential crop shifting to current production regions to account for soil and climatic parameters conducive to crop production. Additionally, we consider type of rice grown in each region as crop physiological requirements may differ based on the rice type. Preliminary results of the optimization framework reveals a rewired domestic rice trade network with a 27% reduction in virtual water use (15.1 billion m^3 reduced to 11.1 billion m^3). However, due to shift in production to more water-efficient states, a 10% increase in marginal land use is required. The land results are significant as they demonstrate the resulting sub-optimal solutions when interdependencies are overlooked. On the other hand, minimizing groundwater usage from water stressed areas also reduces overall GHG emissions demonstrating that these are not mutually exclusive goals in many states. The Pareto optimal solutions showing water-GHG emissions tradeoffs resulting from the model will be presented and their implications for regional FEW nexus sustainability will be discussed.

527168 Food, Energy, Fuels and Chemical Feedstocks from Rice Crops: Multi-Objective Optimisation of Multi-Product Value Chains for the Philippines

Abstract:

Rice is an important staple food in the Philippines such that its 20 M metric tons annual production places the country rank 8th among the rice producers in the world. However, there are many improvements in the efficiency and sustainability of the Philippines rice value chains that could be made by addressing various constraints related to technological, environmental, economic and socio-political factors [1]. Consequently, the Philippine government is striving to achieve Philippine rice competitiveness and to transform its value chains to be more profitable, resilient and sustainable; they recognise the importance of research and development as basis for such a laudable goal [2]. From a recent analysis of the Philippine rice value chain commissioned by the government, by-products like rice straw, rice hull and rice bran can be important sources of raw material for other industry sectors. Ninety-five percent of the rice bran is being used as animal feed component while seventy percent of the of the rice hull is being used as an alternative source of energy and fuel. Moreover, the silica-rich rice hull ash is being used as additive in the cement industry [3]. These could not only provide affordable renewable energy, fuels and materials but also much-needed rural development and employment opportunities. However, there are complex issues that need to be addressed: rice production is a water- and energy-intensive activity. The global water footprint from rice production is 1308 Mm^3/year with irrigated rice cultivation consuming the most water. Energy is required for the production of fertilisers, electricity for pumping water and heat for parboiling, drying and other milling activities [4]. Furthermore, the limited land for rice cultivation is being threatened by the rapid and unregulated conversion of agricultural land to other uses [1]. In this project, a comprehensive optimisation model, based on mixed-integer programming [5-6], is being developed that can support complex decision-making related to multi-product rice value chains and can capture the trade-offs between water, energy and land utilisation. The model is used to identify promising rice value chains for sustainable and efficient production of food, energy, fuels and chemical feedstocks. In this conference, we will present the optimisation model and the key findings from the case studies.

References:

[1] B.A. Excode (2016). Rice production and consumption trends in the Philippines from 2000 to 2015. Yuchengco Center, De La University; Philippines.

[2] C.T. Briones (Ed.) (2016). The PhilRice strategic plan 2017-2022. Philippine Rice Research Institute; Philippines.

[3] A.B. Mataia, A.B. Mataia, R.G. Manalili, J.C. Beltran, B.M. Catudan, N.M. Francisco and A.C. Flores (2016). Analysis of the rice value chain in the Philippines. Philippine Rice Research Institute; Philippines

[4] R. Sims, A. Flammini, M. Puri, S. Bracco (2015). Opportunities for Agri-Food Chains to become energy-smart. Food and Agriculture Organization of the United Nations and the United States Agency for International Development; United States of America.

[5] S. Samsatli, N.J. Samsatli (2018). A multi-objective MILP model for the design and operation of future integrated multi-vector energy networks capturing detailed spatio-temporal dependencies. Applied Energy. DOI: 10.1016/j.apenergy.2017.09.055.

[6] S. Samsatli, N.J. Samsatli, N. Shah (2015). BVCM: a comprehensive and flexible toolkit for whole-system biomass value chain analysis and optimisation - mathematical formulation. Applied Energy, 147, pp. 131-160.

541194 Modeling the Impacts of International Food Trade on Contaminant Transport and Human Exposure

Abstract:

Introduction: Quantifying and addressing the implications of the transfer of resources across borders is a hallmark of research on the food, energy and water (FEW) nexus. International trade in goods can result in severe imbalances in the water stress, green house gas emissions, and energy consumption experienced by or attributed to producing and receiving countries[1,2]. Yet an important component of food trade—that is, the potentially hazardous chemicals contained in foods as a result of accumulation from production environments, processing, transport, or storage—is still poorly quantified[3–5]. Here, we investigate the importance of international food trade on the dietary exposure of the Swiss population to polybrominated diphenyl eithers (PBDEs). PBDEs are lipid soluble compounds used as flame retardants in synthetic fibers and molded plastics. Of major concern is PBDE exposure through the diet. Studies have confirmed that fish, meat, and dairy products contribute most to the daily PBDE intake[6]. Like other countries, Switzerland has reported a gradual shift in dietary patterns over the past few decades, with increasing consumption of fish reported, reaching a peak value of 17.6 kg in 2010, and with a shift in the origin of fish. We therefore constructed a model to quantify the exposure of the Swiss population to PBDEs from local and imported fish and seafood.

Methodology:

Model boundaries: Our study focused on Switzerland as a geographical case study. All calculations referred to fish imports for the year 2016. Fish, shrimp/prawn, and mussels were selected as key fish and seafood types to include in the study, as they are among the most highly consumed.

Data collection: An extensive literature review was conducted to derive useful input data for our model, including PBDE concentrations in fish, shrimp and mussels from different countries, and to estimate daily per capita fish consumption in Switzerland. Trade data (imports, exports, and re-exports) for fish, crustaceans (shrimp and prawns) and mollusks (mussels) was extracted from the UN Comtrade Database. Data were collected for fresh, frozen, fresh fillet, and frozen fillet categories of the respective commodities.

Model Structure: The model consists of four primary components. These are: (i) PBDE concentrations in fish from different origins, resulting from bioaccumulation from their environment and/or food; (ii) food trade data for Switzerland; (iii) local fish production; and (iv) dietary intake to the Swiss population via fish and seafood consumption.

Results and Discussion: Based on the difference between average per capita fish consumption, production, imports and re-exports, we estimate that Switzerland imports approximately 66% of fish consumed, while only 34% is locally caught. Of the imported fish, Norway, Denmark and the UK are the largest exporters of salmon to Switzerland. For shrimp and prawn, the majority is from Vietnam. For other fish, Norway, Denmark, the Netherlands, Vietnam, France, Germany, Italy, and China are the largest exporters. Within the locally produced fish, whitefish (*Coregonus sp.)* is the most common fish contributing to 60% of local the catch, followed by perch (19%), pike (7%) and roach (7%)[7]. Using an average of 23g/day fish consumption, results show that from 70000 metric tons fish consumed in Switzerland in 2016, 66 % (46000metric tons) was imported corresponding to 34 % locally caught fish (24000 metric tons).

Based on published studies of PBDEs in fish, the highest concentrations were detected in whitefish (Coregonus spp.) in Switzerland[8]. Norway also reported high levels of PBDEs in whitefish[9]. The US reported PBDEs as high as 3078 pg/g wet weight in salmon[10]. Compilation of country-specific PBDE body burdens in fish continues. In addition, we are employing modeling approaches to estimate PBDE body burdens in key species from exporting countries where data are lacking. These data will aid in forming a comprehensive source-receptor relationship for Swiss population exposure to PBDEs via seafood and fish.

References Cited

(1) Vora, N.; Shah, A.; Bilec, M. M.; Khanna, V. Food–Energy–Water Nexus: Quantifying Embodied Energy and GHG Emissions from Irrigation through Virtual Water Transfers in Food Trade. *ACS Sustain. Chem. Eng.* **2017**, *5* (3), 2119–2128.

(2) Leck, H.; Conway, D.; Bradshaw, M.; Rees, J. Tracing the Water–Energy–Food Nexus: Description, Theory and Practice. *Geogr. Compass* **2015**, *9* (8), 445–460.

(3) Ng, C. A.; von Goetz, N. The Global Food System as a Transport Pathway for Hazardous Chemicals: The Missing Link between Emissions and Exposure. *Environ. Health Perspect.* **2017**, *125* (1), 1–7.

(4) D'Odorico, P.; Carr, J. A.; Laio, F.; Ridolfi, L.; Vandoni, S. Feeding Humanity through Global Food Trade. *Earths Future* **2014**, *2* (9), 2014EF000250.

(5) Ercsey-Ravasz, M.; Toroczkai, Z.; Lakner, Z.; Baranyi, J. Complexity of the International Agro-Food Trade Network and Its Impact on Food Safety. *PLoS ONE* **2012**, *7* (5), 1–7.

(6) Schecter, A.; Haffner, D.; Colacino, J.; Patel, K.; Päpke, O.; Opel, M.; Birnbaum, L. Polybrominated Diphenyl Ethers (PBDEs) and Hexabromocyclodecane (HBCD) in Composite U.S. Food Samples. *Environ. Health Perspect.* **2010**, *118* (3), 357–362.

(7) Swiss fact: around 1.8 million kilograms of fish are caught in Switzerland every year https://lenews.ch/2017/01/04/swiss-fact-around-1-8-million-kilograms-of-fish-are-caught-in-switzerland-every-year/ (accessed May 31, 2018).

(8) Zennegg, M.; Kohler, M.; Gerecke, A. C.; Schmid, P. Polybrominated Diphenyl Ethers in Whitefish from Swiss Lakes and Farmed Rainbow Trout. *Chemosphere* **2003**, *51* (7), 545–553.

(9) Mariussen, E.; Fjeld, E.; Breivik, K.; Steinnes, E.; Borgen, A.; Kjellberg, G.; Schlabach, M. Elevated Levels of Polybrominated Diphenyl Ethers (PBDEs) in Fish from Lake Mjøsa, Norway. *Sci. Total Environ.* **2008**, *390* (1), 132–141.

(10) Schecter, A.; Päpke, O.; Tung, K.-C.; Staskal, D.; Birnbaum, L. Polybrominated Diphenyl Ethers Contamination of United States Food. *Environ. Sci. Technol.* **2004**, *38* (20), 5306–5311.

523949 Using Agricultural Wastes to Recover Rare Earth Elements from End-of-Life Materials

Abstract:

An economical and eco-friendly bench-scale process that relies on microbial production of organic acids has been developed to solubilize and release rare earth elements (REE) from solid recyclable materials (Reed, Hydrometallurgy, 2016, 166:34-40). Techno-economic analysis (TEA) and life cycle analysis (LCA) for bioleaching of REE from fluid catalytic cracking (FCC) catalysts suggested that a bioleaching plant could be profitable, but could be even more profitable if the cost of the sugar (glucose) used for growing the microbes was reduced (Thompson, ACS Sustainable Chemistry & Engineering, 2017, 6:1602-1609). This presentation will focus on our efforts to improve the economics of REE bioleaching using alternative carbon and energy sources as microbial growth substrates.

In our previous work examining the potential for several organic acid producing microorganisms to leach REE from end-of-life waste materials we identified *Gluconobacter oxydans* B58 as a promising industrial strain. Lixiviant produced by *G. oxydans,* containing predominantly gluconic acid, could recover ~50% of the total REE content of FCC catalyst. The biologically produced lixiviant was more effective than an abiotically prepared solution containing a similar concentration of gluconic acid. TEA and LCA showed that bioleaching could be more profitable and environmentally friendly than industrial standard chemical leaching processes, suggesting commercial viability. However, substantial reductions in operating costs could be achieved by replacing the refined glucose used in the original growth medium with a lower cost substrate. Consequently corn stover and potato wastewater were tested as growth substrates for *G. oxydans*. The initial lixiviants generated were tested against FCC and yielded a leaching efficiency 50-90% of the best case leaching efficiency of glucose-derived biolixiviant. Work is ongoing to optimize biolixiviant production from agricultural wastes. The conditions for improving bioleaching using waste-derived biolixiviants, including TEA and LCA evaluations, will be discussed.

526360 Membranes for Nutrient Concentration, Industrial Separations and Applications Beyond

Abstract:

With the increasing importance of the waterâ€"energyâ€"food nexus, it is critical that membranes can be designed for optimal performance and energy efficiency for large varieties of water and food applications. The cost and energy consumption of milk and whey concentration systems can be minimized when one selects a reverse osmosis element that reduces fluid flow requirements while maintaining optimum dewatering efficiency. In addition, superior performance of RO membranes which address the emerging importance of high rejection will be discussed for nutrient concentration, with a properly designed membrane based on the understanding of membrane structure-property relationships. Potentially there are many more applications where RO technology can be used to meet critical customer requirements with high efficiency and favorable economics.

541565 25 by 25: Chemical Engineering in the Next 25 Years

Abstract:

In 2008, the centennial year for AIChE, we asked 25 chemical engineers for their visions of the profession from around the US and the world: Nine faculty, eight from industry, and eight post-docs and graduate students. We posed four questions related to each individual's industrial sector or area of research, asking for extrapolation, new impacts on existing sectors, new sectors, and an optional, more open-ended comment on the future of the profession as a whole.

In this 110th year of AIChE, we have revisited these questions with many of the same people. Ten years has seen many developments in our profession, in society, and in our thought-leaders' experiences. Here we report on their perspectives, comparing them to the 2008 visions.

541542 The Future of Chemical Engineering Itself

Abstract:

Chemical engineeringâ€™s future is bright because it is vital to so much of society and industry -- and its future is complicated by its rapidly expanding universe of activities. It must represent depth in core skills centered on the chemical sciences including biology and materials. At the same time, it must be transdisciplinary; it must exemplify the need and value of reaching across disciplines in borderless ways.

One way is to create and disseminate a compelling vision and strategy for chemical engineering, research, and education for the future. AIChE and the National Academies are working together to that end. In 1988, the National Academies published "Frontiers in Chemical Engineering: Research Needs and Opportunities.â€ Its committee of authors outlined a wide-ranging roadmap to turn promising research opportunities into reality, while calling for researchers to embrace new frontiers.

Since 1988, that has happened -- along with many other transformative changes like the Web, fracking, personalized medicine, and social media. A new nationwide effort, aided by international contributions, will address both the identity of chemical engineering and its future directions. Its goal is to ensure that the people in chemical engineering are primed to meet society's needs and rapid changes.

544708 Accelerating Innovation through Academic-Industrial Partnerships

Abstract:

In 2011, The Dow Chemical Company launched the University Partnership Initiative (UPI) to sponsor collaborative research at leading U.S. Universities. This 10-year commitment aims to address fundamental questions in industrial chemistry and chemical engineering with alignment to Dowâ€™s next-generation process technology and product development efforts. To date, this portfolio of projects has generated an impressive array of intellectual property, publications, and numerous examples of the collaborative developments implemented at Dow. This presentation will introduce UPI program and highlight several contributions from our collaborators and leaders in the chemical engineering profession.

540559 Maximizing Uptime, Efficiency, and Safety of Industrial Operations through Early Risk Detection

Abstract:

Achieving zero incidents while maximizing process uptime and efficiency is a key vision for operating companies. An important building block towards this vision is establishment of a proactive risk mitigation culture, supported by effective systems and a strong management. Studies have shown that most of the incidents and unexpected process failures can be avoided if the operating teams get timely information about developing risks and take preventive actions early on. While advances have been made in process monitoring and automation in the past decades, there still remains significant technological and behavioral gaps that prevent plant operations from proactively mitigating process risks.

A modern industrial plant monitors thousands of parameters, generating upwards of 50-100 million data points every day. Thanks to recent breakthroughs in machine learning and artificial intelligence approaches, today there are autonomous systems that can sift through this data and point out meaningful and timely insight. This can help operating teams ascertain process issues that are hidden in the data, long before process variables reach critical levels.

In this presentation, a new autonomous system, Dynamic Risk Analyzer™ (DRA), will be introduced that points out early indicators of process issues using proprietary machine learning. The risk indicators enable operations team to respond and make required process changes, days and sometimes weeks ahead of any alarm, avoiding late stage (expensive) fixes. To realize these objectives fully, a proactive management workflow must also be in place. The presentation will discuss new initiatives and workflows developed by plant operations to implement a proactive culture. Real-life case studies from plants using DRA technology will be presented on how this culture change resulted in increased uptime, efficiency and safety of plant operations.

531309 Gaussian Processes for Hybridizing Analytical & Data-Driven Decision-Making

Abstract:

Surrogate models are widely appreciated in process systems engineering [1]. The typical setting focuses on expensive-to-evaluate, possibly uncertain functions. Examples include: modular process simulators [2], integrated gasification combined cycle processes [3], a carbon capture absorber [4], and many other applications, e.g. [5-8]. Resources are typically limited, so effective decision making requires data-efficient learning.

The data science and statistical machine learning communities typically focus on models learned solely from observed data. But chemical engineering applications may also require explicit, parametric models, e.g. modeling known process constraints, operations constraints, and cost objectives [9]. So, work has integrated semi-algebraic functions with those learned from data [10] or developed semi-physical modeling techniques [11, 12].

This presentation surveys the state-of-the-art in hybridizing analytical and data-driven decision making [13]. We consider three probabilistic modeling applications to these hybrid situations:

Design of experiments for model discrimination [14]. We bridge the gap between classical, analytical methods [15] and Monte Carlo-based approaches [16]. Classical methods may have difficulty managing non-analytical model functions and data-driven Monte Carlo approaches come at a high computational cost. We replace the original, parametric models with probabilistic, non-parametric Gaussian process surrogates learned from model evaluations. The surrogates are flexible regression tools that extend classical analytical results to non-analytical models, while providing us with model prediction confidence bounds and avoiding the computational complexity of Monte-Carlo approaches.

Multi-objective optimization [17, 18]. We make novel extensions to Bayesian multi-objective optimization in the case of one analytical objective function and one black-box, i.e. simulation-based, objective function. The resulting method has been applied to a bone neotissue application [19] and a more general test suite.

Scheduling plant operations under uncertainty. For processes with equipment degradation, we use Gaussian processes to approximate large-scale, mixed-integer optimization problems.

We close by offering a broad outlook on applying probabilistic surrogate models to chemical engineering. Statistical machine learning has recently attracted significant interest in process systems engineering [20]. Here we show that state-of-the-art research in Gaussian processes [21, 22] and probabilistic modeling more generally [23] can have a big impact on chemical engineering.

References

[1] Bhosekar A, Ierapetritou M. Advances in surrogate based modeling, feasibility analysis and and optimization: A

review. *Comput Chem Eng*. 108: 250-267, **2018**.

[2] Caballero JA, Grossmann IE. An algorithm for the use of surrogate models in modular flowsheet optimization. *AIChE J*. 54(10):2633-2650, **2008**.

[3] Lang Y, Zitney SE, Biegler LT. Optimization of IGCC processes with reduced order CFD models. *Comput Chem Eng*. 35(9):1705-1717, **2011**.

[4] Cozad A, Sahinidis NV, Miller DC. Learning surrogate models for simulation-based optim-ization. *AIChE J*. 60(6):2211-2227, **2017**.

[5] Boukouvala F, Ierapetritou M. Surrogate-based optimization of expensive flowsheet modeling for continuous pharmaceutical manufacturing. *J Pharm Innov*. 8: 131, **2013**.

[6] Soepyan FB, Cremaschi S, Sarica C, and others. Estimation of percentiles using the Kriging method for uncertainty propagation. *Comput Chem Eng*. 93:143-59, **2016**.

[7] Shokry A, Ardakani MH, Escudero G, Graells M, Espuña A. Dynamic Kriging-based fault detection and diagnosis approach for nonlinear noisy dynamic processes. *Comput Chem Eng*. 106:758-76, **2017**.

[8] Tran AP, Georgakis C. On the estimation of high-dimensional surrogate models of steady-state of plant-wide processes characteristics. *Comput Chem Eng*. DOI 10.1016/j.comp-chemeng.2018.02.014, **2018**.

[9] Boukouvala F, Hasan MF, Floudas CA. Global optimization of general constrained grey-box models: new method and its application to constrained PDEs for pressure swing adsorption. *J Glob Optim*. 67(1-2): 3-42, **2017**.

[10] Boukouvala F, Floudas CA. ARGONAUT: AlgoRithms for Global Optimization of coNstrAined grey-box compUTational problems. *Optim Lett*. 11(5): 895-913, **2017**.

[11] Pearson RK, Pottmann M. Gray-box identification of block-oriented nonlinear models. *J Process Contr*. 10(4):301-315, **2000**.

[12] Cozad A, Sahinidis NV, Miller DC. A combined first-principles and data-driven approach to model building. *Comput Chem Eng*. 73: 116-27, **2015**.

[13] Boukouvala F, Misener R, Floudas CA. Global optimization advances in mixed-integer nonlinear programming, MINLP, and constrained derivative-free optimization, CDFO. *Eur J Oper Res*. 252: 701-727, **2016**.

[14] Olofsson S, Deisenroth MP, Misener R. Design of Experiments for Model Discrimination Hybridising Analytical and Data-Driven Approaches. *arXiv:1802.04170*. **2018**.

[15] Buzzi-Ferraris G, Forzatti P, Emig G, Hofmann H. Sequential experimental design for model discrimination in the case of multiple responses. *Chem Eng Sci*. 39(1): 81-85, **1984**.

[16] Vanlier J, Tiemann CA, Hilbers PAJ, van Riel NAW. Optimal experiment design for model selection in biochemical networks. *BMC Syst Biol*, 8(20), **2014**.

[17] Olofsson S, Mehrian M, Geris L, Calandra R, Deisenroth MP, Misener R. Bayesian Multi-Objective Optimisation of Neotissue Growth in a Perfusion Bioreactor Set-Up. *Computer Aided Chemical Engineering*. 40: 2155-2160, **2017**.

[18] Beykal B, Boukouvala F, Floudas CA, Pistikopoulos EN. Optimal design of energy systems using constrained grey-box multi-objective optimization. *Comput Chem Eng*. DOI 10.1016/j.compchemeng.2018.02.017, **2018**.

[19] Mehrian M, Guyot Y, Papantoniou I, Olofsson S, Sonnaert M, Misener R, Geris L. Maximizing neotissue growth kinetics in a perfusion bioreactor: An *in silico* strategy using model reduction and Bayesian optimization. *Biotechnol Bioeng*. 115(3):617-29, **2018**.

[20] Lee JH, Shin J, Realff MJ. Machine learning: Overview of the recent progresses and implications for the process systems engineering field. *Comput Chem Eng*. DOI 10.1016/j.compchemeng.2017.10.008, **2017**.

[21] Damianou A, Lawrence N. Deep Gaussian processes. In *Artificial Intelligence and Statistics*. 207-215, **2013**.

[22] Deisenroth MP, Ng JW. Distributed Gaussian processes. In *32nd International Conference on International Conference on Machine Learning*. 37: 1481-1490, **2015**.

[23] Ghahramani Z. Bayesian non-parametrics and the probabilistic approach to modelling. *Phil Trans R Soc A*. 371(1984):20110553, **2013**.

533936 Combustion Analysis of Trash to Tank Fuel Derived from Plastic Waste

Abstract:

Trash to Tank is a concept based on the conversion of waste plastic trash into a liquid fuel, suitable for use in diesel or kerosene fuel applications. This contribution focuses on both an experimental and theoretical analysis of combustion emissions of waste plastic derived fuels, comparing them with combustion emissions of petroleum derived diesel and kerosene fuel. Results of experiments to measure the lower heating value (LHV), carbon dioxide emissions and molecular composition of waste plastic derived fuels as well as the corresponding theoretical calculations will be presented. The waste plastic feedstock utilized in this study include, low density polyethylene, high density polyethylene, polypropylene and polystyrene. Thermal decomposition was conducted via slow pyrolysis at 450°C in a batch reactor for each of these materials. Potential benefits to the global environment from the use of Trash to Tank fuels as replacements for petroleum derived fuels will be further discussed.

531638 Atatad Heat Accumulation in Countercurrent Flow, Continuous Phase Sludge Treatment Process to Class a Fertilizer

Abstract:

This document describes reduction to experimental practice, for the first time, for a patented method of pasteurization that converts wastewater sludge into Class A fertilizer using air in a continuous process. Abbreviated atATAD for atmospheric air thermophilic aerobic digestion, the process represents a paradigm shift in sludge treatment from batch to continuous processing that utilizes a multi-stage countercurrent setup to achieve pasteurization temperatures within prescribed residence time. Beyond validating the patentâ€™s concept, atATAD was accomplished with less than half the patentsâ€™ modeled solids content (4-5%). Intensive real-time data collection and post-run sample processing using CHN analysis and bomb calorimetry were used to begin definition of the parameters for a dynamic model. Specifically, a single-stage proxy for the full multi-stage countercurrent system was run as a batch reaction in an attempt to establish the mass and energy balances; the batch runs reached pasteurization temperatures but did not close the mass and energy balance. This initial success of reaching thermophilic temperatures motivated a pilot-scale (1650 gallons) continuous phase reaction with 4-day residence time. This comprehensive pilot-scale demonstration conducted on-site at the local wastewater treatment authority, included far more extensive data sets and verified the use of air and countercurrent flow to yield pasteurization conditions. The goal of biohazard reduction from pasteurization temperatures exceeding 50 °C through aerobic consumption of solids is met.

The rapid transition to a real-world pilot implementation introduced significant operational constraints that prevented steady operation. None-the-less, the resulting dynamic data represents an opportunity which dynamic modeling may be fit to in the future. Dynamic fitting with the continuous temperature data from varying, inevitable condition changes (composition, amount of solids, plugs to the system, energy potential, etc).

538705 Development and Deployment of Advanced Technology for Mswi Fly Ash Stabilization and Utilization Via High-Gravity Carbonation Process

Abstract:

The removal of chlorine and heavy metal in municipal solid wastes incineration (MSWI) fly ash is signiï¬cant necessity before its utilization in cement kiln, ensuring the quality of cement products, maintenance of processing equipment, and the protection of environment. In addition, fly ash could be considered as a suitable material for CO2 mineralization due to high alkalinity, small particle size and high calcium content. In this presentation, we developed an integrated process to reduce the chlorine and heavy metal from MSWI fly ash coupled with the carbonation reaction and water-washing treatment in a high-gravity rotating packed bed. To indicate that the chlorine and heavy metal could be removed from carbonated fly ash, we evaluate the reaction mechanisms at mineral precipitation and leaching kinetic at liquid-solid interface. We also conducted the energy and mass balance to determine the optimal operation condition in this integrated process. Those findings could provide a novel treatment process for MSWI fly ash and broaden the utilization pathway in cement industry.

533373 Investigation on CO_2 co-Gasification of Horticultural Waste and Sewage Sludge for Energy Production: Effects of Temperature, Ash Content and CO_2 flow Rate

Abstract:

In this work, a fixed-bed downdraft gasification system is used to harvest the energy from the municipal solid wastes, i.e. horticultural waste and sewage sludge. In comparison with conventional disposal route of horticultural waste and sewage sludge, e.g. incineration and air gasification, CO_2 gasification is known to be more environmentally friendly with the ability of CO_2 capture, fewer pollutant emissions and the produced synthesis gas in this process can be eventually utilized in boilers, gas engines or turbines to generate electricity or easily upgraded to transportation fuels through some chemical processes. Sewage sludge with high ash content is co-gasified with horticultural waste in this work in order to enhance the overall heating value of the feedstock and avoid agglomeration in real operation. In addition, the alkali metal content in the horticultural waste was reported to act as a catalyst for gasification, thus this co-gasification can help achieve high solid-to-gas conversion efficiency. The effects of operating parameters of a fixed-bed downdraft gasification system, i.e. temperature (700-900Â°C), ash content (5-25%) and CO_2 flow rate (60-100mL/min), on the gasification performance have been comprehensively studied through the Boxâ€"Behnken experiment design.

532732 Waste Management in International Airports: A Case Study of Astana Airport

Abstract:

The gradual increase of airport capacities has lead to an increase of the amount of solid waste. Thus there is a need for sustainable waste management which in turn requires input from experimental studies. The aim of this study was to perform an analysis of the solid waste produced at the Astana International Airport in Kazakhstan. Based on the solid waste charactreristics and amount, different waste management scenarios are compared. The waste management scenarios developed in this work will be a useful tool for decision makers in environmental waste management in airports.

COMPARISON BETWEEN MULTI-STAGE AND SINGLE STAGE MICROALGAE DEWATERING PROCESSES

Fares Almomani, Hazim Qiblawey, Rahul Bhosale, Majeda Khraisheh, Anand kumar, Khaled Al-Jaml

Department of Chemical Engineering, College of Engineering, Qatar University, P. O. Box 2713, Doha, Qatar

falmomani@qu.edu.qa

Abstract

The present work shows a quick, simple and accurate method for algae harvesting with a potential to relate the required polymer dose with in-situ hydrodynamics properties of alga solution. The method also reduces the lag time between the change in influent quality and chemical dosing adjustment during dewatering process. The tests were carried out using single and multi stage conditioning steps. Spectrophotometric measurements can potentially be used to select the best performing polymer and determine the optimum polymer dose for any type of Algae; optimum polymer dose calculated by use of CST and volume of filtrate match with optimum polymer dose achieved with absorbance measurements. Multi-stage conditioning process using ferric chloride and polymer resulted in better settling and algae dewatering when compared to the single stage conditioning using polymer or ferric chloride alone. The amount of conditioner used in multi-stage conditioning processes was less than the doses required for single stage process. The presented method could be useful in monitoring and optimize the polymer concentration and minimize the excessive use of polymers.

Keywords: anaerobic, conditioning, dewatering, polymer dose

INTRODUCTION

The depletion of fossil fuel resources, increase in crude oil prices and the proved evident that the earth's climate is going through global warming as a result of the continuous emissions of greenhouse gases (GHGs) to the atmosphere has resulted in the search for an alternative third-generation fuel[1-4]. Bio-fuels such as biogas[5, 6], biodiesel[7], and bioethanol[8] produced from biological processes are considered promising alternatives to fossil fuels. Bio-fuels have different advantages over fossil fuels such as low toxicity, high biodegradability, low net CO_2 production, are renewable and sustainable, and contains less or no sulfur[9, 10].

Biofuels can be produced from anaerobic digesters, biological stocks, and bio-wastes[11-13]. However, due to the high pre-treatment and operational cost of these processes and the competition with the human food chain, the search for new bio-sources has been a major research endeavor globally[14, 15].

Microalgae which has a rapid growth rate[16-19], high lipid and carbohydrate contents and pose no competition with crops for food is an attractive source for biofuel production[20, 21]. In addition, microalgae have a great potential for carbon capture, wastewater treatment and bio-sequestration. Algal-bacterial ponds or high rate oxidation pond can be used for algae production in addition to efficiently removal of organic matter and nutrients (nitrogen and phosphorus) from wastewater. This dual action consider as a feasible technology for algae production and wastewater treatment that has the ability to remove organic substances and nutrients with reasonable retention time and low operational cost. Different studies have used algal-bacterial interaction for treatment of different types of wastewater. Zhu et al, [22] reported a removal efficiency of 65-76% of chemical oxygen demand (COD), 68-81% of total nitrogen (TN) and 90-100% of total phosphorous (TP) from piggery wastewater with a reasonable algae production rate. Wang et al., [23] reported a reasonable algae growth rate combined with COD removal in the range 50.9% to 83 % from influent of WWTP. The study also showed significant nutrient removals in the range 68.4% to 82.8%. Aslan and Kapdan, [24] found that microalgae *Chlorella vulgaris* can be produced from synthetic wastewater. Zimmo et al., [25] cultivated algae in pilot-scale algae-based ponds (ABPs) using domestic wastewater at cold temperatures (6.5 and 12.7 °C) and warm temperatures (18.4 and 21.3°C). Results showed high algae growth rate and a significant wastewater treatment performance

Although a significant amount of research work has been reported on improving the production of microalgae using different bio-reactors and pond systems, research work that focuses on establishing simple, effective and low-cost downstream dewatering operations is limited. The high cost associated with dewatering large-volumes of microalgae cultures has been the primary hindrance to the development of industrial-scale microalgae biofuels.

The present work focus on improving the separation and dewatering of microalgae after cultivation period in a cost-effective and simple way. The study investigates the conditioning and dewatering of microalgae using single and multiple conditioning stages, determine the optimum conditioner doses, and determine the effect of operational conditions on dewatering performance.

MATERIAL AND METHODS

Chemical

Dewatering experiments were conducting using commercial polymers (Flo Polymer CA 475, CAB 4500, CIBA Zetag, FloPolymer CB 4350 and Flo Polymer CA 4600) manufactured by SNF Inc. and Magnafloc-LT 27AG.. Polymers were selected to represent a wide range of polymers used in algae dewatering. The characteristics of the used polymer are resented in [26, 27]

Algae Sludge samples

Algae strain, spirulina used in the experiments were cultivated in laboratory as pre the procedure reported in [28]

Preparation of polymer solution

Polymers stock solutions was prepared daily in a jar equipment at a concentration of 0.25 % by mixing 1.5 g polymer in 7500 mL of deionized water at a speed of 220 rpm for 5 minutes, the mixture was then mixed for 1.5 hour at a speed of 100 rpm f. then, the mixture was left to sit for 1 hour before use in the experiments.

Dewaterability Measurements

Algae dewaterabilty was tested using Capillary suction time (CST) and filtration volume. Tests were conducted following the procedure stipulated Standard Methods [29]. The detailed procedures for both tests are outline in [30]

RESULTS & DISCUSSION

Figure 1-5 show the optimum polymer dose as determined by CST, filtrate volume, filtrate absorbance for different polymers . Cationic FloPolymer CAB 4500 (Figure 1) with high charge density and medium to high molecular weight showed a significant algae removal at very low optimum dose. The optimum dose was judged at with lowest CST time, and highest filtrate volume. Figure1 presents the change in CST and filtrate volume as function of CAB 4500 polymer doses. ACAB 4500 polymer dose of 7 g/ kg dry solids improved the CST time of sludge from 207 sec for the algae stock to 24 sec. Increasing the polymer dose increased, the CST value showing polymer over dose without any effect on algae harvesting. A similar tendency was observed for filtrate volume; the collected filtrate was 89 mL.

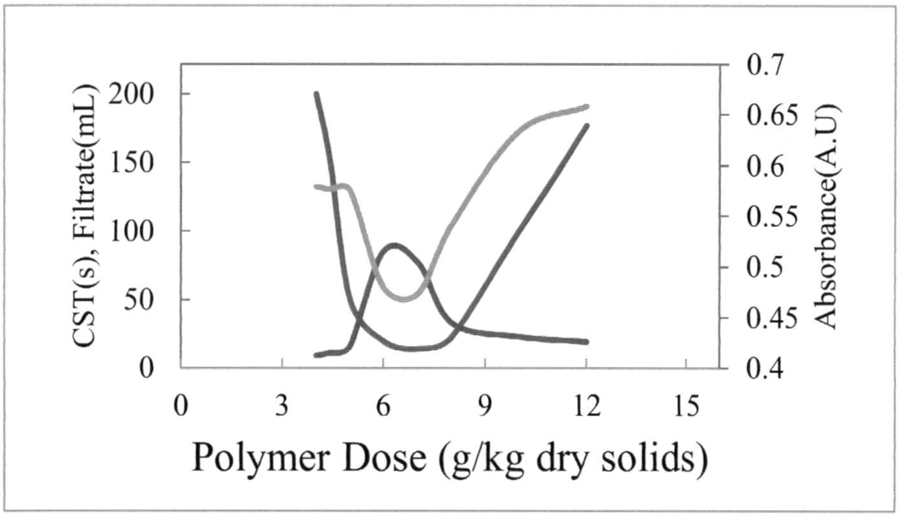

Figure 1: Evolution of CST, filtrate volume and filtrate absorbance as function of CAB 4500 polymer dosage

The cationic polymer FloPolymer CA 475 with high charge density and ultra-high molecular weight polymer showed similar tendency (Figure 2). For under dose region, the collected algae and the volume of filtrate increase with increasing polymer dose until optimum value reach. Using CST, and filtration volume, the optimum dose was found to be 6g/kg dry solids. Extra polymer, overdose, affect negatively the amount of algae harvested

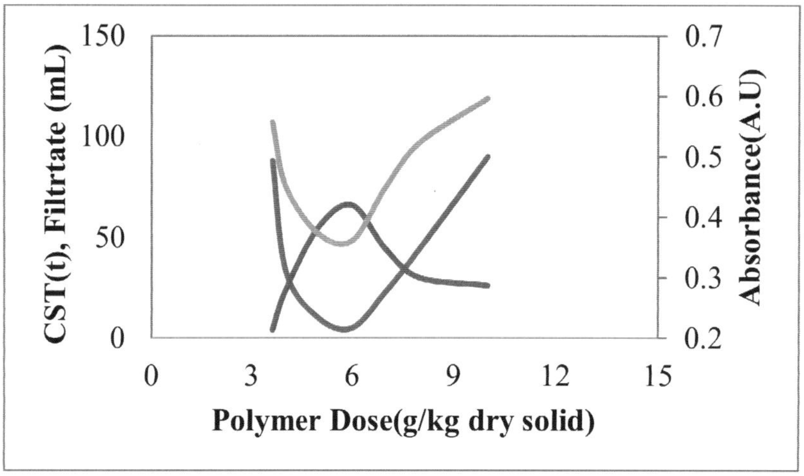

Figure 2: Evolution of CST, filtrate volume and filtrate absorbance as function of SNF Flo Polymer CA475 dosage

Zetag CIBA 7857 polymer was used for algae harvesting. The evaluation of CST and collected volume can be seen in Figure 3. The changes in CST and volume of filtrate follow the previous tendency; the CST value start to decrease as response to polymer action reaching minimum value around 23.4 g/kg DS with and optimum dose of 0.9 g/kg DS. Filtrate volume at the same time increase by increasing the polymer dose reaching a maximum value of 88 mL at optimum polymer dose.

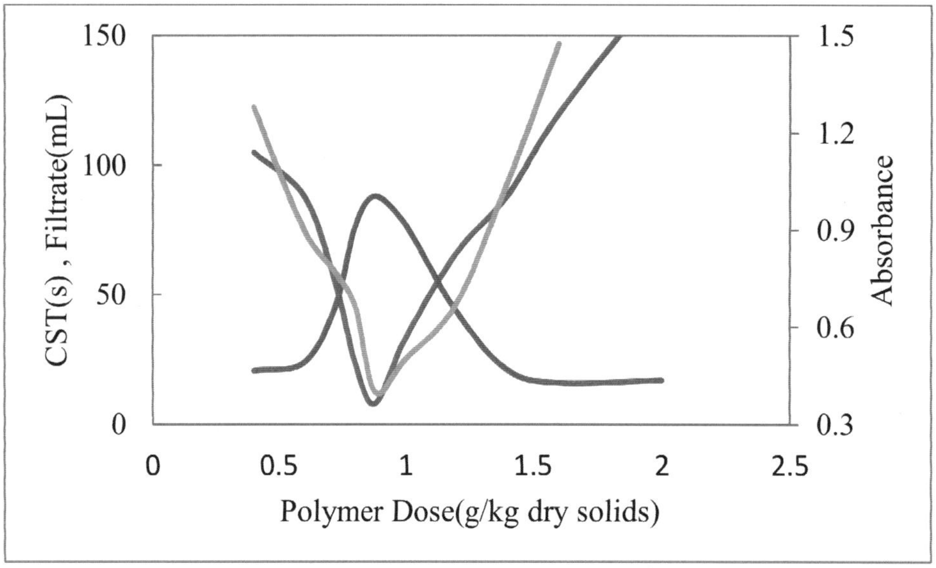

Figure 3: Evolution of CST, filtrate volume and filtrate absorbance as function of CIBA Zetag 7857 polymer dosage

The increase in filtrate absorbance in the under-dose region as results of polymer addition can be attribute the physic-chemical reaction between the filtrate and polymers.

Table 1 shows the optimum polymer dose, optimum CST and collected filtrate for the five studied polymers using single conditioning step. It can be see that for all the studied polymer, the CST value was reduced from 207 sec for the original algae strain to less than 16 sec after conditioning. Moreover, it was observed through all the experiments that the absorbance of filtrate at optimum polymer dose represent the lowest absorbance for each set of experiment. Table 1 also shows the %weight of algae recovered by each polymer suggesting successful harvesting technology. .

Table 1 optimum polymer dose, CST, collected filtrate and % Algae recovery using single conditioning

POLYMER	Optimum polymer dosses	CST at optimum polymer doses	filtrate volume (mL)	% Algae recover
SNF Flo Polymer CA475	6± 0.2	10.1 ± 0.2	88 ± 2	76
SNF Flo Polymer CAB 4500	7 ± 0.5	24.5 ± 0.5	80 ± 1	83
CIBA Zetag 7857	1.5 ±0.2	10.0 ± 0.7	90 ± 4	84
FloPolymer CB 4350	4.2 ± 0.3	11.3 ± 0.3	88 ± 7	81
FloPolymer CA 4600	5.2 ± 0.1	12.6 ± 0.2	85 ± 3	77

Sequential conditioning

Figure 4 shows an example of sequential algae conditioning using CAB4500. Sequential processes showed lower optimum polymer dose compared with single conditioning. Combined it was observed that applying two step conditioning improved algae dewatering, reduce polymer requirement and decrease the CST time. Table 2 shows that for all the tested polymer the optimum dose ranged from 0.5 3.2 (g/kg dry solid).

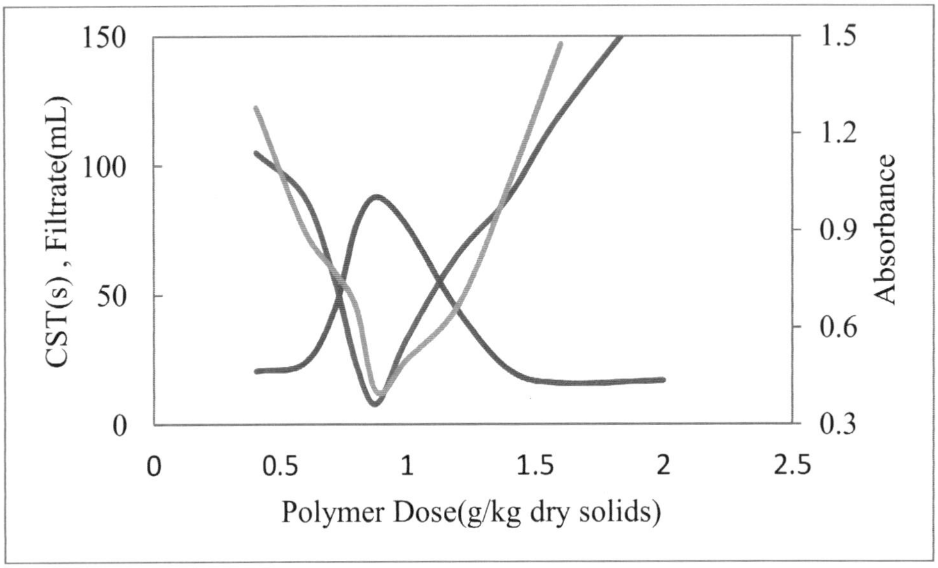

Figure 3: Evolution of CST, filtrate volume and filtrate absorbance as function of polymer dose in sequential conditioning

Table 2 optimum polymer dose, CST, collected filtrate and % Algae recovery using sequential conditioning

POLYMER	Optimum polymer dosses	CST at optimum polymer doses	filtrate volume (mL)	% Algae recover
SNF Flo Polymer CA475	0.5± 0.02	9.1 ± 0.2	83 ± 1	88
SNF Flo Polymer CAB 4500	0.7 ± 0.05	14.5 ± 0.5	86 ± 1	93
CIBA Zetag 7857	0.88±0.02	8.0 ± 0.7	96 ± 4	94
FloPolymer CB 4350	1.2 ± 0.03	7.3 ± 0.3	92 ± 2	91
FloPolymer CA 4600	1.2 ± 0.01	6.6 ± 0.2	89 ± 3	87

Conclusion

The present work provides simple and accurate method to harvest algae strains in water solution and determine the optimum polymer dose for effective conditioning. The method would also be beneficial in preventing the excessive use of polymers during conditioning process.

The method is quick, simple and accurate and has the potential to optimize algae harvesting and conditioning system based on dynamics influent water properties, reduce the magnitude of upset due to the time lag between the change in influent water quality and chemical dosing adjustment and determine the optimum polymer dose based on in-situ hydrodynamics properties of treatment systems. Sequential use of ferric chloride and polymer resulted in better settling and dewatering compared to the single use of polymer or ferric chloride The total conditioner dose used for sequential conditioning was less than the doses required for ferric chloride or polymer conditioning alone The optimum conditioner doses obtained in test tubes were different than the optimum doses obtained from jar tests which can be explained by the differences in mixing. Ferric chloride as well as sequential conditioning resulted in better settling and dewatering at lower pH values.

References

[1] M. Hoel, S. Kverndokk, Depletion of fossil fuels and the impacts of global warming, Resource and Energy Economics, 18 (1996) 115-136.

[2] I.C. Macedo, J.E. Seabra, J.E. Silva, Green house gases emissions in the production and use of ethanol from sugarcane in Brazil: the 2005/2006 averages and a prediction for 2020, Biomass and bioenergy, 32 (2008) 582-595.

[3] T.L. Root, J.T. Price, K.R. Hall, S.H. Schneider, C. Rosenzweig, J.A. Pounds, Fingerprints of global warming on wild animals and plants, Nature, 421 (2003) 57.

[4] P.M. Cox, R.A. Betts, C.D. Jones, S.A. Spall, I.J. Totterdell, Acceleration of global warming due to carbon-cycle feedbacks in a coupled climate model, Nature, 408 (2000) 184.

[5] F. Almomani, R.R. Bhosale, A. Kumar, The effect of intermediate ozonation process on improving biogas production from co-digestion of agricultural waste and manure, Advanced Materials - TechConnect Briefs 2016, 2016, pp. 170-174.

[6] S.J. Judd, F. Al Momani, H. Znad, A. Al Ketife, The cost benefit of algal technology for combined CO2 mitigation and nutrient abatement, Renewable and Sustainable Energy Reviews, 71 (2017) 379-387.

[7] Y. Chisti, Biodiesel from microalgae, Biotechnology advances, 25 (2007) 294-306.

[8] S. Kim, B.E. Dale, Global potential bioethanol production from wasted crops and crop residues, Biomass and bioenergy, 26 (2004) 361-375.

[9] J. Hill, E. Nelson, D. Tilman, S. Polasky, D. Tiffany, Environmental, economic, and energetic costs and benefits of biodiesel and ethanol biofuels, Proceedings of the National Academy of sciences, 103 (2006) 11206-11210.

[10] P.J. Crutzen, A.R. Mosier, K.A. Smith, W. Winiwarter, N 2 O release from agro-biofuel production negates global warming reduction by replacing fossil fuels, Paul J. Crutzen: A pioneer on atmospheric chemistry and climate change in the anthropocene, Springer2016, pp. 227-238.

[11] F.A. Al Momani, N. Jarrah, Treatment and kinetic study of cyanobacterial toxin by ozone, Journal of Environmental Science and Health - Part A Toxic/Hazardous Substances and Environmental Engineering, 45 (2010) 719-731.

[12] N.B. Abderrazik, F. Al Momani, C. Sans, S. Esplugas, Combined advanced oxidation with biological treatment, Afinidad, 59 (2002) 141-146.

[13] F. Al Momani, O. Gonzalez, C. Sans, S. Esplugas, Combining photo-Fenton process with biological sequencing batch reactor for 2,4-dichlorophenol degradation, Water Science and Technology, 2004, pp. 293-298.

[14] S.J. Eichhorn, A. Gandini, Materials from renewable resources, Mrs Bulletin, 35 (2010) 187-193.

[15] H.A. Preisig, B. Wittgens, Thinking towards synergistic green refineries, Energy Procedia, 20 (2012) 59-67.

[16] Y. Liang, N. Sarkany, Y. Cui, Biomass and lipid productivities of Chlorella vulgaris under autotrophic, heterotrophic and mixotrophic growth conditions, Biotechnology letters, 31 (2009) 1043-1049.

[17] H. Znad, A.M. Al Ketife, S. Judd, F. AlMomani, H.B. Vuthaluru, Bioremediation and nutrient removal from wastewater by Chlorella vulgaris, Ecological Engineering, 110 (2018) 1-7.

[18] F.A. AlMomani, B. Örmeci, Performance Of Chlorella Vulgaris, Neochloris Oleoabundans, and mixed indigenous microalgae for treatment of primary effluent, secondary effluent and centrate, Ecological Engineering, 95 (2016) 280-289.
[19] F. Almomani, R. Bhosale, A. Kumar, M. Khreisheh, U. Gosh, The potential use of solar photocatalytic oxidation for removal of emerging pharmaceuticals from effluent of wastewater treatment plant: A pilot plant study, Environmental Division 2017 - Core Programming Area at the 2017 AIChE Spring Meeting and 13th Global Congress on Process Safety, 2017, pp. 50-56.
[20] Q. Hu, M. Sommerfeld, E. Jarvis, M. Ghirardi, M. Posewitz, M. Seibert, A. Darzins, Microalgal triacylglycerols as feedstocks for biofuel production: perspectives and advances, The plant journal, 54 (2008) 621-639.
[21] M.S. Wigmosta, A.M. Coleman, R.J. Skaggs, M.H. Huesemann, L.J. Lane, National microalgae biofuel production potential and resource demand, Water Resources Research, 47 (2011).
[22] L. Zhu, Z. Wang, Q. Shu, J. Takala, E. Hiltunen, P. Feng, Z. Yuan, Nutrient removal and biodiesel production by integration of freshwater algae cultivation with piggery wastewater treatment, Water Research, 47 (2013) 4294-4302.
[23] L. Wang, M. Min, Y. Li, P. Chen, Y. Chen, Y. Liu, Y. Wang, R. Ruan, Cultivation of Green Algae Chlorella sp. in Different Wastewaters from Municipal Wastewater Treatment Plant, Applied Biochemistry and Biotechnology, 162 (2009) 1174-1186.
[24] S. Aslan, I.K. Kapdan, Batch kinetics of nitrogen and phosphorus removal from synthetic wastewater by algae, Ecological Engineering, 28 (2006) 64-70.
[25] O.R. Zimmo, N.P. van der Steen, H.J. Gijzen, Nitrogen mass balance across pilot-scale algae and duckweed-based wastewater stabilisation ponds, Water Research, 38 (2004) 913-920.
[26] F. Al momani, B. Ormeci, In-line and Real-Time Measurement of Polymer Concentration in Water and Wastewater, (2013).
[27] F.A. Al Momani, B. Örmeci, Optimization of polymer dose based on residual polymer concentration in dewatering supernatant, Water, Air, and Soil Pollution, 225 (2014).
[28] S. Mohamad, A. Fares, S. Judd, R. Bhosale, A. Kumar, U. Gosh, M. Khreisheh, Advanced wastewater treatment using microalgae: effect of temperature on removal of nutrients and organic carbon, IOP Conference Series: Earth and Environmental Science, IOP Publishing, 2017, pp. 012032.
[29] A. APHA, WEFF, Standard methods for the examination of water and wastewater, 21 ed., American Public Health Association, Washington, D.C.2005.
[30] F.A. Al Momani, B. Ã–rmeci, Measurement of polyacrylamide polymers in water and wastewater using an in-line UV vis spectrophotometer, Journal of Environmental Chemical Engineering, 2 (2014) 765-772.

IMPROVING BIOGAS PRODUCTION FROM AGRICULTURAL WASTE BY PHOTO-FENTON PROCESS

F. Almomani[1*], R.R. Bhosale[1], A. Kumar[1], M.A.M. Khraisheh[1], M. Shawaqfah[2]

[1] *College of Engineering, Department of Chemical Engineering, Qatar University, P. O. Box 2713, Doha, Qatar.*
[2] *Department of Civil Engineering, Al al-Bayt University, Mafraq, Jordan*

Abstract

The Study investigates the enhancement of biogas production from a mixture of agricultural solid waste, wastewater and manure by Photo-Fenton prices process. The effect of the total solids, organic content, nutrient and the structure of the substrate on the production of the biogas were assisted. The performance of anaerobic digestion (AD) was monitored by following volatile solids removal efficiencies, biogas production rate and biogas yield ands composition. A number of strategies including controlling the composition of the substrate and employing chemical oxidation process where tested to determine if they resulted in an increase in that in amount of biogas production. The cumulative biogas production for 30 h of digestion ranged from 16.2 to 69.8L/kgVS.h, with the highest production rate for substrate with C/NL ratios in the range 20-30., COD in the range 185 to 40 mg/L an TS in the range 100-15%. The production of by biogas increased by increasing substrate VS/TS ratio, reaching a cumulative biogas production rate of 29.8 NL as a VS/TS ratio of 45%, and a 60,5 NL at a VS/TS ratio of 76. Using photo thing as an intermediate oxidation process between two Ads improved biogas production by 19 to 31.3% for the same substrate

1. Introduction

High Energy demand and increase restriction of the use of non-renewable energy resources led researcher to look for alternative energy sources [1-3]. Renewable energy resources draw attention all over the world because they are cheap provide no environmental impact and improve economy[4, 5].

Every year municipalities dispose several million tons of agricultural wastes[6]. Traditional treatment options such as incineration, land up locations and landfilling where resulted in subsequent environmental problems in different parts of the ecosystem; wastes can pollute soil, surface water and underground water[REF]. on other hand, inceneration showed major shortage such as low capacity, low energy output and emission wide varities of poluted gases. Thus, any other solution for solid waste treatment is Highly recommended. A viable solution for this solid waste treatment is the use of anaerobic digestion processs [7, 8]. TRhis process is advantageous for a different reason; it can't be used for treatment of different type of solid wastes, it does not

Corresponding author: Fares Almomani, College of Engineering, Department of Chemical Engineering, Qatar University, P. O. Box 2713, Doha, Qatar. Tel: +974 4403 4140 Fax: +974 4403 4131. E-mail: falmomani @qu.edu.qa

have any environmental impact on the ecosystem, the biogas produced from this process is considered as a valuable energy source which can be use for different applications.

Anaerobic digestion (AD) of agricultural waste and animal manure is any effictive wat to reduce solid waste and the amount of greenhouse gases emission. It was reported that the use of AD can play an important rile in reducing the amounts of ammonia and methane from manure storage facilities [9, 10]. The buffering capacity of manure and the presence of diverse elements have a positive impact on the digestion process stabilities. One of the approach for improving the economics of AD and increase the biogas production rate is by co-digesting different amounts of solid wastes, co-digistion stratigies ami to improve the feed properties, abs consistency, stabilize the feed chemical composition and enhance biogas production. The digested Substrate after AD can be used as an organic fertilizer, rich in nutrients required for the growth of the plants. Utilization of large large amounts of animals manure for bio-energy will reduce the nutrient run off and reduce water resource contaminations.

Different studies reported on improving the production of biogas from AD by using biological additives such as fungi, microbial consortium, and enzymes [11, 12], or chemical reagents such as alkali, acid and oxidative reagents [8]. Advanced oxidation processes were used to improve the biodegradability of organic matter[13, 14], eliminate toxicity[15-17] and increase the production of biogas [18-22]. Ozone and photo-Fenton which are strong oxidizing agents can be used for this purpose [15, 23, 24]

Although AD is very common, there is still needs practical improvement of this part of science, the optimum feed and operating conditions that can lead to high biogas yield still not well identified. Only one work has tested the use of photo-Fenton or the enhancement of biogas production from lignocellulose materials. Improving biogas production for mixed subtract conditions was not tested yet. Accordingly, this work evaluates the impact of using an intermediate oxidation process, photo-Fenton, on biogas production. In addition the study aimed to identify the effect of subtract composition and structure on biogas production from agricultural sold wastes

2.1. Preparation of substrates

The substrate was prepared from a mixture of different agricultural waste and animal manures. The main parameter used in substrate preparation were the solid content, carbon to nitrogen ration (C/N ratio) nitrogen compound content, chemical oxygen demand and water content. The substrate was prepared and immediately stored in a refrigerator at approximately 4 °C. Different waste were collected, during 2010, from municipal wastewater treatment plant and local restaurant.

To study the effect of solid content in the operation of AD was determined by adjusting the solid content by diluting the substrate with tap water. The solid size was classified based on screening the substrate with two different mish sized screens. The characteristics of the substrates tested in this study, are shown in Table 1.

2.2 Experimental set-up

Batch digestion tests were performed on different substrate. The experimental set up used in this work is presented in figure 1. Jacketed bioreactor of total volume of 2 L was used in this work. the reactor is equipped with (1) substrate feeding storage, (2) water bath, (3) biogas collecting bag (4) mechanical stirrer, (5) temperature thermo couple, (6) pH meter, and (8) temperature and pH recorder. For safety and easy operation the main reactor has vent (10) and drainage port (9). The biogas was collected in biogas bag and measured by two converted graduated cylinder filled with water by use of three way value (7).

Table 1: different substrate properties used in this study

Substrate	Substrate	% TS	C/N	COD (mgO_2/L)	NH_3 (mg/L as N)
Sub-1	Straw + cattle manure +water	25	26	220	35
Sub -2	Wood + cattle manure+ water	25	28	240	40
Sub -3	Cow + cattle manure+ water	25	23.5	230	35
Sub -4	Cow + cattle manure+ wastewater	20	20	300	28
Sub -5	Cow + cattle manure+ wastewater+ wood dust	20	35	340	26
Sub -6	Cow + cattle manure+ wastewater+ sawdust	20	33	350	33
Sub -7	Straw + cattle manure +water	10	26	150	20
Sub -8	Wood + cattle manure+ water	10	28	140	16

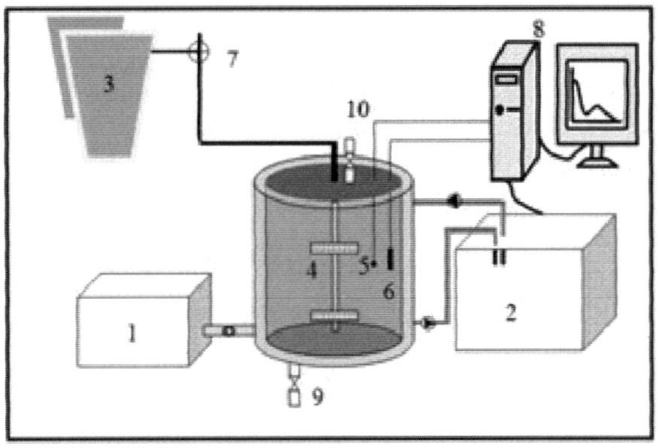

(1) Substrate storage **(2) Water bath** **(3) biogas bag**
(4) Mechanical stirrer **(5) Thermocouple** **(6) pH meter**
(7) Three way valve **(8) PC recorder** **(9) Reactor drain**
(10) Reactor vent

2.3 Biogas measurements and analysis

Biogas production was measured by two converted graduate cylinder filled initially with water. Sliding the upper cylinder upward will create a vacuum space that pull out the biogas from the biogas bag. The biogas is collected between the water surface and the upper biogas bottom.

When all the biogas pulls out of the bag the volume of the gas is recorded. Then, the cylinders were disconnected from the biogas bag and connected to the GC. The collected gas then forced to passed through GC by lowering the upper graduate cylinder and the gas was analyzed.

2.4 Chemical analysis

The TS and VS of the raw samples were measured in triplicate according to standard methods (APHA, 1998).

Kjeldahl nitrogen (TKN) was measured after Kjeldahl nitrogen digestion using a digestion mixture of H_2SO_4, K_2SO_4, and CuSO. Ammonia content of sample was subsequently detected by electrical conductivity cell. Ammonium–nitrogen (NH_4^+—N) was measured by flow injection analysis after steam distillation into dilute sulfuric acid

3. RESULT AND DISCUSSION

3.1 Effect of Total solids and volatile solids

The chemical composition of substrate used in the present study is shown in Table 1. It was observed that the higher volatile solid (VS) content of waste mean the higher energy content, which is desirable from an economic standpoint of view with, regards biogas energy production. Based on chemical analysis it was observed that the agricultural waste has higher VS content in the range (180-260 g/kg), compared with that of manure (90- 110 g/kg). Two values of volatile solid/ total solid (VS/TS) ratios were tested 45% and 86.1%. Figure 2 illustrate the cumulative biogas production for both substrates. It can be seen from the figure that the higher the volatile solid content in the substrate the higher the biogas production. This may be due to the presence of more nutrients and contaminants such as presence of spilled animal feed. It should be mentioned that the TS content of any substrate depends on the amount of water added during dilution. Therefore when using the manure flushing system, the amount of water added should be estimated accurately because it affects not only the digester size but also the volume of the digestate to be disposed of, which consequently affect the costs of treatment[8].

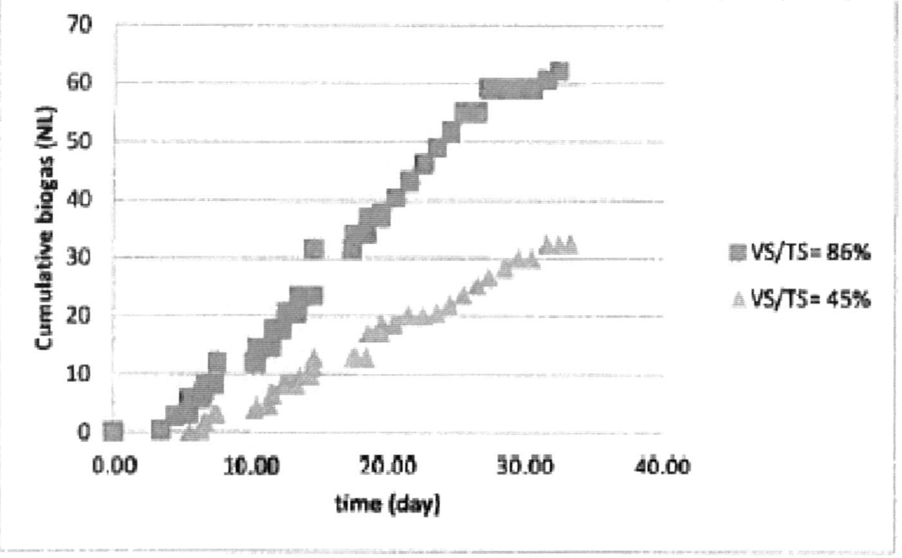

Figure 2 cumulative biogas productions at two different values of (VS/TS) ratios

3.2 Biogas yield

Biogas yields of different substrate presented in table 1 are shown in Fig. 3. Substrate with C/N ration around 25 had the highest biogas yield, compared with substrate with C/N ratio > 30. After 30 days of digestion, cumulative biogas yield was calculated to be 50, 35 and 15 L/kg.VS for Sub -2, Sub -1 and Sub -5, respectively. The higher biogas yield between different substrate may be attributed to the difference in COD value between these substrates and the difference in C/N ratio. Moreover, it was noticed that the higher the nitrogen content of the substrate the lower is the biogas production. The nitrogen effect on biogas production can be explained by the fact that. The presence of high nitrogen content will be converted to ammonia (NH_4). NH_4 will increase the pH value of the content in the digester. A pH higher than 8.5 was found to have toxic effect on bacterial population the biogas production decrease[9].

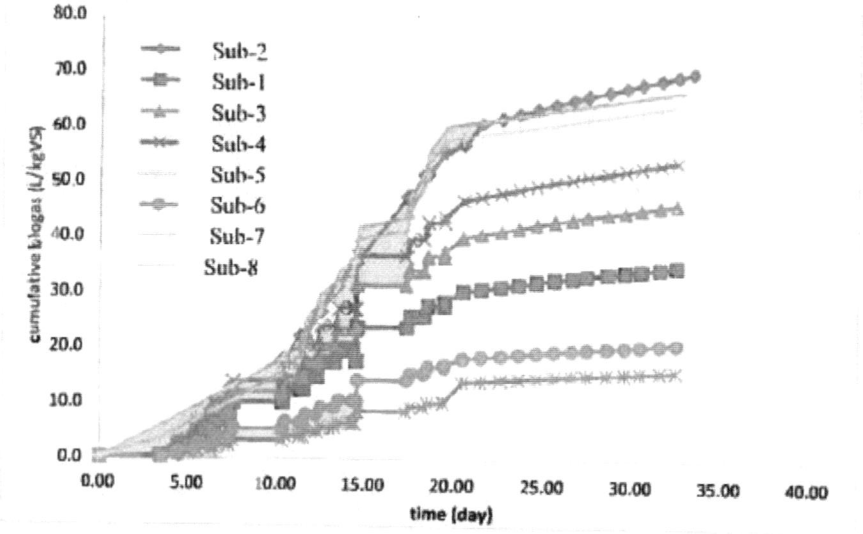

Figure 3: Biogas yields of different substrate presented in table 1

3.3. *Effect of substrate structure*

The effect of substrate structure size was also studied, the substrate was grinding and screens by two different screens 2-mm and 4-mm, the biogas production from fine substrate was higher than that from coarse substrate. The obtained results were attributed to the easy usage of nutrient and organic matter from fine substrate compared with coarse substrate. Similar results were found in different studies. Average daily biogas production rates (L/ kgVS day) for fine and coarse substrate are shown in Fig. 4. For all substrate types, a peak biogas production rate was observed after the second day of digestion. The maximum values of biogas production rate were 47 and 40 L/ kgVS. day for the fine and coarse substrate, respectively. The higher biogas production rate from the fine fraction may be attributed to the smaller particle sizes of this fraction and the presence of easily biodegradable organics compared with both the coarse fraction and the unscreened manure[10].

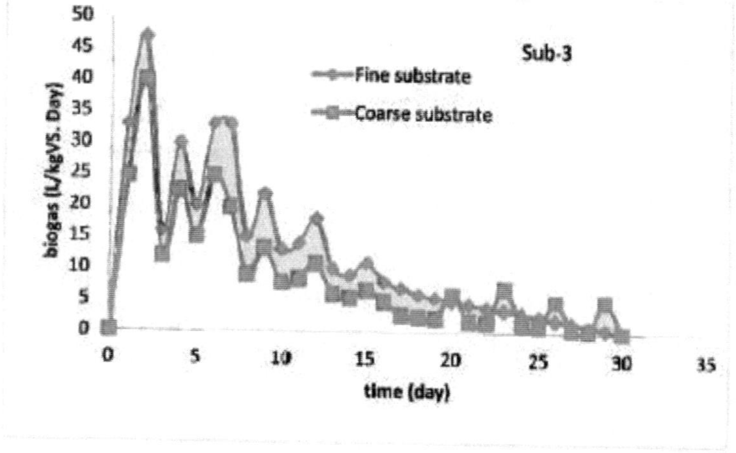

Fig. 4: Average daily biogas production rates for fine and coarse substrate are shown in

3.2 Effect of total solid content

Figure 5 shows the cumulative biogas production for similar substrate but with different total solid (see table 1). It was observed that the biogas production is related to the TS content of the solution. The trend is not clear for Sub -1 and 8 but the tendency was very clear for Sub -1 and 7.Mixing with water is a good method to arrive at the desired consistency of the inputs (e.g. ratio of volume of dung or waste to volume of water could vary from 1:1.25 to even 1:2). The dilution should be made to maintain the total solids from 7 to 10 percent. Experimentally dilution percentage between 10 to 25 % was tested. The relationship between dilution ration and biogas production volume was studied, it was observed that a values between 10 to 15% give optimum values.

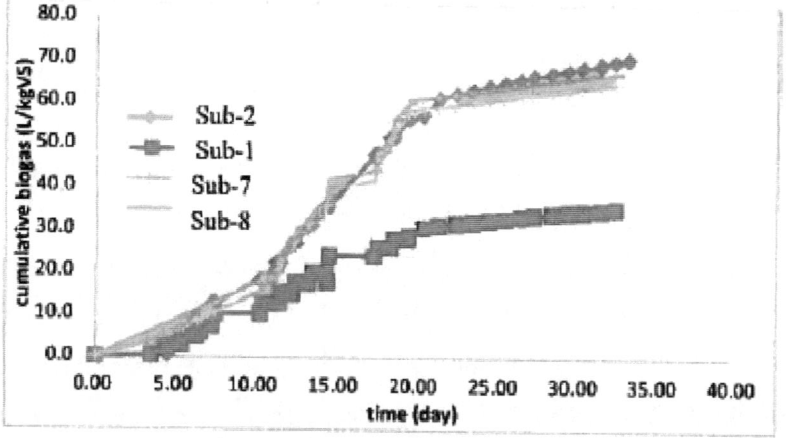

Figure 5: cumulative biogas production at different total solid concentration

3.5 Biogas composition

The measured values of biogas composition are shown in Figure 6. Each data point is the average of duplicate measurements for each substrate. The methane content of the biogas from each of the substrate type was not rather constant over the digestion time. maximum standard deviation of 5% could be estimated over the entire digestion course. The highest average methane content was found to be 64 %. The methane production rate depend on the substrate type, and the total solid composition of the reactor. It is know that methane production is combined with organic matter removal (i.e. volatile solid destruction); The average VS destructions at the end of experiments were directly related to the methane production. Substrates which have high content of organic matter in simple form produce more methane than complex substrate. Table 2 present the average amount obtained during different substrate digestion.

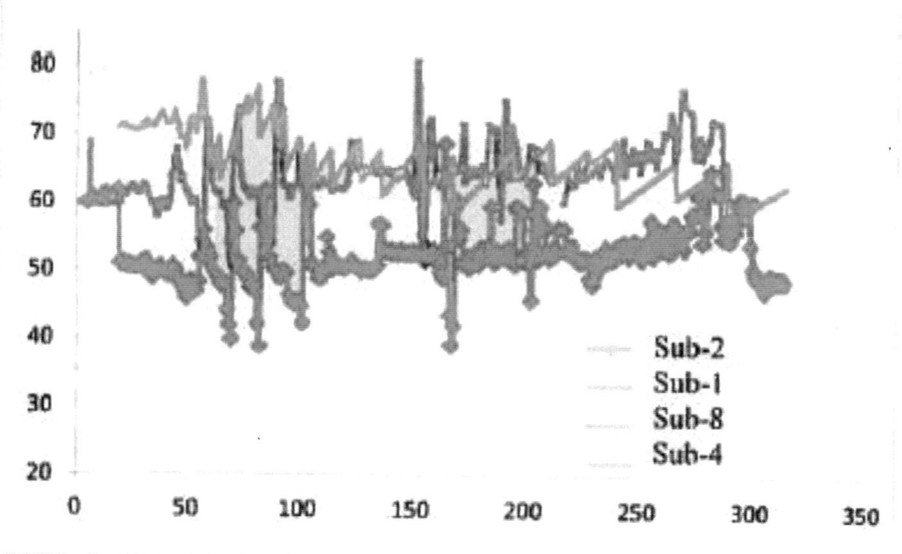

Figure 6: % Methane production at different biogas

It should be reported, here, the obtained methane yields, in this study, were promising for most of feed composition; a methane yield more than 50 % were reported for most processes. Moreover, high methane yields >60 % was obtained in two manures (Sub -1 and Sub -8). Based on our literature review, these values are considered of the highest methane yield reported. The higher % methane yields were achieved by using different substrate of specific characteristics and by co-digestion of different solid waste together instead of one type of solid waste.

Substrate	Biogas yield L/ kgVS. day	%Methane production	%VS removal	pH
Sub -1	35.3	66	42	7
Sub -2	69.8	54	38	7
Sub -3	46.5	50	36	7
Sub -4	54.3	54	42	7
Sub -5	16.2	45	44	7
Sub -6	21.2	43	39	7
Sub -7	64.5	58	45	7
Sub -8	67.0	63	49	7

3.4 Biogas production under intermediate oxidation

An intermediate oxidation step using photo-Fenton was applied to the substrates, tests were conducted by digesting these substrates for 10 days; after which the substrate was oxidized for 10 min and then the substrate was digested again for another 10 days. Table 5 showed the results of biogas production of the oxidized substrate compared with none-oxidized.

Table 3: Cumulative biogas production from two-stages AD with an intermediate chemical oxidation and single stage AD without chemical oxidation.

	AD with intermediate chemical oxidation (two-stages AD)	Single AD process (oxidation)
	Cumulative biogas (L.kg^{-1} VS),	Cumulative Methane (L.kg^{-1} VS),
Sub-1	186	88
Sub-2	280	198
Sub-3	170	55
Sub-4	288	100
Sub-5	190	88

The results in Table 3 indicate that using oxidation process increased the Cumulative biogas compared with single digestion process.

The significant increase in methane production during the two-stage AD (i.e AD an intermediate chemical oxidation) can be explained by the effect of oxidation process on improving substrate hydrolysis. Thus, substrate utilization by microorganism increase and consequently the rate of AD digestion and methane production increase.

3.6 Mathematical Model and biological parameters

The biological parameters of the AD where calculated based on first-order kinetics model that relate the methane yield to these parameters. It is know that the degradation rates of the biodegradable fraction of volatile solids in any AD process are described by Eqs. (1), respectively:

$$\frac{d[sub]}{dt} = \frac{-Ks[sub]}{K+VS} \qquad [1]$$

where:

t = digestion time (days),

Ks = biodegradation kinetics constants for substrate (g/L day),

K = a saturation constant (g/L),

VS = volatile solid content (g/L)

Equation (1) was solved using Matlab software at a time step of 12 hr.

The parameters obtained from the simulation are shown in Table 3. The simulation was based on substrate biodegradability values represented by BOD_5 values, for most substrates presented in table 1 the biodegradability value greater that 40 % of the contained organic matter.

Table 3: Parameters used in the methane production simulations.

Substrate	%Biodegradability	K_s (g/L day)	K (g/L)
Sub -1	45	0.12	0.63
Sub -2	53	0.18	0.76
Sub -3	48	0.09	0.45
Sub -4	50	0.13	0.34
Sub -5	45	0.08	0.44
Sub -6	51	0.13	0.56
Sub -7	60	0.2	0.67
Sub -8	58	0.23	0.65

4 CONCLUSIONS

This study determined the degradation rate and biogas and methane yields of different substrate prepared from agricultural and animal waste using an anaerobic batch digester at 35 °. All the batch digestion tests were performed for 30 days, which was the time determined to achieve the complete digestion of all the substrates studied. It was found that about 90% of the final biogas yield could be obtained after 10 days of digestion, at which point average methane yield over 50 L/kgVS was achieved. There were significant differences in methane yields for different substrat conditions VS, % TS and C/N ratio. A first-order kinetics model was used to model the biodegradation processes. The developed model can be used to predict methane yields for similar substrates

References

1. Herrmann, C., et al., *Optimised biogas production from microalgae through co-digestion with carbon-rich co-substrates.* Bioresource Technology, 2016. **214**: p. 328-337.
2. Gonzalez-Fernandez, C., B. Sialve, and B. Molinuevo-Salces, *Anaerobic digestion of microalgal biomass: Challenges, opportunities and research needs.* Bioresource Technology, 2015. **198**: p. 896-906.
3. Momani, F.A.A. and M. Shawaqfah, *Agricultural Solids Waste in South of Jordan: Facts and Figures.* Journal of Environmental Protection, 2013. **Vol.04No.04**: p. 6.
4. Rodriguez, C., et al., *Pretreatment techniques used in biogas production from grass.* Renewable and Sustainable Energy Reviews, 2016. **in press**
5. Abdeshahian, P., et al., *Potential of biogas production from farm animal waste in Malaysia.* Renewable and Sustainable Energy Reviews, 2016. **60**: p. 714-723.

6. Al-Hamamre, Z., et al., *Assessment of the status and outlook of biomass energy in Jordan.* Energy Conversion and Management, 2014. **77**: p. 183-192.
7. Mao, C., et al., *Review on research achievements of biogas from anaerobic digestion.* Renewable and Sustainable Energy Reviews, 2015. **45**: p. 540-555.
8. Kondusamy, D. and A.S. Kalamdhad, *Pre-treatment and anaerobic digestion of food waste for high rate methane production â€" A review.* Journal of Environmental Chemical Engineering, 2014. **2**(3): p. 1821-1830.
9. Mahanty, B., et al., *Optimization of co-digestion of various industrial sludges for biogas production and sludge treatment: Methane production potential experiments and modeling.* Waste Management, 2014. **34**(6): p. 1018-1024.
10. Nurliyana, M.Y., et al., *Effect of C/N ratio in methane productivity and biodegradability during facultative co-digestion of palm oil mill effluent and empty fruit bunch.* Industrial Crops and Products, 2015. **76**: p. 409-415.
11. Ajeej, A., et al., *An overview of bio augmentation of methane by anaerobic co-digestion of municipal sludge along with microalgae and waste paper.* Renewable and Sustainable Energy Reviews, 2015. **50**: p. 270-276.
12. Zhang, Q., J. Hu, and D.-J. Lee, *Biogas from anaerobic digestion processes: Research updates.* Renewable Energy, 2016. **in press**
13. Al Momani, F., et al., *Combining photo-Fenton process with biological sequencing batch reactor for 2,4-dichlorophenol degradation.* Water Science and Technology, 2004. **49**(4): p. 293-298.
14. Al Momani, F.A. and N. Jarrah, *Treatment and kinetic study of cyanobacterial toxin by ozone.* Journal of Environmental Science and Health - Part A Toxic/Hazardous Substances and Environmental Engineering, 2010. **45**(6): p. 719-731.
15. Al Momani, F., *Degradation of cyanobacteria anatoxin-a by advanced oxidation processes.* Separation and Purification Technology, 2007. **57**(1): p. 85-93.
16. Almomani, F. and E.A. Baranova, *Kinetic study of electro-Fenton oxidation of azo dyes on boron-doped diamond electrode.* Environmental Technology (United Kingdom), 2013. **34**(11): p. 1473-1479.
17. Al Momani, F., *Treatment of air containing volatile organic carbon: Elimination and post treatment.* Environmental Engineering Science, 2007. **24**(8): p. 1038-1047.
18. Cheng, C.-J. and P.K.A. Hong, *Anaerobic digestion of activated sludge after pressure-assisted ozonation.* Bioresource Technology, 2013. **142**: p. 69-76.
19. Chu, L.-B., et al., *Enhanced treatment of practical textile wastewater by microbubble ozonation.* Process Safety and Environmental Protection, 2008. **86**(5): p. 389-393.
20. Carballa, M., et al., *Influence of ozone pre-treatment on sludge anaerobic digestion: Removal of pharmaceutical and personal care products.* Chemosphere, 2007. **67**(7): p. 1444-1452.
21. Almomani, F., et al., *Intermediate ozonation to enhance biogas production in batch and continuous systems using animal dung and agricultural waste.* International Biodeterioration and Biodegradation, 2017. **119**: p. 176-187.
22. Almomani, F., R.R. Bhosale, and A. Kumar. *The effect of intermediate ozonation process on improving biogas production from co-digestion of agricultural waste and manure.* in *Advanced Materials - TechConnect Briefs 2016*. 2016.

23. Bougrier, C., J.P. DelgenÃ¨s, and H. CarrÃ¨re, *Impacts of thermal pre-treatments on the semi-continuous anaerobic digestion of waste activated sludge.* Biochemical Engineering Journal, 2007. **34**(1): p. 20-27.
24. Yan, S.-T., et al., *Analysis of the mechanism of sludge ozonation by a combination of biological and chemical approaches.* Water Research, 2009. **43**(1): p. 195-203.

528041 Measurement of Organic Aerosol Hygroscopicity and Oxidation Level As a Function of Volatility

Abstract:

Hygroscopicity, oxidation level, and volatility are three of the most important properties of atmospheric organic aerosol (OA). Many studies have been dedicated to the determination of these properties individually and chemical transport models (CTMs) use the corresponding results to simulate atmospheric OA, but often struggle to reproduce what is measured in the atmosphere. Cain and Pandis (2017) developed and evaluated a measurement and analysis technique to determine OA hygroscopicity and oxidation level as a function of volatility. Furthermore, Louvaris et al. (2017) combined thermodenuder and isothermal dilution measurements in order to better constrain OA volatility distributions over a large range. This study combines these two techniques and applies them in smog chamber experiments focusing on secondary OA production from the ozonolysis of various volatile organic precursors.

The synthesis of the two techniques resulted in hygroscopicity and oxidation level distributions that spanned a large range of volatility (around 6-8 orders of magnitude) for each OA system. These distributions not only provide a quantitative approach to understanding OA behavior to different perturbations (temperature changes, dilution, etc.), but they can also be converted into the two-dimensional volatility basis set framework to allow for easier implementation into CTMs. The successful combination of these two techniques provides an experimental and analysis framework that can be applied to ambient OA in order to determine a concrete relationship between these three properties, which can be used to update CTMs accordingly.

Cain, K. P., and Pandis, S. N.: A technique for the measurement of organic aerosol hygroscopicity, oxidation level, and volatility distributions, Atmos. Meas. Tech., 10, 4865-4876, 2017.

Louvaris, E. E., Karnezi, E., Kostenidou, E., Kaltsonoudis, C., and Pandis, S. N.: Estimation of the volatility distribution of organic aerosol combining thermodenuder and isothermal dilution measurements, Atmos. Meas. Tech., 10, 3909-3918, 2017.

529795 Secondary Organic Aerosol Formation from Methylfurans By Nitrate Radical Oxidation

Abstract:

An activity that is an important source of both gas- and particle-phase carbon in the atmosphere is biomass burning. Biomass burning contributes to both primary organic aerosol (POA) and secondary organic aerosol (SOA) formation. Recent literature about SOA formation from biomass burning emphasized the importance of the unidentified reaction pathways that contribute to a large fraction of SOA mass. These studies suggested that furan derivatives can be an important contributing factor of nontraditional precursors to SOA formation from biomass burning. One of the main furan derivatives that are emitted by biomass burning is methylfuran isomers. Our current understanding of SOA formation from methylfurans, however, is extremely limited, especially under nighttime oxidation conditions. Here, we investigate SOA formation from 2- and 3-methylfuran by nitrate radical (NO_3) oxidation at the Georgia Tech Environmental Chamber (GTEC) facility. We used High Resolution Time-of-Flight Chemical Ionization Mass Spectrometer (HR-ToF-CIMS) coupled with a Filter Inlet for Gases and AEROsols (FIGAERO) for the real-time measurement of oxidation products. This setup allowed us to measure both gas- and particle-phase composition during the oxidation process. The increase of aerosol mass concentration was measured with a Scanning Mobility Particle Sizer (SMPS) and the methlfuran decay was measured with Gas Chromatography Flame Ionization Detector (GC-FID). Aerosol mass continued to increase after all 2- and 3-methylfuran were consumed, indicating a large contribution of higher-generation oxidation products to SOA mass. Organic nitrate ($RONO_2$) contribution was around 25% to the total organics measured by HR-ToF-AMS. Multiple monomer and dimer species that contain an $-ONO_2$ functional group were detected by FIGAERO-ToF-CIMS. These results indicate that the $RONO_2$ species are also important in terms of SOA formation, which have not yet been considered in 2- and 3-methylfuran SOA studies to date.

535049 Polymorphism of Glutaric Acid Aerosols

Abstract:

Because of the abundance of dicarboxylic acids in the atmosphere, they are commonly studied in the aerosol field, specifically investigating their hygroscopic and cloud condensation nuclei activity properties along with their behavior in inorganic mixtures. Glutaric acid, being a water soluble dicarboxylic acid is commonly used to model water activity in dicarboxylic acid aerosols. Studying trends associated with glutaric acid aerosols in the atmosphere, such as CCN activity and water uptake behavior is common. Glutaric acids behavior in the presence of inorganic particles, such as coating inorganic aerosols and forming co-crystals, is also frequently studied. However, in several circumstances there is some variability within the data associated with glutaric acid, specifically DRH and hygroscopic data, which is hypothesized to be caused by its polymorphism.

Glutaric acid exhibits dimorphism, meaning the molecule can form into two different types of crystal structures, an alpha and a beta polymorph, which can manifest themselves as a bimodal aerosol size distribution when analyzed by SMPS or other mobility sizing methods. Glutaric acid naturally forms in the beta polymorph, which is more stable. The alpha polymorph is the metastable phase, meaning it is less stable. In this work we investigated the formation of the alpha glutaric acid polymorph in aerosols generated from an aqueous glutaric acid solution. In no published research has the formation of both polymorphs from an aqueous solution been documented. Identifying the presence of polymorphism in glutaric acid aerosols will ideally explain the variability present in studies associated with glutaric acid aerosols. Understanding and defining under which conditions the alpha polymorph is present will allow others to avoid it when investigating glutaric acid as a dicarboxylic acid.

Additionally, bimodal distributions have been observed with select mixtures of organic and inorganic substances, however the mechanism creating bimodality appears to be different than polymorphism. These results will be presented and explained.

523792 Investigation of Levoglucosan Decay in Wood Smoke Smog-Chamber Experiments: The Importance of Aerosol Loading, Temperature, and Vapor Wall Losses in Interpreting Results

Abstract:

Levoglucosan has been extensively used as a molecular marker of biomass burning in source-apportionment studies over the last few decades. However, recent studies suggest that the atmospheric lifetime of levoglucosan may be in the order of only 1-2 days under summertime conditions relevant to prescribed/wildfires. Implications of levoglucosan reactivity to wintertime conditions, however, remain uncertain despite significant contributions of domestic wood combustion to wintertime air quality. This study presents smog chamber experiments to investigate levoglucosan decay during photo-oxidation of wood smoke over a temperature range between -10 and 10°C. Significant decay in particle wall-loss corrected levoglucosan is only observed around 10 °C in these experiments. Theoretical analysis shows that apparent chemical lifetime of levoglucosan increases at lower temperatures as well as at higher organic aerosol mass concentrations as a result of smaller vapor fractions. The chemical lifetime of a molecular marker is commonly interpreted by a relationship between relative decay in particle wall-loss corrected marker concentrations versus integrated OH exposure. However, this relationship strongly depends on vapor wall-loss rates in addition to temperature, organic aerosol concentrations, and OH concentrations. Therefore, inferred lifetimes from a small set of experiments from a single chamber cannot easily be generalized for the full range of atmospheric smoke conditions.

537173 CCN Activity and Particle Growth of Aging Diesel Exhaust Particles

Abstract:

Diesel exhaust particles emitted into the atmosphere undergo aging that changes their physical and chemical properties, but these processes are complex and not well understood. Laboratory chamber experiments were conducted to investigate the aging of diesel exhaust in the presence of UV light, ammonium sulfate particles, and/or Î±-pinene vapors. Particle growth and changes in number concentration and size distribution were monitored with a SMPS. Changes in CCN activity were measured with a CCN counter. Results from dark experiments with diesel exhaust alone showed no significant secondary organic aerosol formation and diesel exhaust particles that remained CCN inactive. Aging experiments with UV radiation resulted in a significant increase in particle mass, with deposition-corrected particle mass after 4 hours that was 20-60% higher than the initial mass of freshly emitted diesel soot particles. CCN activity also increased over time indicating that photooxidation of diesel exhaust vapors produces SOA more hygroscopic than the freshly emitted diesel particles. The addition of Î±-pinene to diesel exhaust in UV aging experiments further enhanced both CCN activity and SOA formation. Experiments with ammonium sulfate particles showed no discernible effect on SOA formation for either diesel exhaust alone or diesel exhaust/Î±-pinene mixtures. CCN activity measurements were consistent with a largely external mixture of hygroscopic ammonium sulfate particles and non-hygroscopic diesel soot, coated with moderately hygroscopic SOA. The results from this study underscore the importance of aging via photooxidation of SOA precursors and mixing with water-soluble particles in transforming hydrophobic soot into more CCN-active particles.

526711 Elucidating the Effect of Photons and Aerosols on the Physical and Chemical Transformations of Atmospheric Mercury

Abstract:

The deposition mechanism for mercury from the atmosphere to the Earth's surface largely depends on the chemical form of mercury [1]. Atmospheric mercury exists in three primary forms: Elemental gaseous (Hg(0)), oxidized gaseous (Hg(II)), and particulate-bound (Hg(p)) mercury [2]. Compared to Hg(II) and Hg(p), which deposit to the Earth's surface shortly after emission (within ~100 km), Hg(0) travels globally and remains in the atmosphere for months to years [3]. Therefore, the transformation of mercury between these chemical forms largely dictates the time and length scales for atmospheric mercury transport before eventual deposition. For example, the reduction of Hg(II) to Hg(0) has been shown to occur in the presence of aerosols and photons, which leads to the widespread accumulation of mercury in the environment [4]. Furthermore, Hg(II) species can partition between the gas and particle phase [1], which dictates the mechanism for deposition. By understanding the role of aerosols and photons in the physical and chemical transformations of mercury in the atmosphere, better atmospheric mercury transport models can be achieved.

In this presentation, we show that quantum chemical calculations can be used to understand the physical and chemical transformations of atmospheric mercury. First, we present a computational model to predict the partitioning coefficient for Hg(II) species in the presence of various aerosol surfaces [5]. Second, we show a framework to elucidate the combined role of photons and iron-oxide aerosols in the reduction of Hg(II) species to Hg(0), as this mechanism remains unclear in the literature [6]. We develop the reaction energy surface for the reduction of $HgCl_2$ to Hg(0) to determine potential roles of photons in the chemistry. Through both models, we show how quantum chemical computations can further understandings of atmospheric processes, as well as provide models that can be extended to other pollutants and aerosols.

References

1. A.P. Rutter & J.J. Schauer. *Environ. Sci. Technol.*, **41**, 3934–3939 (2007).
2. S.A. Tacey, L. Xu, M. Mavrikakis, & J.J. Schauer. *J. Phys. Chem. A*, **120**, 2106–2113 (2016).
3. W.H. Schroeder & J. Munthe. *Atmos. Environ.*, **32**, 809–822 (1998).
4. Y. Tong *et al. Environ. Sci. Process. Impacts*, **15**, 1883–1888 (2013).
5. S.A. Tacey, L. Xu, T. Szilvási, J.J. Schauer, & M. Mavrikakis. (Submitted)
6. S.A. Tacey, T. Szilvási, L. Xu, J.J. Schauer, & M. Mavrikakis. (Submitted)

532527 Effective Removal of Pertechnetate from Groundwater By Bimetallic Porous Iron

Abstract:

^{99}Tc is a major long-lived fission product generated during nuclear power generation. Over the years, Tc has been inadvertently introduced into the environment from leaks at waste storage facilities and currently is one of the key risk drivers at the US DOE sites. The most common chemical form of Tc in liquid nuclear wastes or in the environment is pertechnetate (TcO_4^-). TcO_4^- displays limited adsorption to common sediment minerals and is highly mobile making it difficult to be immobilized. As the stockpile of ^{99}Tc-bearing nuclear waste continues to increase rapidly, novel sequestration technologies are needed to reduce its potential contamination of the environment and living organisms.

We developed highly efficient and cost-effective bimetallic porous iron materials that were investigated for TcO_4^- sequestration from groundwater. Catalytic bimetallic nanoparticles were grown onto porous iron by chemical reduction methods. The bimetallic porous iron materials were characterized by several analytical techniques to elucidate their morphology, size, composition, and porosity. Specifically, we used X-ray diffraction, scanning electron microscopy, energy dispersive X-ray spectroscopy, and BET surface area using Kr as the adsorption probe. The bimetallic porous iron materials were evaluated for removal of TcO_4^- and its surrogate (ReO_4^-) from artificial groundwater under ambient conditions using batch experiments. Tc and Re chemical speciation and molecular removal mechanisms were investigated using synchrotron radiation X-ray near-edge structure spectroscopy.

The results show that porous iron is >92% Î±-iron with small amounts of Fe and Al oxides. It has a sponge-like porous structure with a BET surface area of 0.95 m^2/g, pore volume of 0.0068 mL/g, and average pore diameter of 364 Ã…. When evaluated for contaminant removal, the porous iron successfully removed >97% present in an artificial groundwater solution containing 3.2×10^{-6} M TcO_4^-. In contrast, conventional granular zero valent iron, tested as a control, was entirely ineffective with 0% Tc removal. The removal capacity of the porous iron was demonstrated to be 27 mg Tc/g iron. Further, Tc K-edge X-ray absorption spectra of the porous iron samples retrieved from Tc batch experiments showed that both Tc(IV) and Tc(VII) species are present. Linear combination fitting indicated that 50-70% of the Tc was reduced to Tc(IV) that was incorporated into the secondary Fe oxides. In conclusion, these results indicate that porous irons are highly efficient materials for the immobilization of TcO_4^- present in groundwater through both adsorption and chemical reduction mechanisms. They might also be novel materials for effectively immobilizing a wide array of other radioactive contaminants in the environment.

530609 Migrating High Performance Computing to the Amazon Cloud - Methods and Examples Using Environmental Systems Analysis

Abstract:

To adequately analyze a subsurface environmental system, including assessing the inherent uncertainty in the information available to characterize it, and the ability to robustly include the governing physical processes requires numerical computational models. Since these models require inclusion of complex biogeochemical processes and simulate large aerial extents, each model run can require execution times on the order of hours or days to complete.

Adequate analysis of the uncertainty inherent in these subsurface systems performances in their natural state, or due to imposed stresses such as experienced during active dewatering or groundwater remediation, can require 100s or 10,000s of simulations be conducted. Thus, an analysis of complicated subsurface systems can require 100,000s of hours of computer processing time.

With the advent of Cloud computing, computation time on an as-needed basis is readily available. This work discusses implementing large-scale computational analysis on the Amazon Cloud including determining a capacity plan, configuring the project workload, managing the grid engine worker queue, staging and storing results for post processing. Three examples of various size and complexity of actual Cloud computation projects are provided with experience and lessons learned.

544065 Design Optimization for Environmental Projects - A Primer on Optimization Methods

Abstract:

Optimal project solution design for environmental projects involves selecting the appropriate values for any number of design parameters. This presentation provides the practitioner with a general set of tools that are available for use on a wide variety of project types.

536128 Quantifying Shifts in Trace Element Emissions from Coal-Fired Power Plants

Abstract:

Coal contains a variety of trace elements that are volatalized when the coal is combusted in coal-fired power plants (CFPPs). These trace elements are emitted to the environment in the exhaust gas, ashes, or in wastewater effluents. The partitioning of trace elements to the gas, solid, or liquid phase is determined to a large extent by installed air pollution control devices (APCDs). CFPPs have changed their installed APCDs and the coal they combust in response to changes in environmental regulations. We combine data from the U.S. Energy Information Administration and United States Geological Survey to build a model that will estimate the solid, liquid, and gas phase emsisions of seven trace elements (As, Cl, Se, Hg, Br, B, and Pb) from all US CFPPs with a capacity greater than 10 MW. We use this model to quantify how the installation of new APCDs has shifted trace element emisssions from the gas phase to the solid and liquid phase over past decades. We also calculate the impact of the shift towards low-sulfur coals on trace elements in the coal combusted at CFPPs and how alternative, hypothetical environmetnal compliance strategies would have produced different trace element emissions patterns.

527960 Adsorption of Cadmium, Nickel, and Lead on Modified Clinoptilolite: Equilibrium, Kinetic and Selectivity Studies

Abstract:

Environmentally friendly treatment methods for heavy metals prior to discharge are considered favourable so as to protect ecosystems. The performance of modified zeolite from two different sources (South African and the USA) for the adsorption of Ni^{2+}, Cd^{2+} and Pb^{2+} from synthetic industrial effluent contaminated with metal concentration levels at 50, 150 and 500mg/L was assessed. The selectivity of the zeolite for the adsorption of Ni^{2+}, Cd^{2+} and Pb^{2+} was investigated at mixed feed solutions containing all three ions in equal concentrations and single component concentrations containing only one of the ions. The homoionic forms of the zeolite were made Na^{+}, K^{+} and Ca^{2+}. Batch experiments were then conducted to measure the uptake of metals by the zeolites. The zeolites were characterised using SEM, XRD and BET. The South Africa clinoptilolite has a higher surface area and pore volume ($17.52m^2/g$ and $0.047cm^3/g$ respectively) than the USA zeolite ($12.26m^2/g$ and $0.028cm^3/g$ respectively) for the Na^{+} homoionic form. According to the equilibrium studies, the selectivity sequence can be given as Pb^{2+} >Cd^{2} >Ni^{2+}, with good fits being obtained using Langmuir and Freundlich adsorption isotherms for low metal concentrations as compared to higher concentrations. The adsorption was found to depend on the homoionic form of the zeolite and to a lesser extent the source of the zeolite. The selectivity of a particular zeolite for a particular heavy metal can be altered by the homoionic form of the zeolite. Generally, the adsorption capacity of the USA clinoptilolite is higher than the capacity of the SA clinoptilolite revealing the potential of clinoptilolite in industrial effluent treatment.

523475 A Mathematical Programming Model for the Integration of Power Plants Involving Chemical Looping Combustion with Algal Systems Under Carbon Policies Analysis

Abstract:

Nowadays, great amounts of electricity are required. In order to satisfy these demands, power generation plants are commonly used despite of the produced CO_2 emissions. Such tendency has led to several problems, for instance the rise in the atmospheric CO_2 concentration that is considered to contribute to the global warming. Therefore, alternatives to generate electric energy or strategies to reduce the emissions must be proposed and developed. This is not a simple task due to the several issues involved such as economic and technical constraints. The chemical looping combustion can be a potential option since it is a novel technology that allows to capture the carbon dioxide with small penalties in thermal efficiency. Also, once the CO_2 is captured, it can be sent to an algal cultivation system for its utilization. A microalgae system consists of different stages to process the algal biomass and produce biofuels. Accounting for this, the present work proposes a mixed-integer linear programming model to find the optimal integration between power plants and algal cultivation systems for the simultaneous production of electric energy and biofuels. Environmental and economic objectives are taken into account including the inherent separation and mitigation of CO_2 by means of the chemical looping combustion and the algal-to-biodiesel process; as well as the sale of electricity, biodiesel, glycerol, ethanol and proteins and the costs associated with the global system. In addition to this, carbon externalities involving distinct economic penalizations or compensations are evaluated in order to identify their impact on the minimization of emissions and maximization of profit. The results of the proposed approach include the optimal configuration for the system at a macroscopic level which consists of the optimum selection of fuel, combustion system, power cycle, technology for each algal system stage, the amount of CO_2 delivered to the algae and other specific conditions (flow rates and variables involved in the process). Furthermore, the solutions obtained through the case study show different scenarios with carbon taxes and bonuses where tradeoffs between the objective functions are observed. We found that the use of carbon bonus based on avoided emissions provides better compromises for the profit and emissions.

537540 A Framework for Multiscale Consequential Life Cycle Assessment

Abstract:

Background:

A significant drawback of normal Life cycle assessment(LCA) methods is its results sometimes becoming irrelevant on real world application due to unintended effects. Conventional Attributional LCA (ALCA) directly compares a functional unit of new technology with a base technology and determines which one is environmentally superior. It doesnâ€™t consider the introduction of this technology into society, its acceptance and the ripple effects that might occur due to such a technology in the real world. Such a problem can be thought of as being analogous to a chemical engineering design problem in a laboratory (attributional LCA) versus a full-sized industry (real world impacts).

Consequential LCA (2) is one method which is currently used to capture these ripple effects in the society. It considers interaction between markets, looks at knock on effects, and changes in production levels. Due to its relative recent development, this method is not yet well established and different methods are practiced for performing a consequential LCA (CLCA). Broadly, the two methods are Hierarchy based (3) and Equilibrium modeling based (4).

Motivation:

It is observed from CLCA literature that hierarchy based methods are prone to system boundary selection problem due to the subjective nature of selecting marginal technologies. Equilibrium based modeling CLCA literature shows that even though the system expansion is easily obtained through economic model use, significant details are lost due to high data aggregation of such models, rendering them useless for application to specific processes. In this presentation, we propose a novel multiscale solution to this problem by developing a CLCA based optimization framework that combines both these methods. The equilibrium models in this framework will help in system expansion which reduces errors in system boundary selection while the hierarchy based approach can be focused on processes where greater resolution and details are required. Along with that, the presence of third finer engineering scale enables designing of processes while performing CLCA, which is a novel approach in the Sustainable Process Design field that has typically used ALCA for determining environmental impacts.

Objective:

The objective of this research work is to create a consequential LCA based optimization framework that utilizes both the CLCA pathways mentioned in the Background. The advantage of integrating these pathways is that this new framework will be able to derive the advantages of each pathway while addressing each otherâ€™s inadequacies. It is an extension of the P2P framework for process design to consider consequences of the optimization decisions. P2P framework with its multiscale nature serves as a suitable platform for building this CLCA framework. The engineering scale is the scale for determination of industry level design variables for performing process design. The value chain or upstream network scale is suited to model the hierarchy based CLCA based method using a choice of technology approach. The economy scale which previously used a simple economic Input-Output IO model is replaced with an economic general equilibrium model that can explore interaction between markets, price changes, marginal and non-marginal effects.

Results:

A significant challenge for this work is the integration of equilibrium models with the other scales due to its own optimization structure that is separate from the rest of the framework. Obtaining consequential life cycle inventory data and equilibrium models is significantly difficult as well. The framework will be applied to case studies such as but not limited to carbon fiber use in vehicle manufacturing, process intensification for improving the production efficiency of manufacturing processes, decentralization of manufacturing processes to promote better availability of products and services in rural areas and other interesting problems. We also intend to compare with normal ALCA and check differences in results.

References:

1. Hanes, Rebecca J., and Bhavik R. Bakshi. "Sustainable process design by the process to planet framework." AIChE Journal 61.10 (2015): 3320-3331.
2. Zamagni, Alessandra, et al. "Lights and shadows in consequential LCA." *The International Journal of Life Cycle Assessment*7 (2012): 904-918.
3. Ekvall, Tomas, and Bo P. Weidema. "System boundaries and input data in consequential life cycle inventory analysis." *The International Journal of Life Cycle Assessment*3 (2004): 161-171.
4. Earles, J. Mason, et al. "Integrated economic equilibrium and life cycle assessment modeling for policyâ€based consequential LCA." *Journal of Industrial Ecology*3 (2013): 375-384.

531724 Integrated Network Optimization and Uncertainty Analysis Streamlined LCA Method for the Petrochemical Industry

Abstract:

The development of more sustainable products and chemicals has boosted the sustainability evaluation of production systems, ranging from single plants to entire supply chains. However, the complexity and data intensity of chemical processes represent an important burden when trying to apply methodologies such as Life Cycle Assessment (LCA). Environmental databases are generally used to solve these issues, providing results obtained from LCA studies of existing processes. The amount of chemical products and processes present in these repositories, however, is still fairly limited. Additionally, results are obtained considering only fixed mass and energy flows, which contrasts with the dynamic behavior of production processes, particularly for the chemical industry. These differences might lead to severe disagreements between the real case scenarios and environmental repositories, being necessary alternative methodologies to assist in the assessment of environmental impacts.

To this end, we present a novel methodology for the quantification of LCA impacts in the production of chemicals based on the analysis of mathematically modeled networks of chemical processes using linear programming, uncertainty analysis and impact allocation procedures. In order to test this methodology, we assembled a network of the petrochemical industry containing 178 processes and 144 chemical products, being possible to quantify the impacts allocated to different products of the network. This was done for several environmental impact categories such as GWP and ReCiPe 2008. The estimates obtained were fairly similar to those present in popular environmental repositories such as Ecoinvent.

The effect of different network parameters such as final demand or production yield were studied using a sensitivity analysis approach, obtaining the uncertainty for the allocated impact values even for chemicals not previously included in environmental databases. These results demonstrate the capabilities of network approaches to characterize chemical and petrochemical production processes, allowing dynamic and uncertainty analyses of their economic and environmental performance. The presented methodology can be used to improve the transparency and flexibility of the LCA methodology, enhancing the sustainability evaluation of new and existing processes and chemicals.

538431 Study on Regional Air-Quality Impacted By Chemical Plant Shutdown Under the Lower Destruction and Removal Efficiency for Flare Combustion

Abstract:

Flaring is a necessary measure for chemical plant safety. However, the industrial flaring generates large amounts of VOC and NOx, which could transiently aggravate regional ozone concentrations that could affect on human health and the air-quality. Currently, the national standard value of 98% or 99% for destruction and removal efficiency (DRE) is regulated and presumed for industrial flaring operations. In reality, the DRE value, however, could be lower than the national standard under various atmospheric and operating conditions, such as the high wind-speed and flare jet velocity. Thus, it is important to know the quantity and sensitivity of the regional ozone impact due to the lower DREs rather than standard values. In this paper, plant-wide dynamic simulations for plant shutdown flare emissions with regional air-quality modeling were coupled together to quantify the air-quality impact due to the lower flaring DREs caused by wind-speed and flare jet velocity. Firstly, the DRE formulas considering wind-speed and jet velocity were derived based on available references. Secondly, the shutdown flare emissions were estimated based on the DRE formulas and then inserted into emission inventory for air-quality modeling and simulation. After that, comprehensive air quality model with extensions (CAMx) based air-quality modeling was conducted to simulate the regional ozone increment due to the lower DRE values. This study could provide valuable and quantitative support for all relevant stakeholders, including environmental agencies, regional plants, and local communities.

529989 Physical and Chemical Aging of Carbonaceous Aerosol in the Eastern Mediterranean

Abstract:

A month of comprehensive continuous air quality measurements took place at the Finokalia Atmospheric Observatory in Crete, Greece during the summer of 2016. No significant human activities occur within 15 km of the station. The sampled air masses originated from the Balkans or northern Africa or spent several days over the Mediterranean Sea. This study focused on the evolution of organic aerosol in this area of high photochemical activity and its effects on aerosol light absorption.

In addition to a high-resolution Aerosol Mass Spectrometer (HR-ToF-AMS) and a Scanning Mobility Particle Sizer (SMPS) for size distributions and chemical composition measurements, a suite of instruments for the measurement of the light absorption and the refractory black carbon (rBC) mass concentration was deployed. These included an aethalometer, a photoacoustic extinctiometer (PAX_{405}), a Multi-angle Absorption Photometer (MAAP), and a Single Particle Soot Photometer (SP2).

The fine aerosol mass concentration ranged from 0.3 to 8 Î¼g m^{-3}, with the sulfate and organics accounting for almost 80% of the total mass. The organic aerosol was quite oxidized with an average O:C equal to 0.7. The BC concentration ranged from 0.04 to 0.3 Î¼g m^{-3}, representing 2 to 15% of the fine aerosol mass. The measured light absorption was two or more times higher than that of fresh BC. Mie theory was used to evaluate if the effect of the coatings of BC cores by organics and sulfate could explain this absorption enhancement. The role of brown carbon and other non-BC light-absorbing material was also investigated. A thermodenuder was used to link the volatility of the organic aerosol with the observed BC absorption enhancement.

538416 Global Source Apportionment of Atmospheric Particulate Matter

Abstract:

As our society continues to develop and grow, it is necessary to expand our understanding of the impacts of atmospheric pollutant emissions on global scales. Researchers and regulatory agencies alike have used source apportionment as a powerful tool for quantifying and estimating these impacts on the regional scale. Towards expanding these capabilities from regional to global models, we have developed a new source apportionment scheme for the GEOS-Chem global atmospheric chemical transport model. The Global Particulate Matter Source Apportionment Technology (GPSAT) module represents the first incorporation into a global model. We have developed GPSAT within GEOS-Chem so that it will be available to all users in future versions of GEOS-Chem.

GPSAT can estimate contributions from different source classes (including locations and source types) without multiple model iterations or the need to create customized tracers. GPSAT is both computationally efficient and flexible. GPSAT allows us to develop a more complete picture of the impacts of intercontinental transport on global air pollutant concentrations. We will share estimates of the contributions from nine large, continent-scale, regions. We will also share further spatially refined estimates of contributions for the regions with the highest contributions globally. Finally we quantify the contributions of the 20 largest megacities to global air pollution concentrations using the average transport distance. We will specifically discuss the regions of the globe that most disproportionately impact global pollutant concentrations.

Through this project we have demonstrated an enhanced research tool (GEOS-chem with GPSAT) that will allow researchers throughout the globe to address a variety of scientific questions related to our understanding of long-range transport and the climate forcing contributions of different processes throughout the atmospheric component of the Earth System. Our goal in the analysis described here using GPSAT is to demonstrate the power and flexibility of the tool and thereby increase adoption by other researchers. The results from the analysis also provide necessary estimates for international discussions around air pollution. In addition, this analysis identifies areas most in need of focused attention in model improvement to potentially decrease model uncertainties. Finally, we provide a base-framework that other researchers can extend in the future to address the contributions from different processes throughout the entire Earth System and to identifying relative contributions to aerosol-climate feedbacks.

532971 The Impact of Vapor Supersaturation on the Morphology, Mixing State and Optical Properties of Atmospheric Soot

Abstract:

Soot nanoparticles, produced from the incomplete combustion of fossil fuels, absorb solar radiation and contribute to climate change by direct radiative forcing. The magnitude of forcing is strongly influenced by the changes in morphology and mixing state of the soot aggregates as they interact with other aerosols during atmospheric transport. Presently, the pathways leading to these structural transformations and their associated impacts are not well understood. In this study, soot aerosols were exposed to supersaturated vapors of organic and inorganic liquids with a broad range of physical and chemical properties to generate thinly coated soot aggregates. We observed two distinct restructuring behaviors which were governed by the level of vapor supersaturation in our coating chamber. With the aid of a simple analytical model, we show that high vapor supersaturation results in uniform condensation over the aggregates. Conversely, low vapor supersaturation causes preferential condensation in the junctions between contacting primary spheres to form pendular rings which induce structural collapse. We surmise that the degree of vapor supersaturation is the driving factor for the different mixing states and morphological changes observed in lab and field studies. We also discuss the impacts of changes in morphology and mixing state on the optical properties of soot for both condensation behaviors. By incorporating vapor supersaturation in atmospheric models, improvements can be made in simulations of the radiative forcing of soot.

535376 The Competition between Surface and Capillary Condensation of Vapors on Soot Aggregates

Abstract:

Soot is major pollutant and climate forcer [1]. The optical properties of soot that determine its impact on the climate strongly depend on its morphology [2]. When initially generated, soot nanoparticles are branched fractal aggregates of primary carbon spheres. During aging in the atmosphere, soot aggregates often collapse into compact globules, with restructuring driven by a condensate of atmospheric vapors on the aggregate surface. Although there have been several studies of this restructuring effect [3-5], a model which could predict whether the agglomerates will collapse or not is still lacking.

Analysis of the recent laboratory experiments suggested that the spatial distribution of condensate on the soot agglomerates may determine its fate with respect to restructuring [6]. When a vapor condenses in the junction between the spheres, the capillary and disjoining forces are likely to generate mechanical stresses sufficient for plastic deformation of an agglomerate. Therefore, the agglomerate restructuring is directly related to the domination of one of the two scenarios of vapor condensation: if uniform condensation on the surface of primary carbon spheres dominates, the aggregate backbone will remain fractal; if condensation in the junctions between the primary spheres ("capillary condensation") dominates the aggregate will likely collapse.

Here we propose a model describing the vapor condensation on a soot aggregate, consisting of two spherical carbon particles, to elucidate the conditions at which one of the two condensation scenarios is realized. Since the rate of capillary condensation of vapor strongly depends on the curvature of the liquid-vapor interface, we explore three different approximations for the geometry of the droplet forming in the junction, assuming that it has the interface of a cylinder, a globoid, and a catenoid. The former two give the closed-form solutions for the amount of vapor condensed as a function of time, but the latter, which is the most rigorous, does not. However, we show that the difference between the three interfaces is relatively small, and thus, simpler analytical models can be used for semi-quantitative predictions of the vapor condensation on soot and subsequent soot aggregate restructuring.

1. Bond, T. C., Doherty, S. J., Fahey, D., Forster, P., Berntsen, T., DeAngelo, B. J., Flanner, M. G., Ghan, S., Karcher, B., Koch, D., et al. (2013). Bounding the role of black carbon in the climate system: A scientific assessment. J. Geophys. Res. Atmos., 118(11):5380â€“5552
2. Xue, H., Khalizov, A. F., Wang, L., Zheng, J., and Zhang, R. (2009). Effects of dicarboxylic acid coating on the optical properties of soot. Phys. Chem. Chem. Phys., 11(36):7869â€“7875
3. Weingartner, E., Baltensperger, U., and Burtscher, H. (1995). Growth and structural change of combustion aerosols at high relative humidity. Environ. Sci. Technol., 29(12):2982â€“2986.
4. Miljevic, B.; Surawski, N. C.; Bostrom, T. and Ristovski, Z. D., Restructuring of carbonaceous particles upon exposure to organic and water vapours. (2012) J. Aerosol Sci., 47:48-57.
5. Schnitzler, E. G.; Gac, J. M. and Jäger, W. (2017) Coating surface tension dependence of soot aggregate restructuring J. Aerosol Sci., 106:43-55
6. Chen, C., Fan, X., Shaltout, T., Qiu, C., Ma, Y., Goldman, A., and Khalizov, A. F. (2016). An unexpected restructuring of combustion soot aggregates by subnanometer coatings of polycyclic aromatic hydrocarbons. Geophys. Res. Lett., 43(20).

535668 From Nascent to Mature Soot Light Absorption during Agglomeration and Surface Growth

Abstract:

Optical characterization of soot (e.g. Laser Induced Incandescence, LII) and its impact on climate largely depend on light absorption. As soot is incepted by reactive dimerization of small aromatics[1] and grows by acetylene surface reactions[2] and agglomeration,[3] its morphology changes affecting its optical properties. In the early stages of soot formation, small and compact nascent soot agglomerates[2] are made that are almost transparent to visible and near-infrared. Large, ramified mature soot agglomerates formed at longer residence times by coagulation in the absence of surface growth[3] strongly absorb light from the visible to the near-infrared.

Here, the impact of soot morphology on its light absorption is investigated by coupling Discrete Element Modeling (DEM)[2] with Discrete Dipole Approximation (DDA) during soot surface growth and agglomeration. The Mass Absorption Cross-section, *MAC*, of nascent and mature soot agglomerates is estimated by DDA and validated against atomistic point dipole interactions and mesoscale DDA calculations. Using a refractive index, m, for mature soot yields constant average $\langle MAC \rangle$ and absorption function $\langle E \rangle$, that overestimate the nascent soot light absorption up to 60 %. Interpolating m between those of nascent and mature soot for wavelengths, λ = 532 and 1064 nm results in excellent agreement of the DEM-derived evolution of $\langle MAC \rangle$ and $\langle E \rangle$ with LII measurements in premixed flames. The nascent soot mass-mobility exponent, D_{fm}, decreases by agglomeration, doubling the $\langle MAC \rangle$ and increasing $\langle E \rangle$ by 65 %. A quasi-asymptotic D_{fm} of 2.45 is attained faster by larger soot volume fractions, accelerating the increase of $\langle MAC \rangle$ and reaching nearly constant $m = 1.61 - 0.63 \cdot i$ and $1.69 - 0.70 \cdot i$ at λ = 532 and 1064 nm, respectively, for residence time larger than 30 ms. Thus, accounting for the soot morphology dynamics during surface growth and agglomeration are essential to characterize soot by light absorption.

References:

[1] Kholghy, M. R.; Kelesidis, G.A.; Pratsinis, S.E., Reactive polycyclic aromatic hydrocarbon dimerization drives soot nucleation. *Phys Chem Chem Phys* **2018**, *in press*.

[2] Kelesidis, G. A.; Goudeli, E.; Pratsinis, S. E., Flame synthesis of functional nanostructured materials and devices: Surface growth and aggregation. *Proc Combust Inst* **2017**, *36*, 29-50.

[3] Kelesidis, G. A.; Goudeli, E.; Pratsinis, S. E., Morphology and mobility diameter of carbonaceous aerosols during agglomeration and surface growth. *Carbon* **2017**, *121*, 527-535.

sustainability

MDPI

Article

A Questionnaire Case Study of Chinese Opinions on the Haze Pollution and Economic Growth

Frederick Qiu [1], Ethan Wang [2], Matthew Fan [3], Hong Liao [4], Litao Wang [5] and Zuyi Huang [6,*]

1 Central Bucks High School East, Doylestown, PA 18902, USA; frdrckq@gmail.com
2 Central Bucks High School South, Warrington, PA 18976, USA; mrethanhw@gmail.com
3 Conestoga High School, Berwyn, PA 19312, USA; matthewfan20@gmail.com
4 School of Environmental Science and Engineering, Nanjing University of Information Science and Technology, Nanjing 210044, China; hongliao@nuist.edu.cn
5 Department of Environmental Engineering, Hebei University of Engineering, Handan 056038, China; wanglitao@tsinghua.org.cn
6 Department of Chemical Engineering, Villanova University, Villanova, PA 19085, USA
* Correspondence: zuyi.huang@villanova.edu; Tel.: +1-610-519-4848

Received: 23 May 2018; Accepted: 4 June 2018; Published: 12 June 2018

check for updates

Abstract: Haze pollution in China has reshaped daily life for the Chinese and led to serious health issues. At the same time, the Chinese have enjoyed the rapid economic growth that has contributed to this pollution. While questionnaire-based studies have been conducted within certain regions of China to learn the public's opinions of haze pollution, little work has been done to understand how Chinese citizens value haze treatment in relation to their nation's economic growth at a nationwide scale. To fill this knowledge gap, this project conducted a nationwide investigation of Chinese opinions on the benefits of economic growth versus the disadvantages of haze pollution, as well as their responses to efforts by the Chinese government to combat haze and to the influence of haze on Chinese daily life and on personal health. The study also sought suggestions for combatting haze. In particular, an anonymous questionnaire consisting of 29 questions was given in the summer and fall of 2017 to 1233 people of different genders, ages, child statuses, educational backgrounds, occupations, living areas (rural, suburban, and urban), and living regions. The statistical Chi squared test was then used to identify the demographic group of respondents supporting the economic slowdown policy or requesting more efforts from the Chinese government to combat haze pollution. A multivariate statistical approach—principal component analysis—was further applied to visualize respondents' feedback on the impact of haze on their daily life and personal health, as well as the change of environment and economic conditions in the last 10 years. The results show that more than 50% of respondents, especially those with children, those between the ages of 31 to 50, and those living in high-pollution regions, supported the economic slowdown policy. Totally 40.63% of the entire group of respondents believed the government's efforts to control haze were small or very small. Only 27.84% of respondents held the opposite opinions. In total, 72.91% of respondents believed the environment in China became worse or much worse in the past 10 years; however, most responded positively to the idea of resolving the haze issue within 15 or more years. Haze has caused health issues in and around half of the respondents and has significantly reshaped their outdoor activities.

Keywords: haze pollution; statistical survey; principal component analysis; Chi squared test

1. Introduction

The economic boom in China following the reforms introduced by Deng Xiaoping during the 1980s turned the country into a global powerhouse [1]. Specifically, while 27.6% of China was urbanized in 1992, this percentage increased to 52.6% in 2012 (cited from China Statistical Yearbooks). The Gross domestic product (GDP) grew by an average of 9% or more each year from the late 1980s to 2014 [2]. This economic growth in China is largely reliant on the large industries powered by fossil fuels, including steel factories. For instance, the 6.9% GDP growth in the second quarter of 2017 was largely due to the growth of the steel, coal, and cement industries, which are regarded as some of the most damaging industries for the environment [3]. China has developed as the top producer of coal in the world, producing 3.46 billion tons in 2016 alone, according to the Chinese National Bureau of Statistics. Coal is proven to be harmful to the environment: emissions due to coal often release mercury upon combustion [4–6]. Additionally, coal has contributed to 40% of all fossil fuel CO_2 emissions [7]. The rapid infrastructure construction and urban expansion in China impose the highest demand for cement in the world. In particular, China produced around 57% of global cement production (i.e., 1.87 billion metric tons of cement) in 2010, with an average annual growth rate of 11.6% in cement production over the past two decades. China's recent rapid industrial and urban growth has resulted in environmental complications such as the haze that is the focus of this study. A slight obscuration of the atmosphere by small solid or liquid particles, haze represents a form of air pollution that has become a prevalent concern in China, especially near city clusters [1]. Particles in the air with a diameter of less than 2.5 micrometers (so-called $PM_{2.5}$) has caused 652,000 premature deaths in 2015 in 161 cities within nine regions across China, accounting for a total of 6.92% of all deaths that year [8]. The air quality in China as of the winter of 2016 has far exceeded dangerous levels, causing cancelled flights, closed schools, and poor visibility. While a concentration of 300 micrograms per cubic meter of $PM_{2.5}$ particles is considered hazardous, 74 Chinese cities exceeded this number in 2013, with some even reporting numbers as high as 700 [9]. The problem posed by Chinese air pollution can no longer be ignored. The rapid economic growth that is driven by coal, steel, cement and other industries presents a growing contrast between the country's economic and environmental goals. It is thus crucial to understand where the public stands on the issue because it can reveal to government policy makers and industrial owners the best course of action for implementing changes that will most effectively improve the quality of life for the Chinese populations. It may influence future economic endeavors that China can take on to maintain an environmentally sustainable but also economically successful nation.

To address the aforementioned issues, this project presents a questionnaire to survey the opinions of the Chinese public on how they prioritize controlling haze pollution versus encouraging economic growth and their thoughts on efforts that might be made by the Chinese government to reduce haze pollution. Questionnaire survey approaches have been widely implemented as a means to obtain Chinese opinions on topics that are related to China's haze pollution problem: (1) the annual consumption of various energy varieties (e.g., coal and fuel gas) in rural areas of northern China [10]; (2) tourists' perception of haze pollution and the potential impacts on travel to Beijing in China [11]; (3) public awareness of smog pollution in China [12–16]; (4) the public's willingness to pay (WTP) for reducing or tracking haze pollution [17–22]; (5) the public's adaptive behavior responses to haze pollution in urban China [23,24]; (6) personal protection strategies from haze pollution [23,25]; and (7) consumers' purchase intentions related to products that deal with haze pollution [26].

These studies indicate that the questionnaire-based survey is an effective approach for obtaining the public's opinions on haze pollution in China. However, few of these studies have covered the topic of how far people are willing to go in sacrificing economic growth and halting the exacerbation of the environmental situation in China. This project aims to fill this knowledge gap. While several studies have been conducted to evaluate the government's performance in responding to and controlling haze pollution [12,19], these were limited to either local areas (e.g., Xinjie, a rural area in central China) or specific populations (e.g., skilled workers). We revisit this question here by exposing it to more general populations with various demographic backgrounds (i.e., gender, age, child status, educational background, occupation, living area (rural, suburban, and urban), and living region in China). In addition, this work implements advanced multivariate statistics methods to extract important patterns from the wealth of data obtained through the 29-question survey, which includes seven demographics questions concerning gender, child status, age, education background, occupation, population density of the living area, and living region in China; 18 multiple choice questions surveying opinions regarding haze pollution and economic growth; three checklist questions (select all applicable answers); and one open-ended question asking for suggestions on battling the haze in China. The findings from this work can provide Chinese policymakers with information such as: (1) which demographic groups are satisfied/dissatisfied with the government's performance in reducing haze pollution; (2) which demographic groups want to encourage/discourage slowing down economic growth to treat haze pollution; (3) a nation-wide investigation of the impact of haze pollution on Chinese daily life and public health; and (4) respondent's suggestions on how to reduce haze pollution.

2. Materials and Methods

2.1. Questionnaire Design and Sampling

The questionnaire designed for this project relied on psychometric paradigm methods [27,28]. The detailed decription of the questionnaire is provided as follows. The questionnaire includes the following topics: (1) respondents' demographic information; (2) respondents' opinions on the relationship between haze pollution and economic growth in China; (3) respondents' comments on the treatment of haze pollution in China; and (4) the impact of haze pollution on respondents' daily life and personal heath. The demographic parameters included in the questionnaire have been found important for studying people's opinions on haze pollution. In particular, the respondent's age, gender, education [24], living regions [14], and occupation [29] were found to play an important role in determining respondents' opinions on haze pollution. The study presented in Reference [16] considered all these demographic parameters. The aforementioned demographic parameters were investigated in Questions 1–7. Specifically, Questions 1–3 surveyed general information of the respondents including their gender, age, and child status; Questions 4 and 5 ascertained the level of knowledge of the respondent by questioning his/her education and occupation; and Questions 6 and 7 determined the location and living area of the respondent. These questions allowed us to analyze the severity of haze pollution in different parts of China and analyze the concern and awareness of the problem by those who experience it firsthand.

Questions 8–29 surveyed the respondent's opinions on the experience with several issues regarding haze pollution versus the economic growth, haze pollution treatment, impact of haze pollution on the respondent' daily life and personal health. In particular, Questions 8–12 studied the respondent' opinions on quality of life and the environment pollution in China, including: whether the haze had negative impact on respondents' life in the past 10 years (Question 8); how the environment changed in the past 10 years in China (Question 9); how the environment will change in China in the future (Question 10); how long the respondent thinks it will take for China to resolve the haze problem (Question 11); and how the respondent' quality of life (with the consideration of both environment and economic conditions) has changed in the past 10 years (Question 12). Questions 13–17 further surveyed the respondent's willingness to sacrifice economic growth in favor of controlling haze pollution: Question 15 asked about whether the government had done enough to combat haze; Question 16 was about whether the respondents supported potential policies that might slow down economic growth to work toward reducing haze pollution; and Question 17 listed options respondents could select to prioritize economic growth over reducing haze pollution or to prioritize reducing haze pollution over economic growth. Questions 18–22 were focused on haze pollution and its treatment, including: approaches for reducing haze pollution (Question 18); determinations of the global warming effect of haze pollution (Question 19); daily assessments by respondents to check the daily haze value, i.e., PM2.5 (Question 20); existence in the respondent's city of haze pollution (Question 21); and possible troubles the haze may have caused in the respondent's daily life or work (Question 22). Questions 23–28 collected information on the respondents' personal experiences with haze pollution and the effects it had on the respondents' families. Question 23 asked how much respondents had reduced their outdoor activities. Question 24 surveyed whether respondents changed their habits (e.g., taking buses instead of driving personal cars) due to the haze pollution. Question 25 studied whether respondents intended to leave if their city had haze. Question 26 investigated the approaches respondents used to protect themselves. Question 27 asked how often the respondents or their immediate families had contracted haze related health complications. Question 28 further studied the affected area in the respondents' bodies if they got sick during periods of high haze pollution. Question 29 was an open-ended question that allowed respondents to input comments on haze pollution and thoughts on moving toward a sustainable environment in China.

The questionnaire survey was mainly conducted through one of the most popular online surveying websites in China (www.diaochapai.com) and distributed through the most common social media apps (e.g., WeChat) used by Chinese. To make as many Chinese aware of this survey as possible, the survey was disseminated by local Chinese people from different educational backgrounds throughout different regions of China and to as many people as they knew. The questionnaire ultimately surveyed 1233 people in total, gathering the demographics of each respondent and their opinions on many different subjects concerning haze pollution and the Chinese economy. The samples were mainly collected in the summer and fall of 2017. Since the policies in China may change with time, the survey results may not reflect the latest opinions on haze pollution and economic growth. To get more complete (e.g., more samples) and timely Chinese opinions on these issues, questionnaires need to be manually and frequently handed to people living all over China. With the consideration of time, the legal permit, and the cost for conducting such a survey in China, we conclude that a more complete and timely survey can only be done by the Chinese government. Nevertheless, this work aims to provide a reference that reflects the opinions of some Chinese people on the haze pollution.

2.2. Statistical Analyses

The survey data were analyzed to get the histogram for each of Questions 1–28. This can indicate the preference of respondents for the options listed for each question. Once the result for each question was shown, multivariate statistical methods, including Chi-squared test (CST) and principal component analysis (PCA), were further implemented to analyze respondents' opinions on the key problems that this work was focused on. These problems are mainly related to respondents' satisfaction on the government's effort in preventing the haze pollution, respondents' preference on slowing economics to prevent the haze pollution, and respondents' opinions on the change in living environment, quality of life, and daily activities caused by haze pollution.

In this work, the CST is mainly used to determine whether respondents' opinions on Question 15 (i.e., whether the government had done enough to combat haze,) and Question 16 (i.e., whether respondents supported the policy to slow down economic growth for reducing haze pollution) are independent of their demographic backgrounds (i.e., gender, child status, age, educational background, occupation, population density of the living area, and living region in China). For example, Table 1 shows the results of respondents' opinions on Question 16 based on their ages. The null hypothesis is that respondents' opinions are independent of their ages. As for the respondents between the ages of 41 and 50 years, they take 35.5% (calculated by 438/1233) of the whole 1233 respondents. It is thus expected that 35.5% of the 669 respondents who support the policy should be between the ages of 41 and 50 years. This expected number, represented by $n_{\text{expected},i}$ in Equation (1), equals 237 (calculated by 35.5% $\times$ 669). It is observed from the table that there are actually 273 people between 41 and 50 years supporting the policy. The difference between this observed number (represented by $n_{\text{observed},i}$) and the expected number $n_{\text{expected},i}$ contributes one term in Equation (1) in the CST, i.e., $(273 - 237)^2/237 = 5.47$. The same approach is used to quantify the difference between the observed and expected numbers for other age categories so that all age categories are considered in the Chi-squared value (i.e., χ^2). The Chi-squared value, along with the number of age categories (i.e., five age groups) and opinion categories (i.e., four different opinion, is then used as the input of the Chi-squared distribution to determine whether the difference is statistically significant.

Typically, a p-value less than 0.05 is used to reject the null hypothesis (i.e., respondents' opinions are independent of ages in this case). The command *chisq.test* in R was used in this work to perform the CST. Since the calculated p-value is 2.0×10^{-5}, much less than 0.05, it is concluded that the alternative hypothesis (i.e., respondents' opinions depend on ages) is correct. A CST was performed for all demographic categories for Questions 15 and 16 in this project.

$$\chi^2 = \sum_{i=1}^{k} \frac{\left(n_{observed,i} - n_{expected,i}\right)^2}{n_{expected,i}} \quad (1)$$

Table 1. The opinions of respondents to policy that would slow down economic growth to reduce haze pollution.

		Respondents' Opinions of Policy That Would Slow down Economic Growth in Order to Reduce Haze Pollution				Total
		No Support	Neutral	Support	No Opinion	
Ages	Below 21 years	10	23	18	1	52
	21 to 30 years	39	81	93	3	216
	31 to 40 years	33	57	86	4	180
	41 to 50 years	62	92	**273**	11	**438**
	Above 50 years	55	82	199	11	347
	Total	199	335	**669**	30	**1233**

Note: the highlighted numbers in the table are used to illustrate the principal of the CST.

While the CST is used to identify the demographic groups who expect more efforts from the government toward haze prevention (Question 15) and who prefer to slow down economic growth for haze prevention (Question 16), principal component analysis is further used to visualize respondents' opinions on the impact of haze pollution in respondents' living environment, quality of life, and daily activities over the last few years. Principal component analysis (PCA) is one of the most commonly used multivariate statistical tools to analyze and visualize high-dimensional datasets [30,31]. The idea of PCA is to represent the original data using fewer new orthogonal variables called principal components that are obtained as linear combinations of the original coordinate variables. As shown in Figure 1, PCA can use a two-dimensional space that is represented by the first two principal components (i.e., PC1 and PC2) to present the original three-dimensional data. Each of individual principal components is essentially a linear combination of original coordinate variables. The essential idea of PCA is to rotate the original coordinate system so that the projections of the original data onto PC1 have the largest variance, which means that the PC1 can retain the largest portion of information from the original data. PC2 is orthogonal to PC1 and retains the second largest portion of information from the original data. Typically, the original data can be shown by the reduced dimensional space represented by the first few principal components. This explains why PCA has been widely used in high-dimensional data reduction and visualization. In this work, PCA was used to visualize the opinions of 1233 respondents on Question 8 (i.e., whether the negative impact on daily life was large), Question 9 (i.e., whether the environment became worse and worse in the last 10 years), Question 12 (i.e., whether the living quality, with the consideration of both economic and environment situations, became worse in the last 10 years), Question 22 (i.e., whether the haze pollution caused troubles in respondents' work and daily life), and Question 23 (i.e., whether respondents reduced outdoor activities due to the haze pollution). The command *prcomp* in R, which is a popular programming language widely used in statistics analysis of statistical data [32] and data in engineering [33], was used to perform principal component analysis.

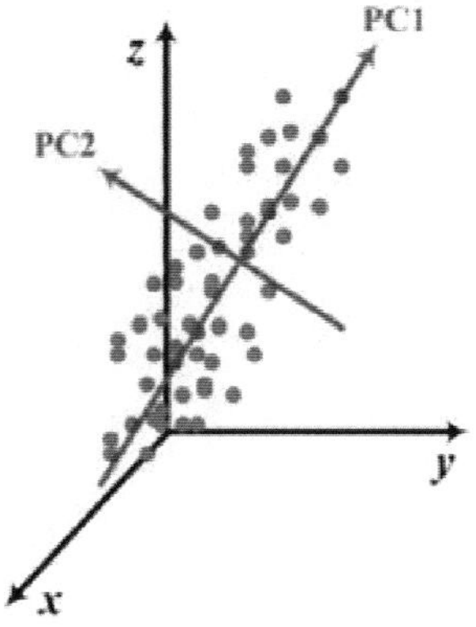

Figure 1. An illustrative diagram for principal component analysis. The red color points show the three-dimensional data, which can be represented by their projections onto a two-dimensional space coordinated by the first two principal components (i.e., PC1 and PC2). In this figure, the projects onto PC1 has higher variance than those onto PC2.

3. Results

After analyzing the statistics, an overview of the results for all questions in the survey is shown below. This is followed by the analysis of respondents' opinions on Question 16 to study the influence from demographic parameters like age and educational background on respondents' opinions on the policy that would slow down the economy to combat haze. This is followed by respondents' satisfaction with government efforts in haze prevention (i.e., Question 15). While the CST can be applied to study other questions, this work is mainly focused on Questions 15 and 16, which were seldom studied in a nationwide scale in China. To get a comprehensive viewpoint of respondents on the change in their living environment, the life quality, and daily activities in the last few years, principal component analysis was used to project respondents' feedback for Questions 8, 9, 12, 23 and 24 onto a two-dimensional space. Finally, results from the impact of haze on respondents' daily life and personal health and from respondents' suggestions for combating haze were presented.

3.1. Overview of the Survey Results

Questions 1–7 were designed for the demographic analysis of gender, child status, age, educational background, occupation, population density of the living area, and living region in China. Questions 8–29 then showed respondents' opinions on the haze pollution and economic growth. Figures 2–29 show Questions 1–28 and the survey results one by one. Both the question and the discussion of the result are given in the title of each figure. The result for Question 29, which was an open-ended question for respondents to input their comments on the haze pollution, is shown in the Section 3.5.

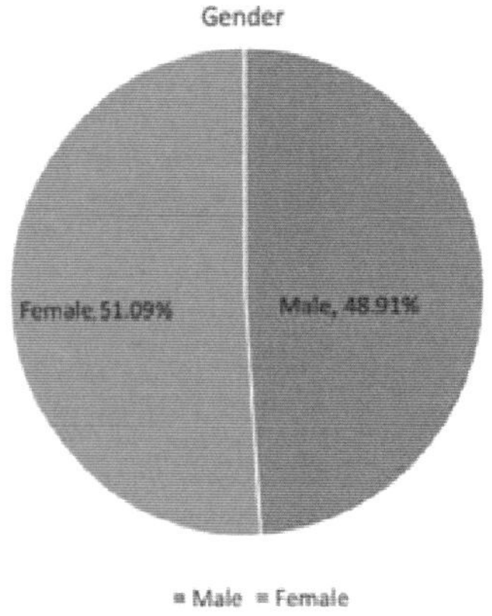

Figure 2. The result for Question 1 "What is your gender?": The number of males and females that responded to the survey were roughly equal.

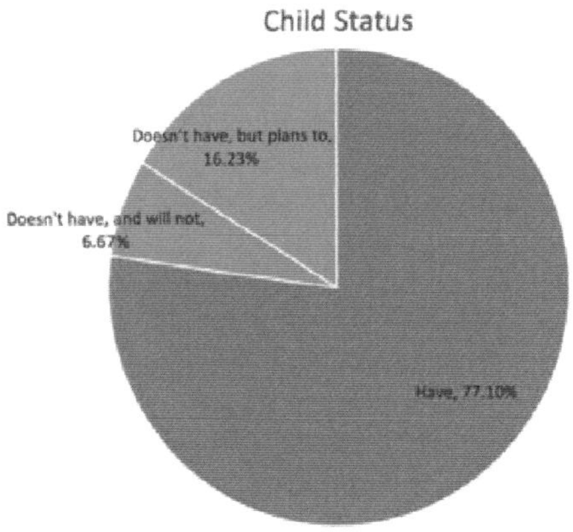

Figure 3. The result Question 2 "Do you have children?": Most respondents had children, and only 6.67% of respondents did not plan to have children.

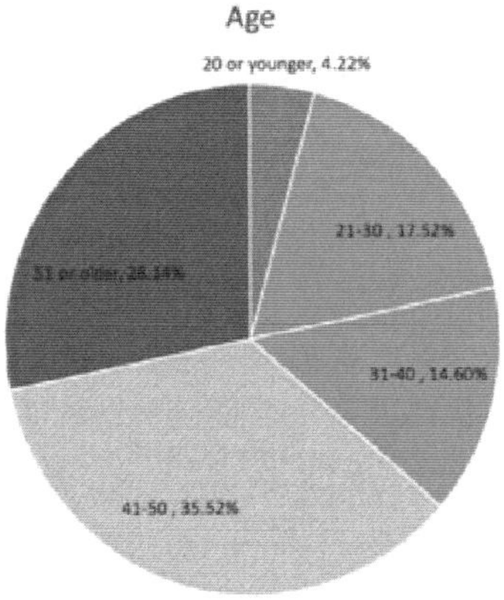

Figure 4. The result for Question 3 "How old are you?": There were more older people than younger people, with 63.66% of the respondents being above 40.

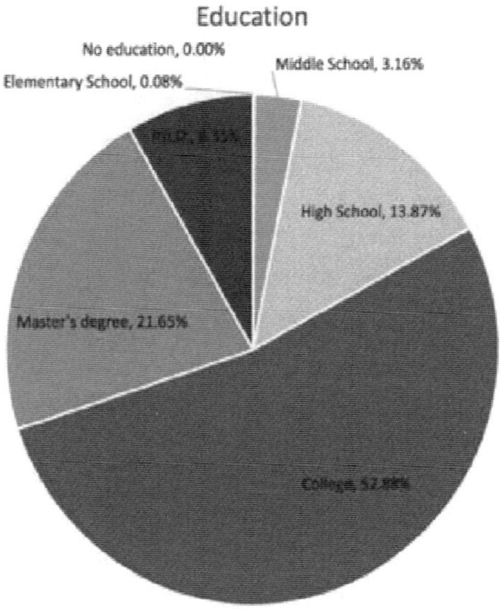

Figure 5. The result for Question 4 "What is your education background?": The respondents were generally well-educated, with 82.88% having college diplomas or higher.

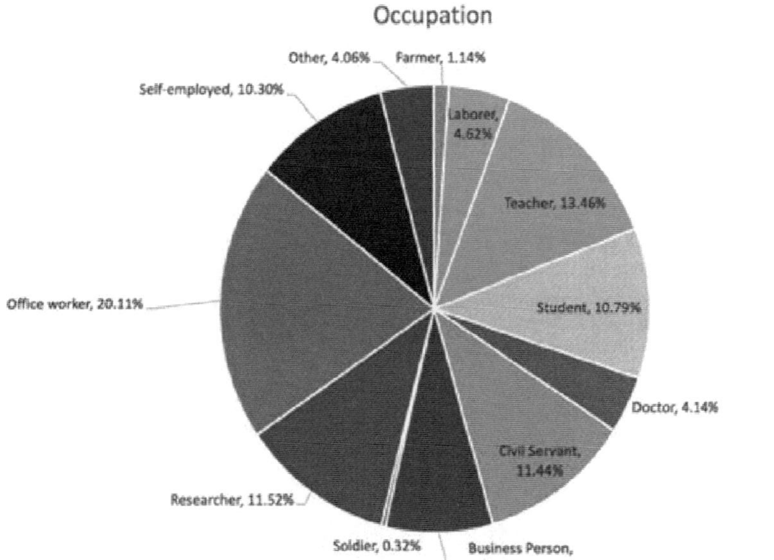

Figure 6. The result for Question 5 "What is your occupation?": Office workers pulled ahead numerically, making up 20.11% of the participants.

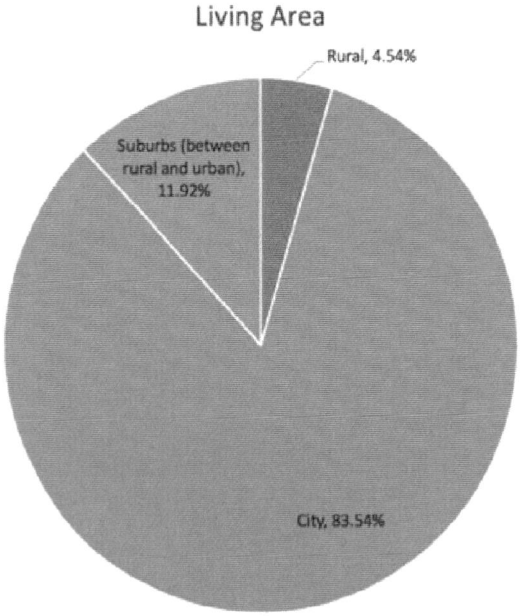

Figure 7. The result for Question 6 "What area (population density) do you live in?": Most of the people surveyed live in cities.

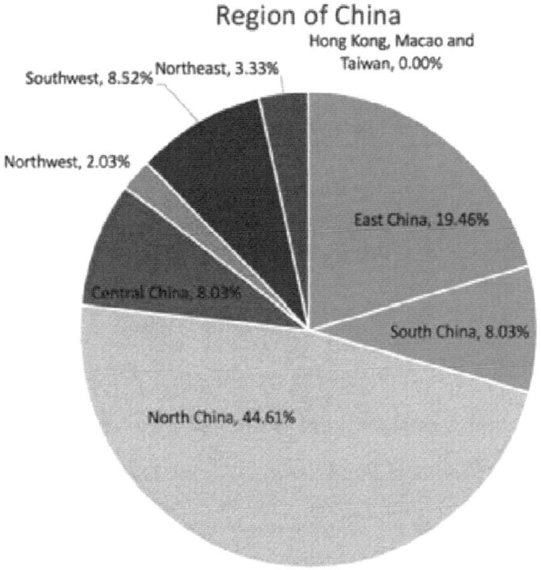

Figure 8. The result for Question 7 "What region of China do you live in?": It seems that 67.40% of the respondents lived in northern, eastern, or northeastern China.

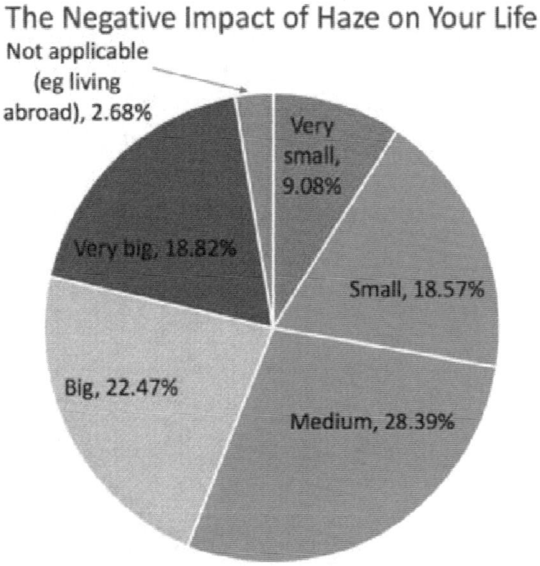

Figure 9. The result for Question 8 "What impact did haze have on your life in the past 10 years?": 28.39% of respondents stated that haze had a medium level negative impact on their lives. Overall, 22.47% of respondents said haze had a big impact on their lives, and 18.82% stated it had a very large impact. A small impact was selected by 18.57%, and very small impact was selected by 9.08%.

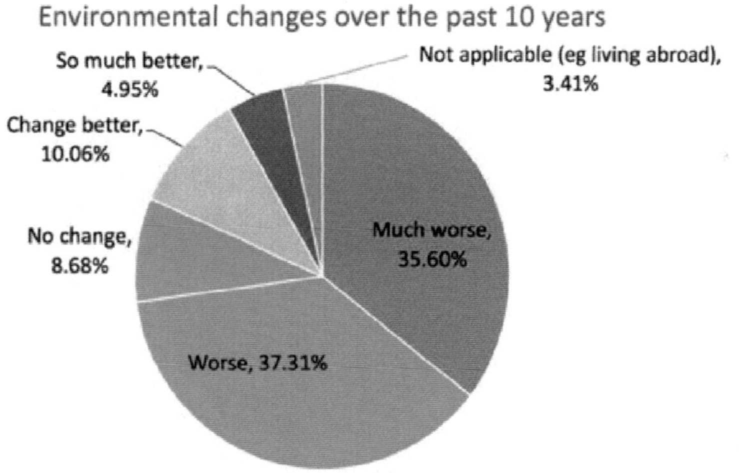

Figure 10. The result for Question 9 "How did the environment change in the past 10 years?": 72.91% of those surveyed stated that the environment had gotten worse or much worse in the past ten years.

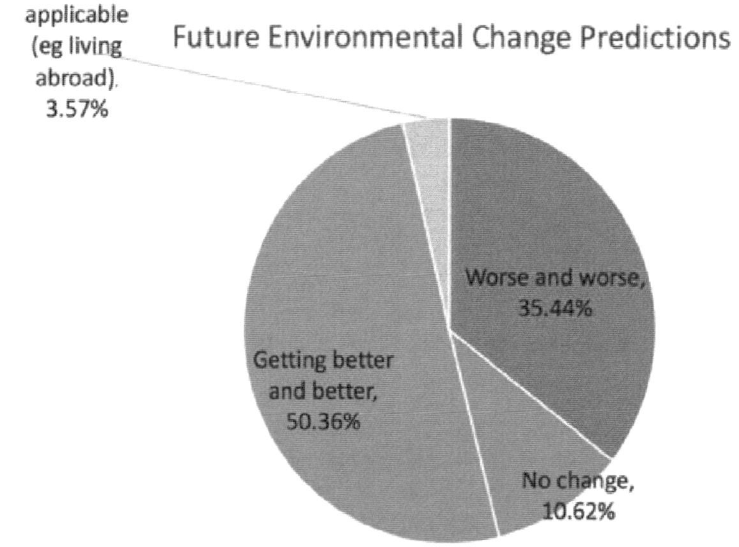

Figure 11. The result for Question 10 "How will the environment change in China in the future?": It is noted that 35.44% of people had a negative prediction on the potential for improvement of the environment China, but 50.36% believed the environment would get better.

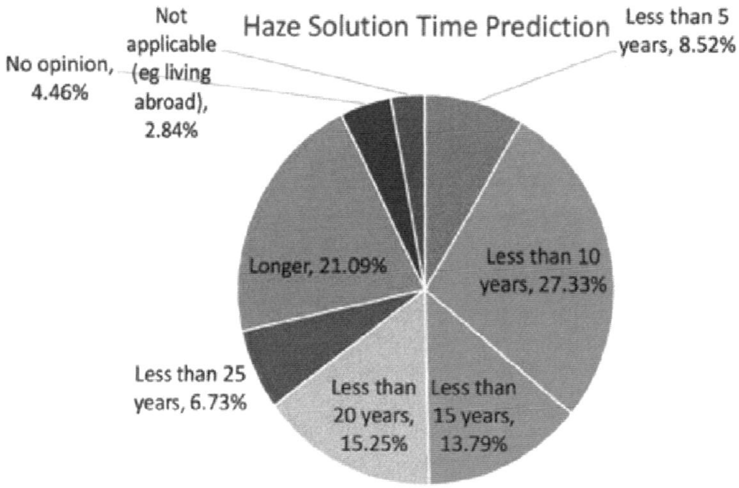

Figure 12. The result for Question 11 "How long do you think Chinese can get haze resolved?": 27.33% of people believed that haze pollution in China would be solved in 5–10 years, with 8.52% of people choosing less than 5 years; 13.79% of people chose 10–15 years; 15.25% of people chose in 15–20 years; and 27.82% of people chose 20 years or longer to solve haze.

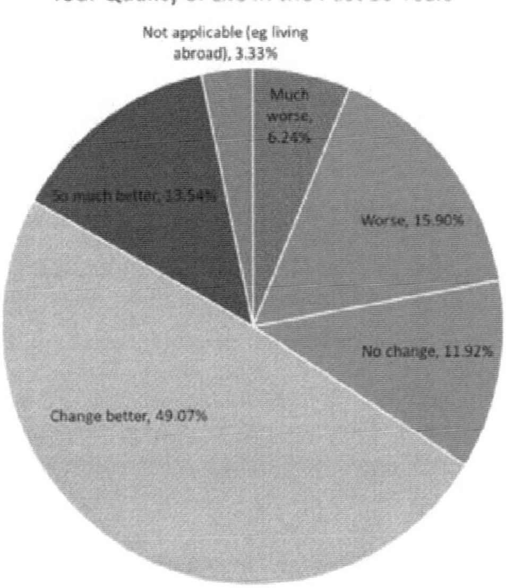

Figure 13. The result for Question 12 "How did your life quality (with the consideration of both environment and economic conditions) in the past 10 years?": 49.07% and 13.54% of people agreed that quality of life (with the consideration of economic conditions) in the past years for the general population changed for the "better" and "much better", respectively. On the contrary, 15.9% said quality of life got worse, and for 6.24%, it got much worse, while 11.92% said their life quality did not change.

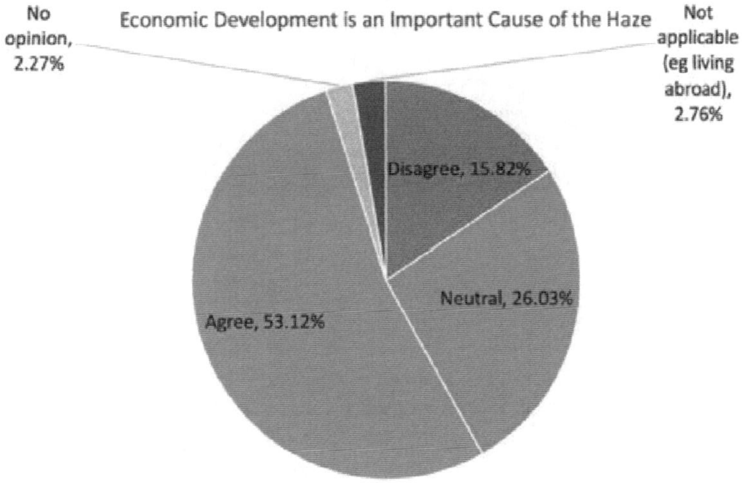

Figure 14. The result for Question 13 "Do you think the economic development is a prominent cause of haze?": 53.12% of respondents agreed that economic development and haze are connected by causation. In contrast, 15.82% disagreed that economic development has led to haze pollution, and 26.03% were neutral.

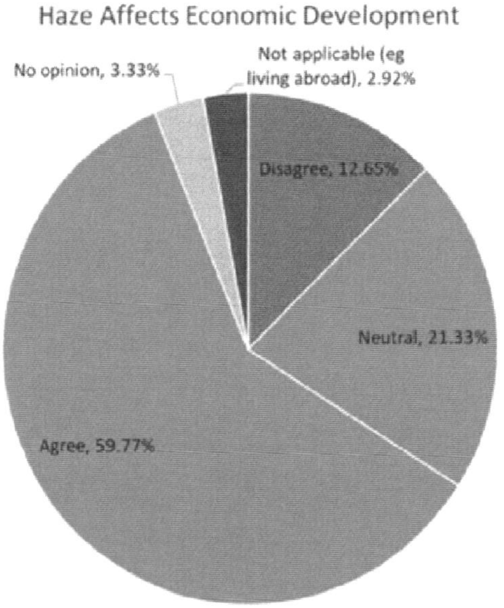

Figure 15. The result for Question 14 "Do you agree the haze pollution impair the economic growth?": 59.77% of respondents agreed that haze affected economic development, 12.65% disagreed that haze affected economic development, and 21.33% were neutral.

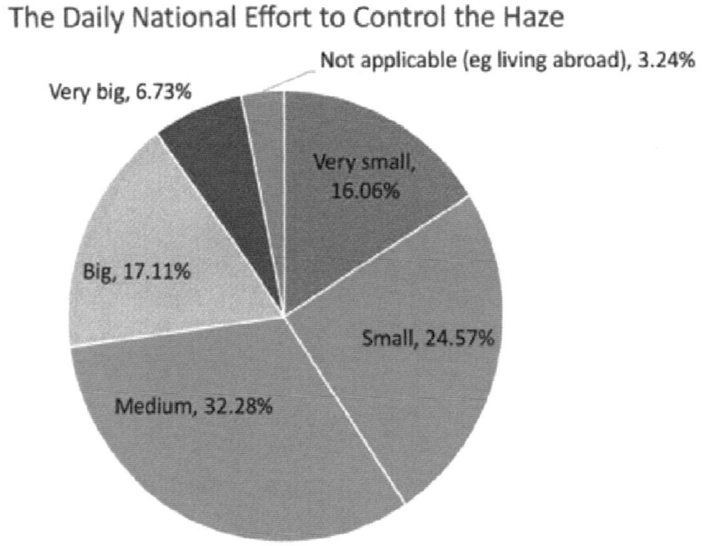

Figure 16. The result for Question 15 "How do you rate the government's effort in handling the haze pollution?": 24.57% believed national efforts to control haze were small, and 16.06% believed the efforts were very small, totaling 40.63% who believed efforts were small or very small. Overall, 17.11% believed the efforts were big, 6.73% believed the efforts were very big, for a total of 23.84% who believed there was a large effort. Overall, 32.28% believed there was medium national effort.

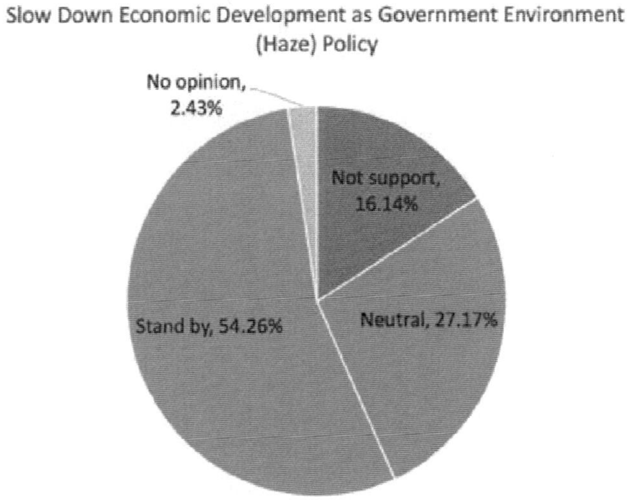

Figure 17. The result for Question 16 "Do you support the policy of slowing down the economic growth to reduce the haze pollution?": 54.26% of respondents stood by a government policy that would slow down economic growth to focus on haze policy and prevention. Overall, 27.17% were neutral, and 16.14% disagreed with such a policy.

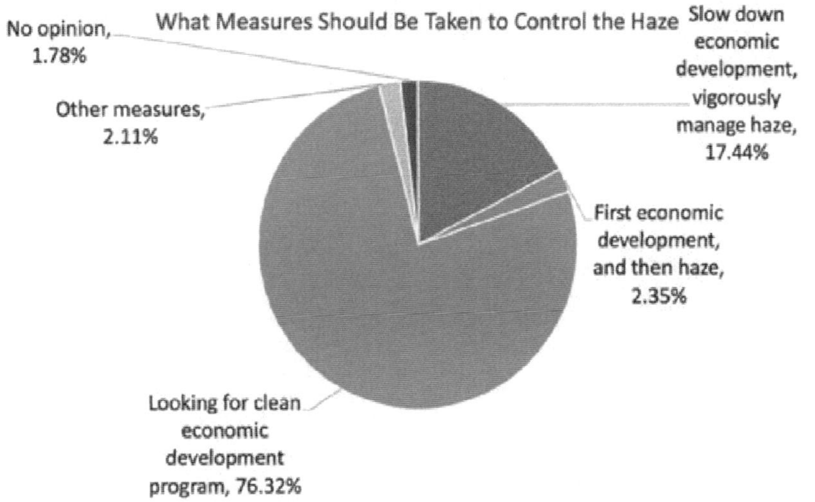

Figure 18. The result for Question 17 "Which option do you agree to handle the economic growth versus the haze pollution?": 76.32% of people agreed that a clean economic development plan would be the best course of action as a measure to control the haze. This is opposed to 17.44% who believed that haze prevention should be vigorously managed before economic growth is considered, and only 2.35% who believed in an economy-first plan over environmental protection.

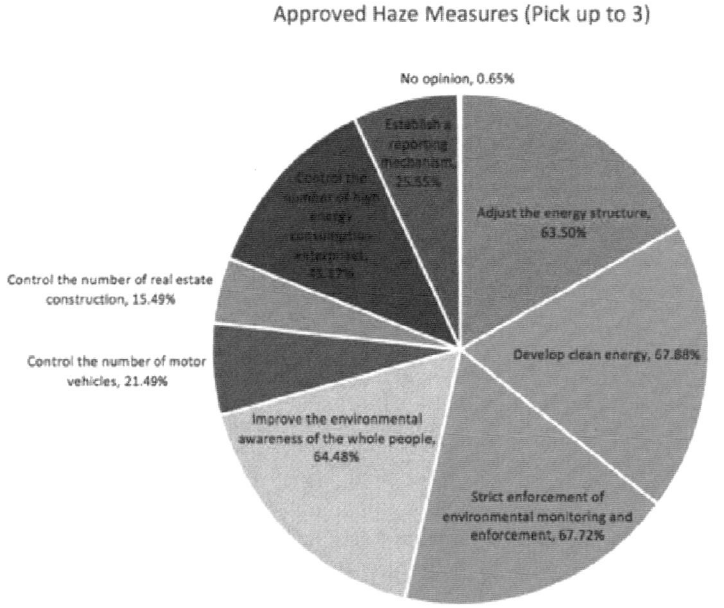

Figure 19. The result for Question 18 "Which three approaches do you choose to handle the haze pollution?": A majority of people supported the need for an increase in environmental awareness, strict environmental monitoring, the development of clean energy, and the adjustment of energy structure. Specifically, 67.88% of those surveyed preferred the development of clean energy.

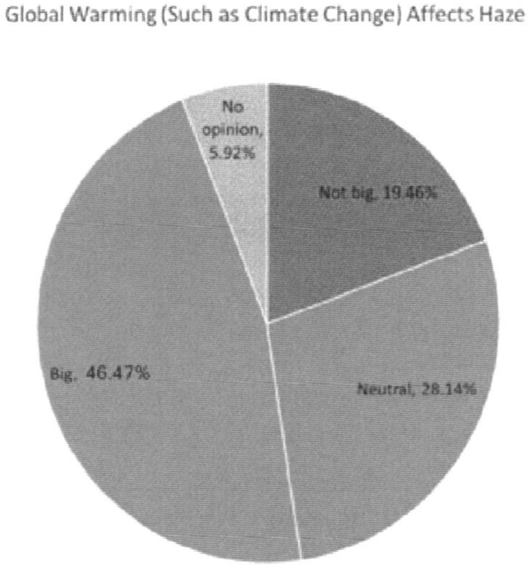

Figure 20. The result for Question 19 "Does the global warming affect the haze pollution?": Almost half the people (46.47%) thought that climate change affected haze.

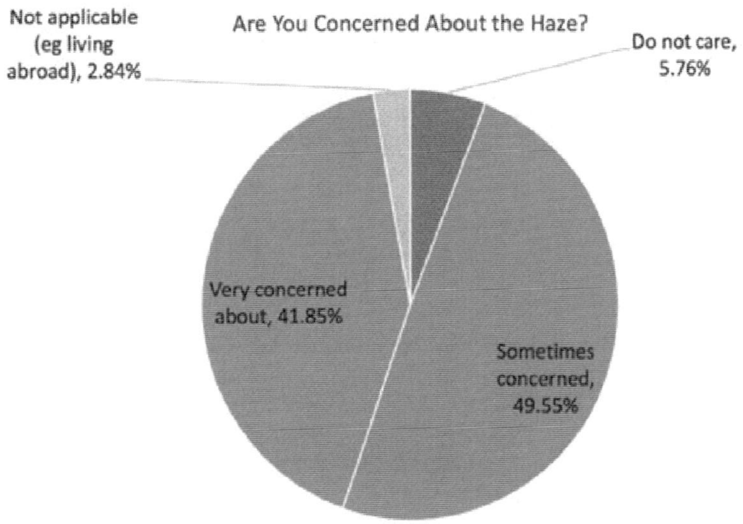

Figure 21. The result for Question 20 "Are you concerned about the haze pollution (e.g., check daily PM2.5)?": 91.40% of the Chinese people were sometimes or very concerned about the haze and checked daily $PM_{2.5}$ value.

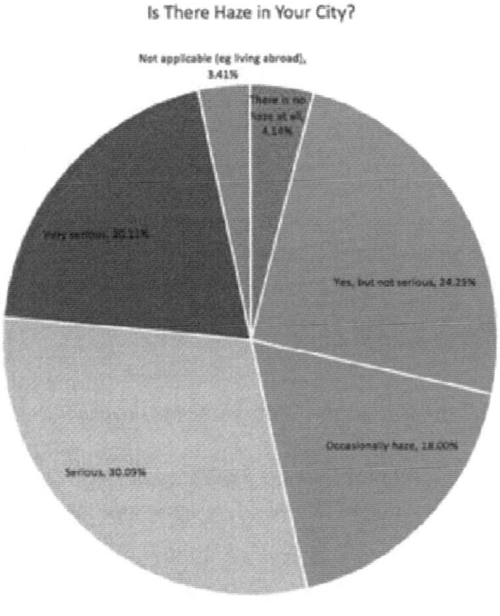

Figure 22. The result for Question 21 "Is there haze in the city you live in?": Only 4.14% said that there was no haze in their city, while 50.20% had serious or very serious haze in their cities.

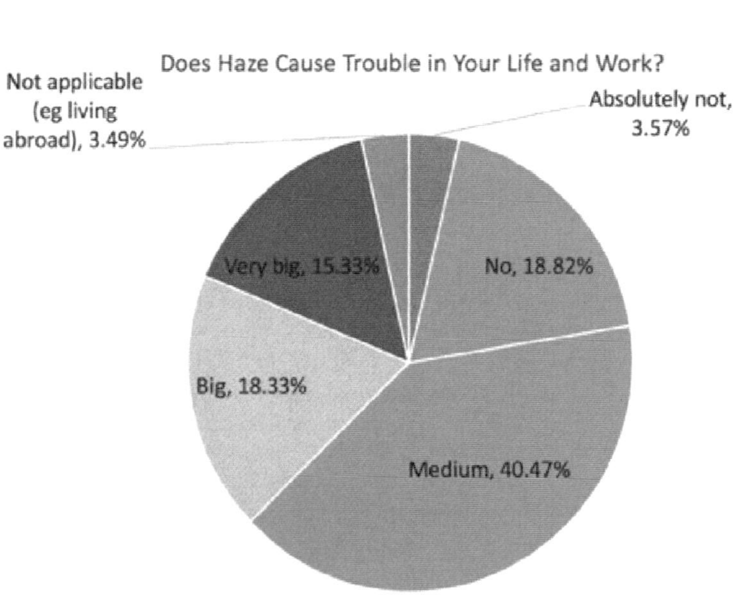

Figure 23. The result for Question 22 "Does haze cause trouble in your daily life or work?": 74.13% of those surveyed were affected in life and at work.

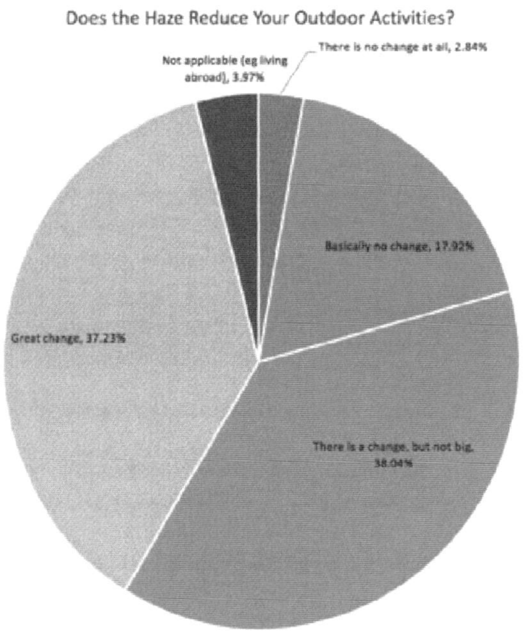

Figure 24. The result for Question 23 "Does haze change your outdoor activities?": 75.27% reduced outdoor activity because of the haze.

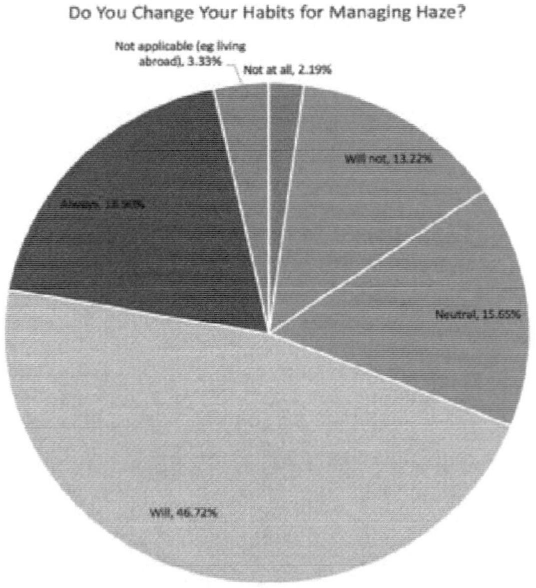

Figure 25. The result for Question 24 "Do you change your habits (e.g., taking buses instead of driving personal cars) due to the haze pollution?": 65.62% changed their habits to adjust to the haze.

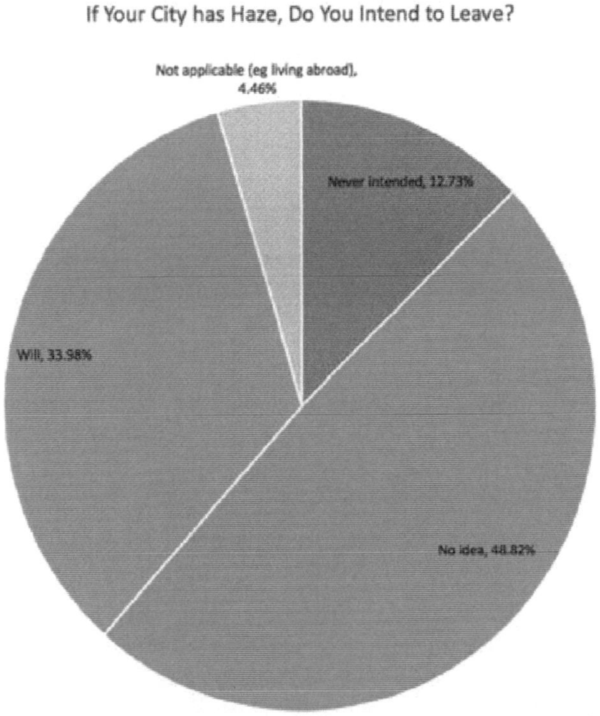

Figure 26. The result for Question 25 "Do you intend to leave if your city has haze?": 48.82% of the people did not know if they would leave their city because of the haze, but 33.98% of them planned to leave.

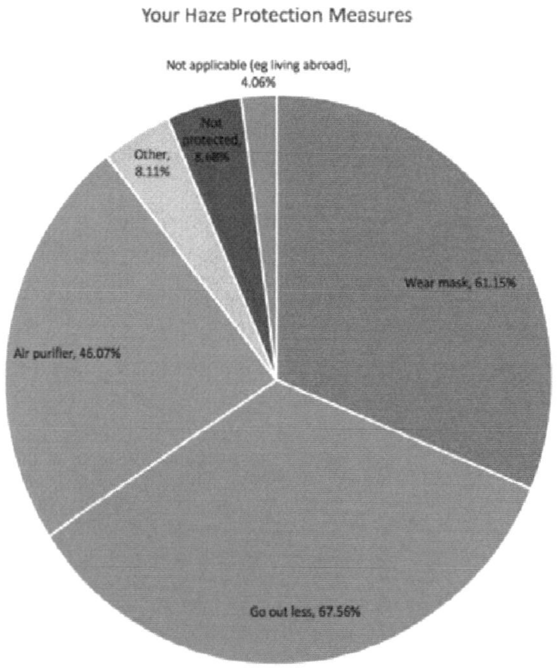

Figure 27. The result for Question 26 "Which approaches did you use to protect yourself (multiple options can be chosen)?": Many people said they used an air purifier (46.07%), went out less (67.56%), and wore masks (61.15%) to protect themselves from haze.

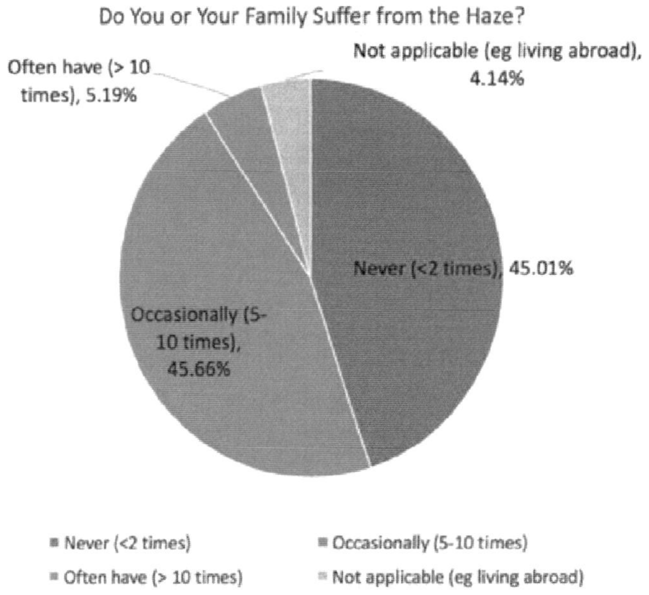

Figure 28. The result for Question 27 "Did you or your family members get sick during the haze period?": 45.66% of respondents had family that suffered from haze or suffered from it themselves, but another half did not.

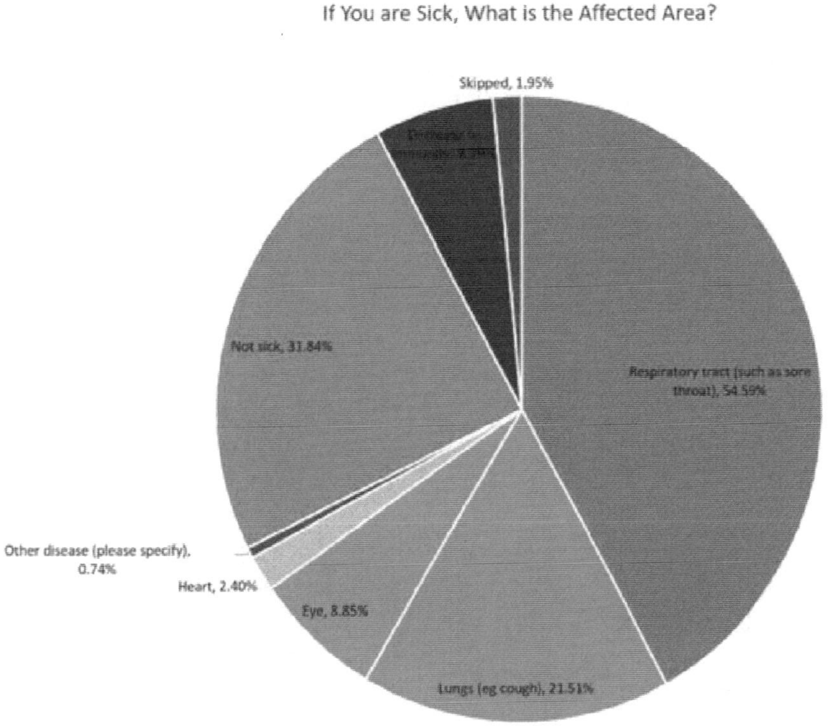

Figure 29. The result for Question 28 "If you are sick, what is the affected area in your body?": 54.59% of respondents were affected in their respiratory tract by haze. The lungs and eyes were the other major areas where respondents got sick.

3.2. Respondents' Opinions on Haze versus the Economic Growth

Respondents' feedback on Questions 13 and 14 indicated that more than 50% of respondents agreed that the economic growth and the haze pollution were highly correlated, and that the rapid economic growth was one of the major factors leading to the haze pollution. Here, we further investigated respondents' preference on rapid economic growth versus lower haze pollution (i.e., Question 16). The CST indicated that respondents' opinions on the policy to slow down economic growth to combat haze were dependent on their genders (p-value of 1.06×10^{-3}), their child statuses (p-value of 1.37×10^{-4}), their ages (p-value of 1.98×10^{-5}), and the regions they lived in (p-value of 3.52×10^{-4}). Since the CSTs on respondents' child statuses and ages were of the smallest p-value, detailed analysis of the results on these two demographic categories is given below. The results on respondents' genders and living regions were then briefly summarized.

As shown in the previous section, 54.26% of respondents generally supported the policy to slow down economic growth to combat haze (i.e., Question 16). This implied that Chinese people may be willing to sacrifice economic growth for a better living environment. We further investigated respondents' opinions from different demographic groups (i.e., gender, child status, age, educational background, occupation, population density of the living area, and living region). As shown in Figure 30, 57.48% of those currently with children stated that they would support government policy that slowed down the economy to prevent haze from worsening. This is compared to the weaker support coming from those planning to have children, only 39.51% of whom support the policy, and those not planning to have children, 45.69% of whom support the policy. For all three demographic groups, the percentage of people supporting the policy was much larger than the one opposing the policy and the one without preference (i.e., neutral). Among these three groups, the respondents with children were more supportive of government policies against haze. This is likely because the parents were more stressed by the health issues caused by haze pollution on their young children. We further analyzed respondents' opinions to Question 16 based on respondents' ages (Figure 31). A majority of respondents between the ages of 31 and 50 believed that the economic development should be slowed down to control the haze. However, for the older group (above 50 years old) or younger group (below 30 years old), people were more reserved and were neutral about economic slow-down as a government policy. This may be because these respondents need more stable employment or financial help. While economic slowdown may be beneficial for middle aged respondents in stable lifestyles, those just entering or about to leave the workforce may experience adverse results.

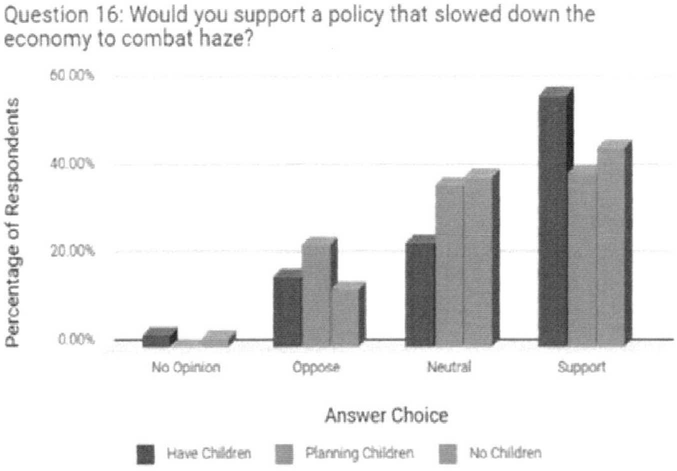

Figure 30. Grouping of support for potential government policy by respondents' child status. The percentage on the Y-Axis represents the percent of respondents in the specific group. For example, summing the percentages for all the bars representing people with children yield 100%.

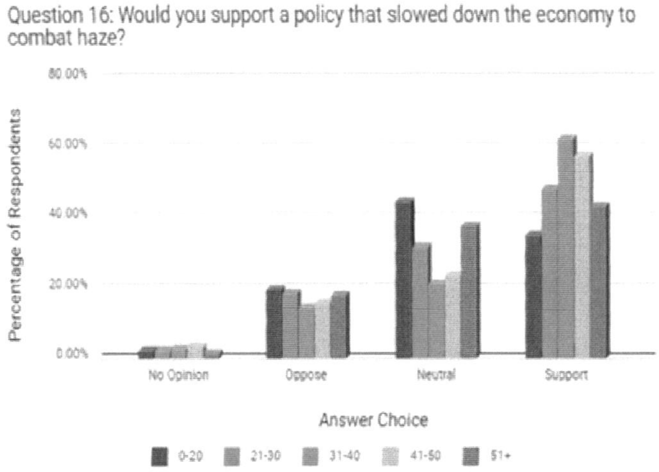

Figure 31. Grouping of support for potential government policy by ages. Respondents were categorized into five age groups: 0–20, 21–30, 31–40, 41–50, and above 50.

Similar analysis was performed on respondents' opinions on the economic-slowdown policy according respondents' genders and living regions. While more than 53% of respondents in each of the female and male categories supported the economic-slowdown policy, 19.2% of the male respondents, versus 13.8% of female respondents, opposed the policy. That may be because husbands contribute to the major financial income in most Chinese families. As for living regions, respondents from East China (55.8%), North China (56.4%), Northwest China (64%), and Southwest China (59.0%) showed stronger support of the economic-slowdown policy than other regions. On the other hand, respondents from Northeast China (34.14%) showed the least support on the policy. Since economic growth in Northeast China has slowed down recently, it is understandable that people living there are against the economic-slowdown policy.

3.3. *Respondents' Opinions about the Government Effort in Haze Prevention*

A CST was performed for the seven demographic categories in Question 15 (i.e., whether the government's effort in handling haze pollution was significant). It indicated that respondents' opinions depended on their genders (p-value of 5.45×10^{-3}), ages (p-value of 5.07×10^{-6}), educational backgrounds (p-value of 1.45×10^{-5}), occupations (p-value of 2.89×10^{-6}), living area population density (p-value of 1.07×10^{-11}), and living region (p-value of 2.2×10^{-16}). As mentioned above, 40.63% of all respondents believed the government's efforts to control haze were small or very small, which represented the largest portion of respondents in the five options of "very small effort", "small effort", "medium effort", "large effort", and "very large effort". A sum percentage of larger than 40.63% in the options of "very small effort" and "small effort" thus indicates the corresponding demographic categories of respondents desired more government effort to combat haze pollution. This may indicate that Chinese people are expecting more effort from the Chinese government to control haze. The following demographic categories of respondents requested more efforts from the government in handling the haze pollution: female respondents in the gender category, respondents below 40 in the age category, respondents with bachelor or high-school degrees in the education category, respondents working as farmers, doctors, students, and regular company employees in the occupation category, respondents living in rural areas in the category of population density, and respondents from Northeast China, South China, and Southwest China in the category of living region. It is interesting to see in Figure 32 that people with a Bachelor's degree or lower mostly said that the government was not doing much about the haze problem, while people with Master's degrees were less inclined to respond that way. People with PhDs were evenly split in their responses. The respondents with higher education levels may know more about the (technical) challenges involved in reducing haze pollution in a short time frame and thus appreciate the government's effort. While at first glance it may seem that the national environmental effort is small, a deeper understanding of the laws and regulations being implemented may reveal stronger efforts by the government to treat haze. People from areas with less severe haze pollution, such as rural residents and people from Northeast China, South China, and Southwest China, requested more government effort in handling the haze pollution. A potential reason for this is that most of the government's policies on haze pollution were applied to those areas with severe haze pollution. Respondents not living in those areas thus may not feel the government's effort in handling the haze pollution.

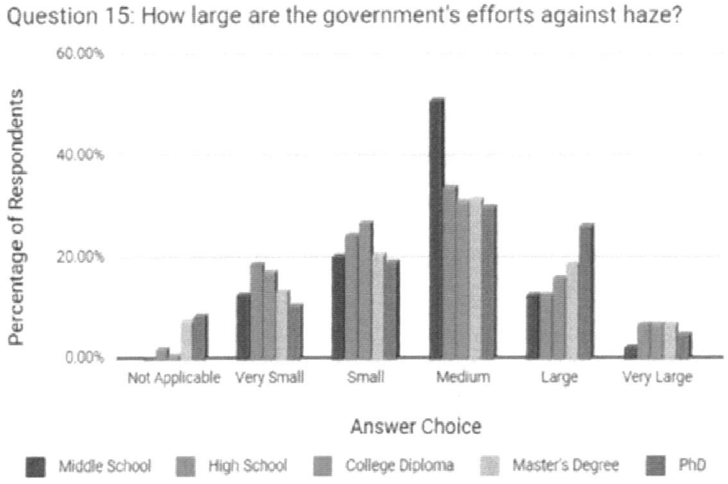

Figure 32. Respondents' opinion on government's current policies against haze according to their educational backgrounds.

3.4. The Impact of Haze on Respondents' Daily Life and Personal Health

The following questions in the survey are related to the influence of haze on respondents' daily life: Question 8 (i.e., whether the negative impact on daily life was large), Question 9 (i.e., whether the environment worsened progressively over the last 10 years), Question 12 (i.e., whether living quality—considering both economic and environmental situations—worsened over the last 10 years), Question 22 (i.e., whether the haze pollution caused troubles in respondents' work and daily life), and Question 23 (i.e., whether respondents reduced outdoor activities due to haze pollution). Respondents' opinions on these five questions may be correlated to each other. For instance, respondents agreeing with that the haze pollution caused troubles in their work and daily life (i.e., Question 23) may also feel that the negative impact of the haze pollution on their daily life was large. To project the opinions of respondents on these five questions, principal component analysis was implemented to identify a two-dimensional space defined by the first two principal components PC1 and PC2 that are orthogonal (or uncorrelated). Both PC1 and PC2 are the linear combination of the variables representing the respondents' answers to the five questions. To facilitate the understanding of the projections of original data onto the PC1-PC2 space, vectors representing respondents' opinions of the five questions were also plotted in Figure 33. Vectors in similar directions indicate that respondents' answers to those questions are correlated. For instance, respondents whose daily life was impaired by haze (Question 8) tended to agree that haze caused troubles in their work and daily life (Question 22). Similarly, respondents who thought the environment got worse or much worse in the last 10 years (Question 9) tended to agree that life quality got worse (Question 12) and tended to reduce outdoor activities (Question 23). The power of principal component analysis is in its ability to visualize five-dimensional data so that we can get a more comprehensive view. In Figure 33, bubbles were used to represent an assembly of respondents, with larger sizes indicating more respondents. For example, the large bubble along the direction of Question 8 indicated that many respondents agreed that the negative impact on their daily life was large. Similar, the respondents' opinions on other questions can be visualized in the figure. Figure 33 also shows that the large bubbles are mainly located in the sector from vector Q#12 clockwise to Q#22. This means that respondents generally agreed that the environment became worse and that haze caused trouble in their daily life. This resulted in reduced outdoor activities. Figure 34 shows the reduction of outdoor activities for respondents living in areas with different population densities (Q#23). Those living in rural districts did not reduce their outdoor activity (due to haze) as much as those living in urban and suburban zones.

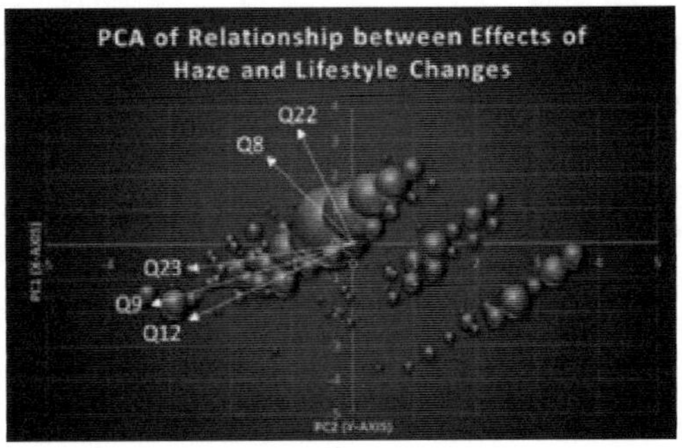

Figure 33. Projections of respondents' opinions on Q#8 (i.e., whether the negative impact on daily life was large), Q#9 (i.e., whether the environment became worse and worse in the last 10 years), Q#12 (i.e., whether the living quality, with the consideration of both economic and environment situations, became worse in the last 10 years), Q#22 (i.e., whether the haze pollution caused troubles in respondents' work and daily life), and Q#23 (i.e., whether respondents reduced outdoor activities due to the haze pollution) onto a two-dimensional space coordinated by the first two principal components, i.e., PC1 and PC2. Bubbles represent an assembly of respondents, with larger sizes indicating more respondents. Vectors representing the five questions are shown in White color. The bubbles show the distribution of respondents' opinions among the five questions.

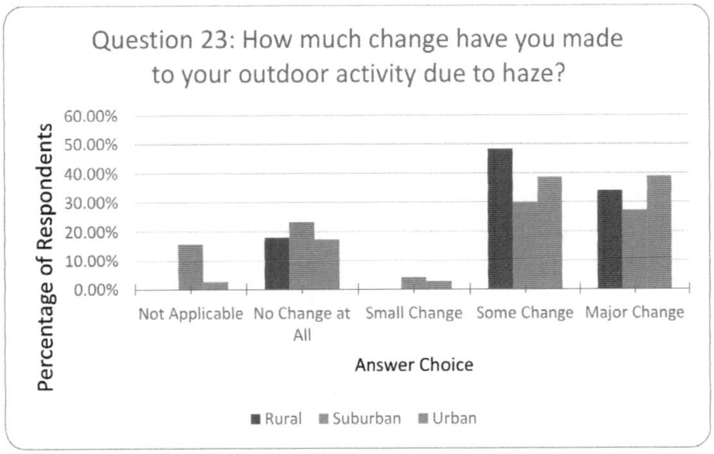

Figure 34. The change of outdoor activities in respondents living in rural, suburban, and urban areas.

While Figures 5 and 6 show that respondents' daily life was significantly influenced by haze, we further studied the impact of the haze pollution on respondents' personal heath. Since most respondents were adults, the results shown below were not applied to young children who generally have weaker immune systems to defend against diseases. As mentioned before, 45.66% of respondents had family members suffering from haze, but another half did not (Question 27). A CST indicated that respondents' feedback was mainly dependent on the regions they lived in (Figure 35). Specifically, respondents living in Northeast China, South China, Southwest China, and East China were mostly unaffected or affected relatively little by haze in terms of health. On the other hand, respondents living in the more polluted regions such as Northern, Central and Northwestern China were more likely to be afflicted by health problems. Among those respondents having health issues during the haze periods, a majority (54.59%) of respondents had complications with their respiratory tracts (Figure 36). In addition, 21.51% of them had lung problems.

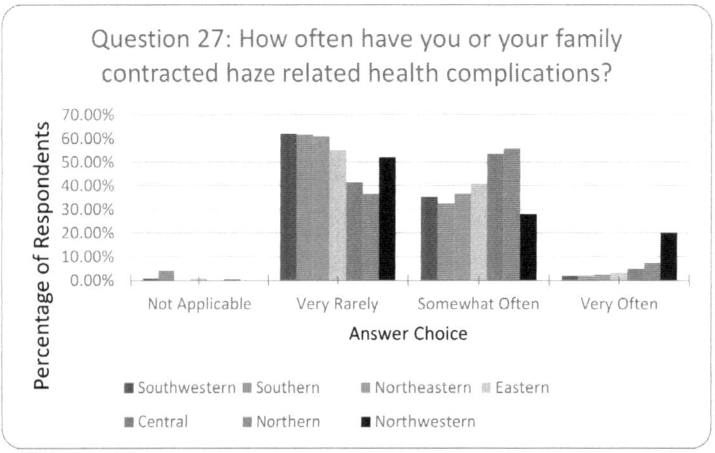

Figure 35. Grouping of frequency of haze related sickness by respondents' living regions. Respondents living in North China, Central China, and Northwest China suffered more in their health.

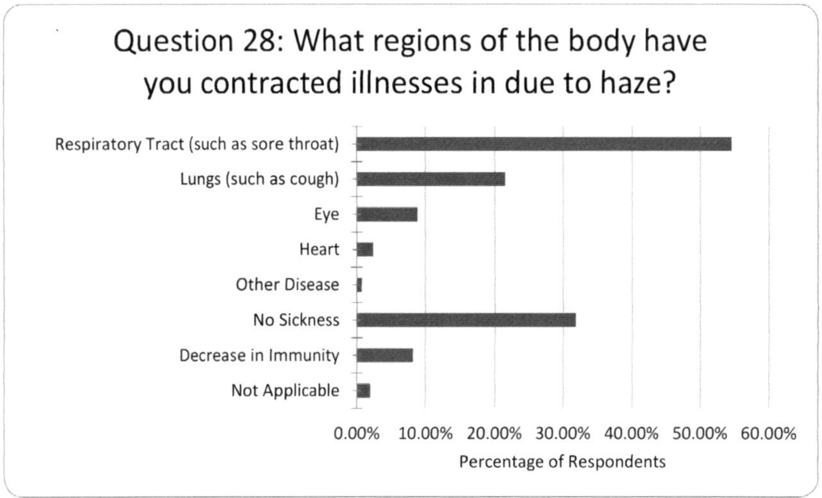

Figure 36. Areas of the body of respondents having health issues likely caused by haze. Multiple choices could be selected by respondents. No sickness option shown in the figure means the respondents did not have any illness during the haze periods.

While haze has reshaped Chinese daily life and caused health issues, people in China have found or created different solutions for protecting themselves during the haze attack (Question 26). Figure 37 shows that 67.56% of the people surveyed combatted haze by going out less and another majority (61.15%) wore masks. Another common haze protection approach used by 46.07% of people was installing air purifiers in homes. In the survey, respondents were allowed to choose multiple approaches they used to protect themselves from haze. Therefore, respondents might simultaneously reduce outdoor activities, wear masks, and install air purifiers at home.

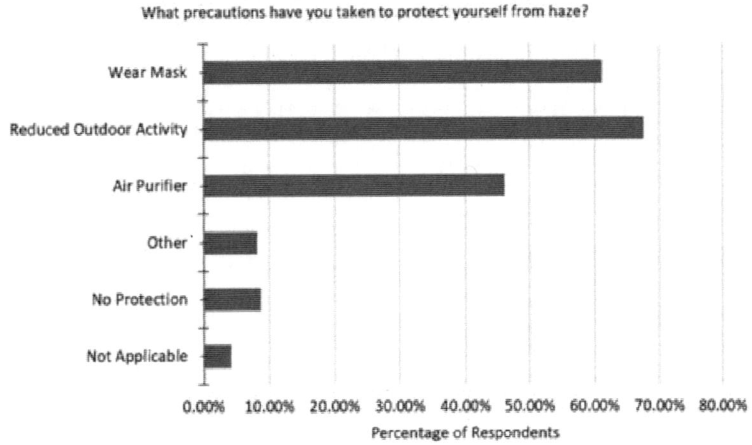

Figure 37. Actions taken or items used by respondents to protect themselves from haze pollution. Multiple choices could be selected by respondents.

3.5. Respondents' Suggestions for Combating Haze

Although more than 35% of respondents did not think the environment in China would be better in the coming 10 years, more than 50% of respondents still expressed positive feelings about the improvement of the environment in China (Question 10). On the other hand, more than 43% of respondents agreed that it will take 15 years or longer to resolve the haze pollution issue in China (Question 11). This section shows the suggestions provided by respondents for handling haze pollution. Question 18 asked respondents to select all applicable solutions for reducing the haze pollution (Figure 38). The most popular solutions chosen by respondents were as follows: develop new forms of clean energy (67.9% of respondents); design and strictly enforce environmental law (67.7% of respondents); increase the use of existing clean energy (63.5% of respondents); and prioritize haze prevention policy (64.5% of respondents). In total, 45.2% of respondents thought that companies with high energy consumption should face penalties. Only 25.5% thought that establishing organizations similar to the United States Environmental Protection Agency in China would be beneficial for haze prevention, with even fewer people supporting strict regulation of the number of motor vehicles or the development of real estate (21.5% and 15.5%, respectively).

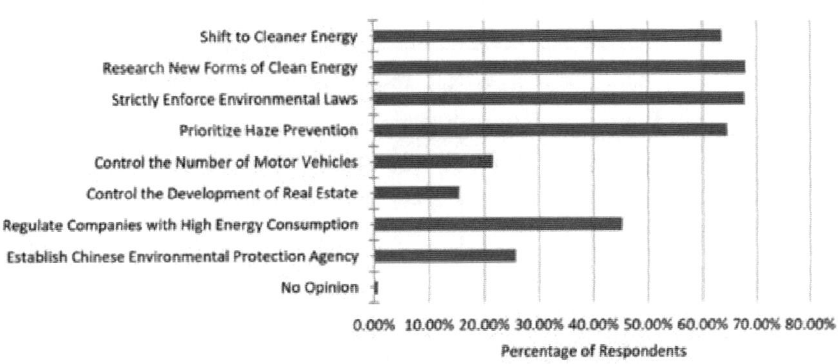

Figure 38. Strategies selected by respondents for combatting haze. Multiple choices could be selected by respondents.

While Question 18 allowed respondents to select among several suggested strategies for combatting the haze pollution, Question 29, which was an open-ended question, offered a chance for the respondents to input their own suggestions for haze prevention/protection (Figure 39). The most popular comments given by respondents focused on the use of alternative vehicle transportation or the reduction of automobile exhaust. This was found in 39 responses. Many who were polled felt that industrial emissions were too high. Suggestions included supervision and reduction of polluting enterprises (36 mentions), complete shutdown of those polluting enterprises (13 mentions), and stronger governmental legislation and enforcement of pollution-preventing policies (30 mentions). The following suggestions/comments were mentioned more than eight times by respondents: a need for public awareness (25 mentions), decreasing coal usage (18 mentions), afforestation (17 mentions), use of clean/renewable energy (15 mentions), reducing construction/infrastructure projects (11 mentions), decentralizing overcrowded urban population (10 mentions), enforcing effective waste treatment (9 mentions), and reshaping economic structure (8 mentions). In terms of general solutions to the haze issue, it was unsurprising that most respondents endorsed the search and production of clean energy, which, when replacing traditional energy gradually, can maintain the nation's needs while fighting haze. Controlling the number of cars or the development of real estate was not as popular as expected, as this approach would require direct sacrifice on the part of respondents. Another trend observed from respondents' comments was the willingness of respondents to support a shift from traditional fossil fuel emissions toward new renewable and efficient energy methods. All of these suggestions show that respondents were willing to seek change, minor and drastic, in order to suit the needs of haze prevention, through decreased industrial pollution.

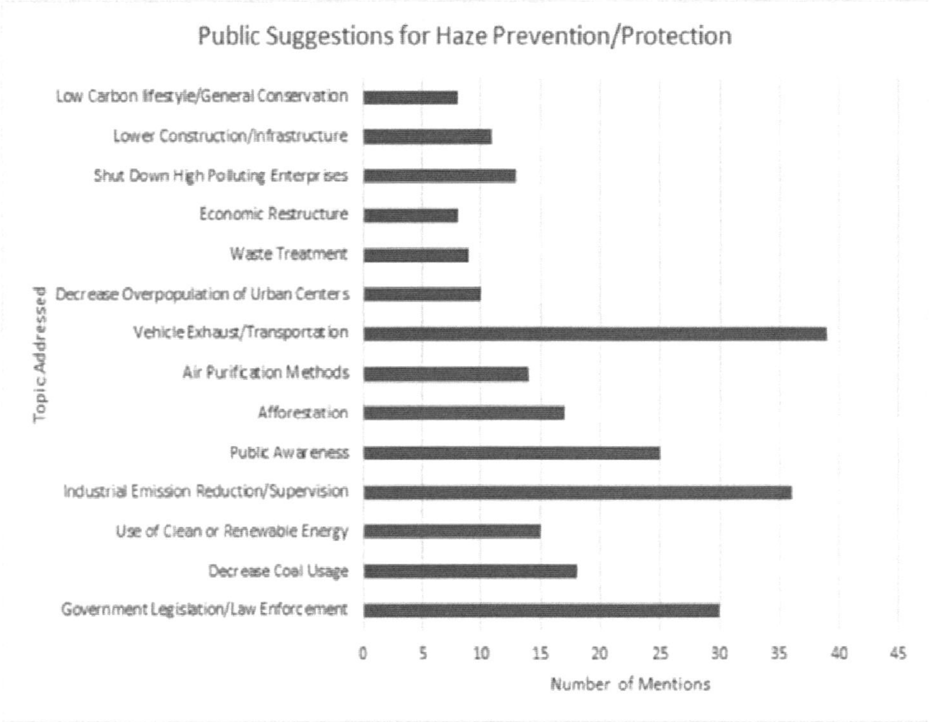

Figure 39. Suggestions from the public for haze prevention or protection.

4. Discussion

It is interesting to see from the results that most respondents supported the policy of slowing down economic growth to prevent the deterioration of the haze pollution. Certain respondents also suggested switching coal energy to clean energy so that China can still keep the rapid economic growth without causing more pollution. In addition, more than 40% of respondents expected more effort from the government in treating the haze pollution. All of these are helpful for the policy makers in China to make the right regulation on their economic and environmental policies. While tremendous effort was spent in outreach to as many respondents as possible in China, the sampled respondents may not completely reflect the opinions of most Chinese on the haze pollution and economic growth issues in China. In addition, questionnaires were mainly distributed via social media, and people living in remote areas with limited access to internet may not have been able to participate in the survey. As shown in Figure 34, urban residents changed their outdoor habits the most compared to their suburban and rural counterparts. This is because haze is most densely concentrated in industrial zones, which are often located near cities. Another issue is that the samples were mainly collected in the summer and fall of 2017. Thus, the results shown in this work mainly reflect the opinions of respondents in that period. To obtain the public's opinion in a timely manner, the survey should be provided more frequently to more people in various locations across China. This can only be done by the Chinese government. Nevertheless, the obtained results can still provide a reference that reflects the opinions of some Chinese people on the haze pollution. Relating to the collected data itself, more analysis still needs to be performed. While a thorough breakdown of Questions 1–7 was conducted along with specific cases pertaining to the questions addressed in this study, the influence of the combination of these seven demographic questions on respondents' opinions has not been thoroughly explored, which would be time-consuming and thus delay the publication of these timely survey results.

5. Conclusions

This study presented the first nationwide investigation of Chinese opinions on economic growth in relation to haze pollution. It applied multivariate statistical methods to analyze questionnaire data from 1233 respondents. Most respondents agreed that rapid economic growth was one of the main causes of haze pollution. More than half of respondents supported a policy that would slow down economic growth to combat haze, especially the respondents that had children at the time of the study, that were aged 31–50, or that were living in regions with high haze pollution. The respondents who believed the government's efforts to control haze to be small or very small outnumbered those with opposite opinions almost two times. CST showed that the respondents below 40 years of age, with Bachelor's degrees or high-school diploma, working as farmers, doctors, students, and regular company employees, desired more effort from the government to control haze. Most respondents thought the environment became worse or much worse in the past 10 years. The haze pollution also significantly influenced their daily life and personal health, especially for the respondents living in the more polluted regions such as Northern, Central, and Northwestern China. Most respondents protected themselves from the haze pollution by going out less, wearing masks, and installing air purifiers in their homes. Most respondents were positive about the potential improvement of the environment in China, although most of them thought it would take more than 10 years to resolve the haze issue. Suggestions from respondents for reducing pollution included, but were not limited to: developing new forms of clean energy, designing and strictly enforcing environmental law, using alternative vehicle transportation to decrease automobile exhaust, prioritizing haze prevention policy, and regulating highly-polluting enterprises. There are many elements to explore that can develop from this project. The most obvious would be to obtain a larger sample. Certain demographics were not fully represented because there were not enough of respondents. This would need the support of the Chinese government. Nevertheless, this work serves as a starting point for getting a more timely and complete survey of Chinese opinions on haze pollution and its relation to economic growth.

Author Contributions: Conceptualization, H.L., L.W. and Z.H.; Formal Analysis, F.Q., E.W., M.F. and Z.H.; Data Curation, all authors.; Writing-Original Draft Preparation, F.Q., E.W. and M.F.; Writing-Review & Editing, H.L., L.W. and Z.H.; Supervision, Z.H.

Funding: This research received no external funding.

Conflicts of Interest: The authors declare no conflict of interest.

References

1. Shao, M.; Tang, X.Y.; Zhang, Y.H.; Li, W.J. City clusters in China: Air and surface water pollution. *Front. Ecol. Environ.* **2006**, *4*, 353–361. [CrossRef]
2. Wang, S.J.; Li, Q.Y.; Fang, C.L.; Zhou, C.S. The relationship between economic growth, energy consumption, and CO_2 emissions: Empirical evidence from China. *Sci. Total Environ.* **2016**, *542*, 360–371. [CrossRef] [PubMed]
3. Tan, H. China Reports 6.9% Second-Quarter GDP Growth on Year, Topping Expectations. CNBC: 2017. Available online: https://www.cnbc.com/2017/07/16/china-q2-gdp-second-quarter-data-as-stability-reigns-ahead-of-party-congress.html (accessed on 20 December 2017).
4. Kolker, A.; Senior, C.L.; Quick, J.C. Mercury in coal and the impact of coal quality on mercury emissions from combustion systems. *Appl. Geochem.* **2006**, *21*, 1821–1836. [CrossRef]
5. Baba, A.; Gurdal, G.; Yucel, D.S. Enrichment of trace element concentrations in coal and its combustion residues and their potential environmental and human health impact: Can Coal Basin, NW Turkey as a case study. *Int. J. Environ. Technol. Manag.* **2016**, *19*, 455–480. [CrossRef]
6. Carlson, C.L.; Adriano, D.C. Environmental Impacts of Coal Combustion Residues. *J. Environ. Qual.* **1993**, *22*, 227–247. [CrossRef]
7. Le Quere, C.; Raupach, M.R.; Canadell, J.G.; Marland, G.; Bopp, L.; Ciais, P.; Conway, T.J.; Doney, S.C.; Feely, R.A.; Foster, P.; et al. Trends in the sources and sinks of carbon dioxide. *Nat. Geosci.* **2009**, *2*, 831–836. [CrossRef]
8. Maji, K.J.; Dikshit, A.K.; Arora, M.; Deshpande, A. Estimating premature mortality attributable to PM2.5 exposure and benefit of air pollution control policies in China for 2020. *Sci. Total Environ.* **2018**, *612*, 683–693. [CrossRef] [PubMed]
9. Xu, N.; Zhang, C. How Did China's Air Pollution Get This Bad? China Dialogue, 2013. Available online: https://www.chinadialogue.net/article/show/single/en/5604-How-did-China-s-air-pollution-get-this-bad- (accessed on 20 December 2017).
10. Zhi, G.R.; Zhang, Y.Y.; Sun, J.Z.; Cheng, M.M.; Dang, H.Y.; Liu, S.J.; Yang, J.C.; Zhang, Y.Z.; Xue, Z.G.; Li, S.Y.; et al. Village energy survey reveals missing rural raw coal in northern China: Significance in science and policy. *Environ. Pollut.* **2017**, *223*, 705–712. [CrossRef] [PubMed]
11. Zhang, A.P.; Zhong, L.S.; Xu, Y.; Wang, H.; Dang, L.J. Tourists' Perception of Haze Pollution and the Potential Impacts on Travel: Reshaping the Features of Tourism Seasonality in Beijing, China. *Sustainability* **2015**, *7*, 2397–2414. [CrossRef]
12. Jiang, L.; Hiltunen, E.; He, X.L.; Zhu, L.D. A Questionnaire Case Study to Investigate Public Awareness of Smog Pollution in China's Rural Areas. *Sustainability* **2016**, *8*, 1111. [CrossRef]
13. Lan, G.L.; Yuan, Z.K.; Maddock, J.E.; Cook, A.; Chu, Y.Y.; Pan, B.B.; Tu, H.; Fan, S.; Liao, X.; Lu, Y.N. Public perception of air pollution and health effects in Nanchang, China. *Air Qual. Atmos. Health* **2016**, *9*, 951–959. [CrossRef]
14. Yang, S.C.; Shi, L.Y. Public perception of smog: A case study in Ningbo City, China. *J. Air Waste Manag. Assoc.* **2017**, *67*, 219–230. [CrossRef] [PubMed]
15. Huang, L.; Rao, C.; van der Kuijp, T.J.; Bi, J.; Liu, Y. A comparison of individual exposure, perception, and acceptable levels of PM2.5 with air pollution policy objectives in China. *Environ. Res.* **2017**, *157*, 78–86. [CrossRef] [PubMed]
16. Qian, X.J.; Xu, G.Z.; Li, L.; Shen, Y.P.; He, T.F.; Liang, Y.J.; Yang, Z.Y.; Zhou, W.W.; Xu, J.Y. Knowledge and perceptions of air pollution in Ningbo, China. *BMC Public Health* **2016**, *16*, 1138. [CrossRef] [PubMed]
17. Sun, C.W.; Yuan, X.; Yao, X. Social acceptance towards the air pollution in China: Evidence from public's willingness to pay for smog mitigation. *Energy Policy* **2016**, *92*, 313–324. [CrossRef]

18. Wang, Y.T.; Sun, M.X.; Yang, X.C.; Yuan, X.L. Public awareness and willingness to pay for tackling smog pollution in China: A case study. *J. Clean. Prod.* **2016**, *112*, 1627–1634. [CrossRef]
19. Lu, H.; Yue, A.L.; Chen, H.; Long, R.Y. Could smog pollution lead to the migration of local skilled workers? Evidence from the Jing-Jin-Ji region in China. *Resour. Conserv. Recycl.* **2018**, *130*, 177–187. [CrossRef]
20. Tang, C.X.; Zhang, Y.C. Using discrete choice experiments to value preferences for air quality improvement: The case of curbing haze in urban China. *J. Environ. Plan. Manag.* **2016**, *59*, 1473–1494. [CrossRef]
21. Sun, C.W.; Yuan, X.; Xu, M.L. The public perceptions and willingness to pay: From the perspective of the smog crisis in China. *J. Clean. Prod.* **2016**, *112*, 1635–1644. [CrossRef]
22. Chu, Y.Y.; Liu, Y.S.; Lu, Y.N.; Yu, L.C.; Lu, H.S.; Guo, Y.L.; Liu, F.F.; Wu, Y.Y.; Mao, Z.F.; Ren, M.; et al. Propensity to Migrate and Willingness to Pay Related to Air Pollution among Different Populations in Wuhan, China. *Aerosol Air Qual. Res.* **2017**, *17*, 752–760. [CrossRef]
23. Wei, J.C.; Zhu, W.W.; Marinova, D.; Wang, F. Household adoption of smog protective behavior: A comparison between two Chinese cities. *J. Risk Res.* **2017**, *20*, 846–867. [CrossRef]
24. Ban, J.; Zhou, L.; Zhang, Y.; Anderson, G.B.; Li, T.T. The health policy implications of individual adaptive behavior responses to smog pollution in urban China. *Environ. Int.* **2017**, *106*, 144–152. [CrossRef] [PubMed]
25. Johnson, T.; Mol, A.P.J.; Zhang, L.; Yang, S. Living under the dome: Individual strategies against air pollution in Beijing. *Habitat Int.* **2017**, *59*, 110–117. [CrossRef]
26. Wu, X.; Qi, W.; Hu, X.; Zhang, S.S.; Zhao, D.T. Consumers' purchase intentions toward products against city smog: Exploring the influence of risk information processing. *Nat. Hazards* **2017**, *88*, 611–632. [CrossRef]
27. Huang, L.; Zhou, Y.; Han, Y.T.; Hammitt, J.K.; Bi, J.; Liu, Y. Effect of the Fukushima nuclear accident on the risk perception of residents near a nuclear power plant in China. *Prac. Natl. Acad. Sci. USA* **2013**, *110*, 19742–19747. [CrossRef] [PubMed]
28. Slovic, P. Perception of Risk. *Science* **1987**, *236*, 280–285. [CrossRef] [PubMed]
29. Cheng, P.; Wei, J.C.; Ge, Y. Who should be blamed? The attribution of responsibility for a city smog event in China. *Nat. Hazards* **2017**, *85*, 669–689. [CrossRef]
30. Wold, S.; Esbensen, K.; Geladi, P. Principal Component Analysis. *Chemometr. Intell. Lab. Syst.* **1987**, *2*, 37–52. [CrossRef]
31. Jolliffe, I.T.; Cadima, J. Principal component analysis: A review and recent developments. *Philos. Trans. R. Soc. A* **2016**, *374*. [CrossRef] [PubMed]
32. Virag, K. A Handbook of Statistical Analyses Using R. *Acta Sci. Math.* **2015**, *81*, 356–357.
33. Nadarajah, S. Simple R Programs for Statistical Distributions Most Commonly Used in Engineering. *Comput. Appl. Eng. Educ.* **2011**, *19*, 826–834. [CrossRef]

541944 Activation of Persulfates by Catalytic Nickel Nanoparticles Supported on N-doped Carbon Nanofibers for Degradation of Organic Pollutants in Water

Abstract:

Peroxydisulfate (PDS) and peroxymonosulfate (PMS) as the active superoxides have been extensively studied to generate highly reactive species for wastewater decontamination during the past decade. Since PDS is stable, soluble, cheap (<$2.65/kg for $Na_2S_2O_8$) and owns high redox potential (2.01 V), it is widely used as a source of radicals. A series of methods have been used to activate PDS, including UV irradiation, heat, ozone, quinone and transition metal catalysis. In this paper, an N-doped carbon nanofiber cloth (CC) with anchored nickel nanoparticles (Ni@N-CC) was synthesized from a facile pyrolysis process and employed as a catalyst to oxidize target contaminants using PDS as both radical precursors and electron acceptors. An effective strategy was developed to control the porous structures and catalytic performances by optimizing the precursor weights and pyrolysis temperatures for Ni@N-CC preparation. The optimal temperature is 700 °C, and the best dicyanodiamine mass is 1.0 g. Ni@N-CC was found to be superior for PDS activation to CC and nickel nanoparticles (NPs), ascribing to highly active sites, intimate connection between the nickel NPs and highly conductive N-doped CC, as well as the formed three-dimensional architecture. The oxidation rates were influenced by the oxidant loading (0.185–1.11mM), initial organics concentration (10–50mg/L), temperature (5–45°C), pH (2.65–10.47), and inorganic anions. Furthermore, mechanistic investigations using various probe reagents and spin trapping technique identified the generation of several active species for oxidation. The reaction was found to proceed via the electron transfer mediation from organics to PDS on N-doped CC and one electron reduction of PDS on Ni^0 NPs. This study highlights the design of highly active and reusable heterogeneous carbon/metal hybrids for more efficient PDS activation in environmental remediation.

530187 Synthesis and Characterization of Nitrogen-Dope $Sr_4Nb_2O_9$ by Mechanochemical Method

Abstract:

Perovskite-type oxynitride often exhibit intense colors and have potential for applications such as visible light driven photocatalyst. Conventionally, these oxynitride have been synthesized by the direct nitridation method which require very severe reaction conditions such as high temperature (900-1100â„ƒ) under ammonia flow for long time (e.g., 10 h). Nitrogen-doped $Sr_4Nb_2O_9$ is one of them, and Wu et al. reported its high oxygen generation efficiency for water splitting under visible light irradiation.

Therefore, in this study, we attempted to develop a more safer and simple synthesis method of perovskite type oxynitride such as nitrogen-doped $Sr_4Nb_2O_9$ by using the mechanochemical method. under various conditions and evaluated characteristics of the resulting samples. The mechanochemical method does not need to treat sample under ammonia high temperature under ammonia flow for long time to dope nitrogen, and instead, solid materials such as $CO(NH_2)_2$ can be utilized as nitrogen sources.

At first, $Sr_4Nb_2O_9$ was synthesized by a solid-state reaction method using $SrCO_3$ and Nb_2O_5 as raw materials at 1000â„ƒ for 9 h in air. Then, nitrogen-doped $Sr_4Nb_2O_9$ was synthesized by the mechanochemical treatment. In the experiment, a planetary ball mill (Frisch, P-7) was used to grind $Sr_4Nb_2O_9$ and $CO(NH_2)_2$ (15wt%) at a speed of 700 rpm for 180 min and then the obtained samples were heated at 400â„ƒ in air for 1 h to remove the unreacted $CO(NH_2)_2$. In order to determine the crystalline structure of the obtained sample, the X-ray diffraction pattern was measured. The result showed that the structures of both before and after the mechanochemical treated samples were $Sr_4Nb_2O_9$, however, the peak of the mechanochemical treated sample shifted to a higher angle by 0.1°than the original. In addition, the absorption spectrum was shifted to the visible light region. These results demonstrated that nitrogen-doped $Sr_4Nb_2O_9$ was successfully synthesized by using the mechanochemical method with much safer reaction condition than the conventional method.

530318 Effect of Power Frequency on Various Organic Compounds(VOCs) Decomposition Using Nonthermal Plasma Reactor Combined with Ceramic Filter

Abstract:

Recently, volatile Organic Compounds(VOCs) emissions have been controlled in various countries in the world. There are two major methods of VOCs processing methods such as incineration and adsorption. These methods are suitable for VOCs with high concentration levels and flow late, however, these are not suitable for medium- and small-scale emission sources. Therefore, the nonthermal plasma decomposition method has been drawing more attention these days as an alternative approach because of its advantages over conventional methods. In our previous work, we developed a new type electrode system for gas treatment in which nonthermal plasma and a ceramic filter were integrated and applied it for VOCs treatment. Configuration of the reactor is as follows; a copper wire electrode is wound in a coil form on the surface of a tubular porous ceramic filter as a discharge electrode. Another copper rod electrode which is insulated by a quartz tube is inserted inside the tube and is used as a counter electrode. The ceramic tube is made of alumina.

ã€€In this study, in order to investigate the effect of both power frequency and chemical structure of VOC on the decomposition efficiency, discharge decomposition experiment of various VOCs was carried out using our nonthermal plasma reactor. In the experiment, trichloroethylene (TCE), n-hexane, benzene, n-pentane, 1-hexene, 1-clorohexane were used as VOCs to be decomposed, and the power supply frequencies were 0.5, 1, 3, and 5 kHz to perform discharge decomposition. As a result, in the case of 1-hexene and TCE, very high decomposition efficiency was shown, and at the same time, the decomposition efficiency was significantly influenced by the power frequency. On the other hand, the decompositions of 1-clorohexane, n-hexane, n-pentane and benzene were influenced by the power frequency slightly, and the decomposition efficiency was remarkably lower than the case of 1-hexane and TCE. Furthermore, when the decomposition efficiency of n-hexane was compared to 1-clorohexane, which has almost the same structure as n-hexane and just replacing one hydrogen to chlorine, the efficiency in 1-clorohexane was higher at any frequency. In addition, when the decomposition efficiency was compared between n-hexane and n-pentane, no significant difference was observed. It indicates that only one carbon number difference does not affect discharge decomposition largely in the range of this study.

532993 Carbon Capture and Utilization Using Metal Cation in Seawater-Based Wastewater

Abstract:

Many nations in the world are concerning the climate change caused by greenhouse gases emission. There are some types of greenhouse gases and the level of contribution for each gas is all different. However, for carbon dioxide, the global warming potential (GWP) is the lowest but it contribute to the climate change quite seriously due to the amount of its emission. Also, carbon dioxide is produced in the fossil fuel combustions and the amount of fossil fuel usage is gradually increasing. Many industrial processes and the powerplants are operated based on the fossil fuel and the carbon dioxide should be captured before they are emitted to the atmosphere. In order to prevent carbon dioxide emission, the CCS (Carbon Capture and Storage) technology was introduced. The main concept of the technology is to separate the carbon dioxide contained in the flue gas and store the carbon dioxide in the safe palce such as deep ocean or the underground. However, the places which was thought to be safe can be changed to the most unstable place due to crust activities and the ocean ecosystems may affected when carbon dioxide is injected in the ocean. With this drawbacks of CCS technology, the new and novel technology that is called CCU (Carbon Capture and Utilization) have got much attention. In ths CCU technology, the carbon dioxide can be converted into useful materials that is organic or inorganic. In this research, inorganic carbon dioxide utilization was dealt with. Carbon dioxide was captured using liquid absorbent comprised of alkanolamine type absorbents. After its separation or capture, industrial seawater containing high concentrations of metal ions was added to the saturated absorbents. When carbon dioxide is dissolved at the amine type absorbents, they are converted to ionic state such as carbonate, bicarbonate or carbamate. Since the reactivity of ionic substances are extremely higher than any other substances, metal carbonate salt which can be used for various purposes can be produced. The merits of this indirect aqueous carbon fixation is that they consume less energy compared to the direct carbonation reaction. The carbonation reaction was taken place under ambient pressure and room temperature. But for direct carbonation, the carbonation reaction can be occurred under high pressure and temperature. As a result of carbon fixation, calcium carbonate salt was produced. In order to investigate the physical properties of the product, XRD (X-ray diffractogram) and SEM (Scanning Electron Microscopy) were used. The final products can be used for construction materials, additives or fillers, and pharmaceutical purpoes and so on.

537818 Sustainability Assessment By Emergy Approach: Gold Mining Extraction in Colombia

Abstract:

Due of its great climate diversity, Colombia has an abundance of natural resources and possesses a wide range of mining products, including coal, gold, platinum, nickel, emerald, limestone, among others that are extracted at a smaller scale. This extraction is carried out in two ways: a formal largescale extraction and other smallerscale, traditional and artisanal; the latter lacks adequate and defined technology, usually in some cases it is considered an illegal activity, making it unsafe, unprofitable, uncompetitive and environmentally unsustainable.

In the last decade, there has been a significant increase in both the volume and value of production of primary goods, particularly in the mining and hydrocarbons sector, which has positively and significantly affected the country's gross domestic product (GDP) (FEDESARROLLO, 2012), currently being one of the most dynamic of the Colombian economy. The GDP from this sector has grown from 5.2 to 10.3 trillion pesos between 2000 and 2011 (Colombian Ombudsman, 2010), although in the last two years (2013 2014) this activity has gone into recess.

Thus, according to the largescale mining sector (LSM), in 2011 mining represented 24.2% of exports; 2.4% of GDP; 20% of the total foreign direct investment; 650 billion pesos in construction of infrastructure; 2.6 trillion pesos in purchases from domestic suppliers, 65 billion pesos of investment in social responsibility and 178 billion pesos in environmental responsibility (LSM, 2012). This constitutes evidence for the mining sector in Colombia to be considered one of the main economic engines, since it does not only contribute with employment creation (there were 836000 in 2011, between direct and indirect jobs), but also with investments in infrastructure, public utilities, and social and environmental management (LSM, 2012). In royalties, the mining sector contribution to the country is 16.5%; 13% of it comes from coal companies (1.2 trillion pesos a year).

This constitutes evidence for the mining sector in Colombia to be considered one of the main economic engines, since it does not only contribute with employment creation (there were 836000 in 2011, between direct and indirect jobs), but also with investments in infrastructure, public utilities, and social and environmental management (LSM, 2012). However, it is impossible to avoid discrepancies and political, economic, environmental and social polemics around this productive activity in the country, since mishandled mining practices lead to: social impacts, since many of these unformalized activities allow child labor, make use of intensive techniques in unskilled labor, generating poor quality jobs with low levels of industrial, social, and health security for miners, it is usually developed in remote and deprived areas; furthermore, it generates conflicts between large mining industries due to concession contracts awarding. Regarding political and economic impacts, a minimum working capital and scarce financial resources for investment are evidenced, and therefore an improper handling of inputs and royalties from this activity is presented. In a greater relevance, there are the irreversible environmental impacts such as destruction of ecosystems, damage to protected areas, air pollution, and *pollution and impact on the availability of water resources.* Globally, water demand is driven primarily by the population and economic growth, and it is anticipated that in the coming decades, it will increase considerably, while supplies will remain almost constant or decreasing by factors such as climate change, over pumping and pollution of aquifers (Drelich, 2012). It is not unknown that the primary input in gold extraction is water; the water that is used in this sector comes from a variety of sources, and sources and quality are the main areas of controversy and debate in this productive sector. *The main problem of the mining industry is to generate confidence in the development of a responsible, sustainable and transparent strategy in the use of water* (Drelich, 2012).

Undoubtedly, the flow of resources from this activity is increasing and projections show that the growing trend will continue, since the international scene shows an important dynamism of the mining market. This situation could be exploited by Colombia, given its geographical location and its geological potential (FEDESARROLLO, 2012); it is estimated that US$250 billion will be invested in mining projects in Latin America between now and 2020. Thus, the mining sector gains importance among public policies, not only due to the high revenues generated by its operation, but also because of the social and environmental impact of the sector activity. Hence the importance of seeking for solutions, methodologies and sustainable alternatives to assimilate the challenge of turning the mineral wealth in an opportunity for development, and thus respond to future generations about how resources from nonrenewable assets were invested without affecting renewable resources.

This highlights the fact that, despite the importance that the mining sector has taken, the debate about the true benefits and costs, or the environmental, economic, energy and social problems of the sector's activity has intensified. It is misleading to decide whether mining offers a significant potential or a threat to society without a methodological tool that allows predicting how sustainable is this productive activity, which represents a profitable business for many, which provides social and economic welfare (ultimate goal of the implementation of development projects of this magnitude), and for other sectors of the society symbolizes the opposite, considering as ineffective and undervalued all corrective and preventive measures that have been implemented by the government agencies or those regulatory entities that have some kind of control in the field of sustainable management of the processes of extraction and use of nonrenewable resources, including products that are generated in such activity, in response to the environmental awareness that has been growing in recent years.

Based on the concept of sustainability as "the compatibility between energy, economics (maximum performance) and environmental aspects" (Redclif, 1987), all development projects especially those that threaten the environmental integrity such as the exploitation of natural resources and mining processes, should be focused on being an economic use alternative that provides an energy yield with acceptable environmental burden. That is, this highly demanding production activity in the country (including byproducts generated in it) should be given a particular management by means of methodological tools to harmonize the economic with the energy and environmental aspects; since these three factors cannot be assessed independently, but they must be analyzed together and in an interrelated way, in order to give management on a sustainable basis, which avoids the discrepancies noted above. In addition to these tools, it is possible to identify where the weaknesses lie, which apparently make this activity a threat to some sectors of the society.

It is then necessary to establish, among the many valuation methodologies available, those that allow a sustainable assessment of the current management and mining process in Colombia in both artisanal and large-scale mining; likewise identify the gaps in it, whose purpose is to make mining a lever for development not only with good labor, industrial and environmental safety practices, but also going further in its commitment to the country and the city that hosts it. That is, for the management of mining resources, a valuation method should be used thereof, which is bearable and balanced from the economic, energy and environmental levels, as in the current regulations governing this sector a large imbalance is evident between these aspects. That is to say, in many situations mining is presented with a great economic benefit, but it entails irreversible environmental damage or major social problems that trigger collective chaos; or on the contrary, many mining licenses are denied impeding economic growth because of an environmental impact that may be acceptable, discrepancies caused by no uniformity of a regulatory framework for the mining sector and the lack of methodological tools to assess the sustainability of different productive sectors of society, which and/or balance economic, energy and environmental aspects.

Globally, the possibility of quantifying the sustainability of different production sectors (agriculture, energy, legislative, sanitation) has been subject of study using the method of emergy which allow not only to quantify parameters, but also to analyze the implications and interrelationships of economic and environmental aspects of these long-term processes.Thus, this research aims to develop a methodological tool; emergy analysis to assess and compare the sustainability of the current small and large scale mining systems in Colombia, including byproducts generated in this process, emphasizing on the use of water resources, gaps in these systems which may not make them sustainable, and a uniform methodology that can be used as a road map to regulate that productive sector in a sustainable way, because without doubt, regulation constitutes the source of many of the limitations of the mining sector

532994 Investigation on CO_2 Capture and Utilization Using Simulated Bio-Gas and Extracted Metal Cations

Abstract:

Since Industrial Revolution in 18th century, the amount of energy usage by human-being have been drastically increased. The energy source that mostly used was fossil fuel. When fossil fuel is combusted, carbon dioxide is produced as a product. When carbon dioxide is emitted into the atmosphere in large amount, it may absorb the radiation energy emitted towards space from earth and reflect it towards the surface of the earth. In fact, there are six representative types of greenhouse gases (GHGs) including carbon dioxide (CO2), methane (CH4), nitrous oxides (NOx), hydrofluorocarbons (HFCs), perfluorocarbons (PFCs), and sulfur hexafluoride (SF6). Among them, the global warming potential (GWP) of carbon dioxide is minimum, however, considering the portion in the total emission of GHGs, carbon dioxide is thought to be the main cause of greenhouse effects and climate change, furthermore. Hence, the technology to reduce carbon dioxide have been studied over a decade. One of the problem in common carbon reduction technology is that there are not enough spaces for its storage. So, the technology to utilize carbon dioxide after its capture had come arise. These technologies are commonly called Carbon Capture and Utilization (CCU). One of the method in this CCU technology is to convert carbon dioxide in to metal carbonate salt that can be used as raw materials in various industrial fields such as cement manufacturing industries, paper making industries and so on. The main research topic regarding this inorganic CCU is to secure metal cation supplying source in order to convert carbon dioxide into metal carbonate precipitate. The metal cation can be obtained from various sources: natural minerals, industrial wastes and seawater and so on. In this research, carbon dioxide is captured using alkanolamine based absorbents followed by carbon fixation using metal cation extracted from desulfurization ash produced from coal-fired powerplants. It is known that the main constituent of desulfurization ash is calcium sulfate. In common coal-fired powerplant which supplies electricity to metropolitan area, very large amount of desulfurization ash is produced every day and most of the facilities are suffering from the treatment of the ash. Using the calcium component contained in the desulfurization ash, calcium cation can be extracted using standardized method. As a result of the experiments, calcium carbonate precipitate was produced as final product. In order to investigate the type of substances, X-Ray Diffraction (XRD) analysis was performed. Also, to support its data, Scanning Electron Microscope (SEM) images were also obtained. Throughout the XRD and SEM analysis result, it is proven that the calcium carbonate in the crystal structure of calcite was dominant. Also, the absorption and conversion characteristics were demonstrated by suggesting carbon dioxide loading curve during absorption stage. The result and data obtained from this research may useful to process designing using industrial waste.

534110 Selective Lithium Recovery from Brine Using $Li_{1-x}Ni_{0.5}Mn_{1.5}O_4$ /Ag Battery System

Abstract:

Lithium shortage is inevitable given the massive growth on its demand for use in electric vehicles (EVs) and various technologies. Consequently, recovery of lithium (Li) from other sources such as seawater, brine, and Li containing wastewater has been recently gaining attention in energy related fields. Discussions are directed towards exploitation of these aqueous sources to potentially alleviate the impending Li supply shortage. Yet, the most promising aqueous source for economical Li extraction are brine-like solutions. However, the challenge is to find an environmentally benign and more energy-efficient processes with fast Li^+ production rate for sustainable Li^+ supply. Electrochemical Li recovery from aqueous solutions is an attractive method as it provides fast recovery rate. However, several studies on this process are challenged with the stability of the materials in aqueous environment and large amount of energy needed entailing high production cost. Until recently, electrochemical related Li extraction has yet to realize its feasibility for industrial-scale application. The success of electrochemical method relies on utilizing highly effective electrodes that can selectively capture Li^+ at a fast rate and at a competitive uptake capacity with minimal energy requirement.

In this study, $LiNi_{0.5}Mn_{1.5}O_4$ (LNMO) was investigated for its feasibility as a selective Li^+ capturing electrode for electrochemical Li recovery in brine solution. The LNMO is paired with Ag as a chloride capturing counter electrode. LNMO was first oxidized (delithiated) to facilitate its Li preference during capture phase. By applying a negative current, the delithiated $Li_{1-x}Ni_{0.5}Mn_{1.5}O_4$ or (NMO) paired with Ag (NMO/Ag) was able to capture Li^+ (i.e. discharge mode). The electrochemical system was proved to be stable in aqueous environment. The captured ions were then released by reversal of the polarity of the current (charging) in a separate cell containing a recovery electrolyte solution. The operating conditions were optimized through response surface methodology utilizing the central composite design. The determined optimum conditions were used for the repetitive Li capture-release process. Cycling (discharge-charge) experiments and cyclic voltammetry showed high selectivity and stability of the electrode in the presence of other competing cations (Na^+, K^+, Ca^{2+}, Mg^{2+}). Overall, the results demonstrate the feasibility of the electrochemical system (NMO/Ag) for high-throughput selective Li^+ mining process with low energy requirement.

This research was supported by the National Research Foundation of Korea (NRF) under the Ministry of Science and ICT (No. 2017R1A2B2002109) and the Ministry of Education (No. 22A20130012051(BK21Plus).

537141 Environmental Impacts of Pavement Rejuvenators

Abstract:

The research conducted in this project investigated the potential environmental impacts of three types of pavement rejuvenators: asphalt-based, bio-based, and coal tar-based. Pavement rejuvenators are applied to pavement surfaces to â€œrevitalizeâ€ old asphalt pavements and are thereby expected to extend their useful life. Applying rejuvenators may allow municipalities to maintain roadway pavements at a lower cost than completely renovating (removing and repaving) roadways. However, there is little published research on the extent contaminants may be released from rejuvenated pavement materials and as a result, little is known about the impacts of rejuvenators to the environment. This project looked into the environmental concerns of three types of pavement rejuvenators and analyzed the chemicals present to determine any potentially harmful environmental impacts. The three rejuvenators evaluated were an asphalt-based rejuvenator, a coal tar-based rejuvenator and a bio-based rejuvenator.

To conduct this evaluation, a PaveTesting Model Mobile Load Simulator (MMLS) was used to simulate surface challenges that roadway pavements are expected to withstand. Pavement samples coated with the different types of rejuvenators were subjected to testing in the MMLS to simulate roads being driven on through repeated tire-pavement contacts. The MMLS simulation dislodged particulate matter that was subsequently chemically analyzed through gas chromotography analysis. Concentrations of PAHs in each sample were determined.

To assess the environmental impacts, particulate matter detached from pavements treated with each type of rejuvenator was collected, extracted with methylene chloride, and injected into a gas chromatograph with a flame ionization detector (Agilent Technologies) to quantify polycyclic aromatic hydrocarbons (PAHs) present in the particulate matter. Many PAHs are a concern to public health and aquatic life, with some considered carcinogens. With significant quantities of PAHs present in this particulate matter detached during simulation testing, it can be expected that they will be present in particulate matter released from rejuvenated pavements in actual application. The mobilization of this particulate matter could cause problems in the environment because of the presence of PAHs.

Results indicated that the pavement samples treated with coal tar-based rejuvenator produced the least amount of particulate material during MMLS testing (0.2458 g), yet that particulate material had the greatest PAH concentrations, 3,440 ± 380 mg total PAH/kg sample. While the coal tar-based rejuvenator performed best with respect to limiting the mass of particulate that was mobilized from the car tire abrasion, the total amount of PAH mobilized from the pavement was the greatest. The pavement samples treated with asphalt-based rejuvenator resulted in the greatest mass of particulate matter released, 0.8315 g, and the total PAH concentration of that material was found to be 364 ± 26 mg PAH/kg sample. The pavement samples treated with the bio-based rejuvenator contained 582 ± 57 mg total PAH/kg sample, and resulted in 0.5559 g of detached material.
The use of coal tar-based rejuvenator is of concern due to its potential for release of PAH contaminants through detached particulate material. The bio-based rejuvenator appear to have promise due to the low potential for PAH mobilization. It is recommended that additional research be undertaken to evaluate the performance of rejuvenators in extending the useful life of asphalt pavements. Of particular interest is further evaluation of the environmental impact of these products.

537292 Microencapsulated Fluorescent Gold for Ppb-Level Chromium(VI) Sensing

Abstract:

Fluorescent gold nanoclusters (Au NC) are a promising probe material for selective chemical sensing. However, low fluorescence intensity and an incomplete understanding of the mechanistic origin of the fluorescence limit their practical implementation. We induced glutathione-capped Au NCs to aggregate within silica-coated microcapsular structures using polymer-salt aggregate (PSA) self-assembly chemistry. The encapsulated NCs can detect Cr(VI) at concentrations as low as 5.8 ppb (= 0.05 ÂµM CrO_4^{2-}) through fluorescence quenching, compared to free Au NCs which have a QY of ~15% and a limit of detection (LOD) of 116 ppb (= 1 Î¼M CrO_4^{2-}). This LOD is 15Ã— lower than the US EPA maximum contaminant level for total chromium (Cr(III) and Cr(VI), 100 ppb) in drinking water. No pH adjustment is needed using the encapsulated Au NCs, unlike the case for free Au NCs. This material may serve as a basis for a practical Cr(VI)-sensitive sensor at parts-per-billion concentration levels.

535745 Benzene Methylation Catalysed By Hierarchically Porous Zeolite: An Effective Way to Promote Xylene Selectivity and Catalyst Lifetime for Large Scale Commercial Use

By Author

Abstract:

Benzene Methylation Catalysed by Hierarchically Porous Zeolite: An Effective Way to Promote Xylene Selectivity and Catalyst Lifetime for Large Scale Commercial Use

Transformations of aromatic hydrocarbons represent one of the most important groups of reactions in petrochemical process. In this respect, methylation of benzene is a promising alternative to produce xylenes, by converting the carcinogenic benzene which is almost restricted for its use in gasoline. Also, for the purpose of our study, benzene makes monitoring of this alkylation reaction over zeolitic catalysts easier compared to other aromatics. In this work, a combined dry-gel-conversion and steam-assisted-crystallization method is used to synthesize TNU-9 zeolites with varied crystal size, which in turn show different accessibility of active sites and confinement effect. N_2 physisorption, NH_3- TPD, SEM and FT-IR are employed to characterize the texture and acidity of both hierarchically porous zeolites and their conventional counterparts. Catalytic properties are test on benzene methylation at 460 °C to investigate the effect of additional porosity. It is found that hierarchical porosity markedly enhances the utility of zeolite and selectivity towards xylenes via improved mass transport. Under optimized hierarchical TNU-9 catalyst, xylene selectivity reaches 36.9 %, a comparable result than that of ZSM-5, the best catalyst tested for this reaction to date.

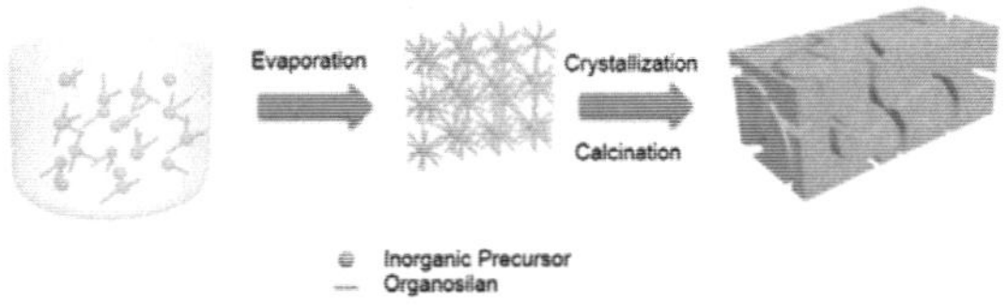

Fig.1 Crystallization mechanism on hierarchical porous ZSM-5 in this work

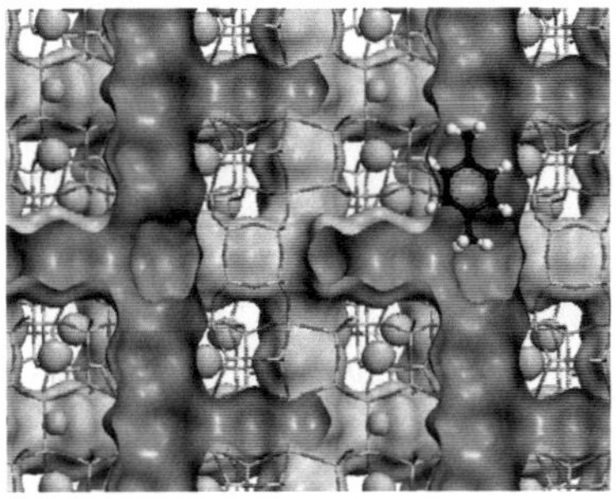

Fig.2 TUN topology and para-xylene

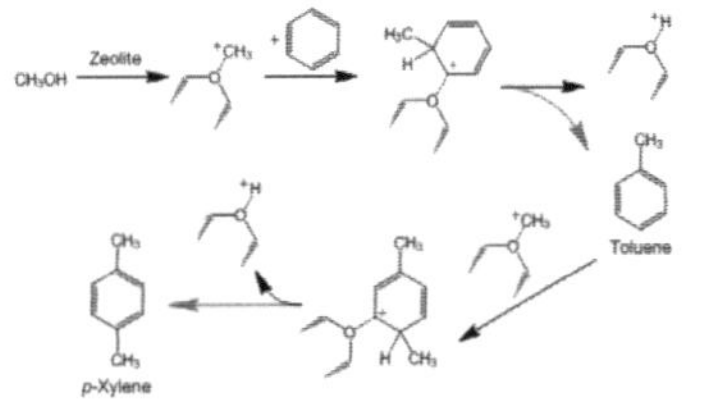

Fig.3 Mechanism on benzene alkylation with methanol

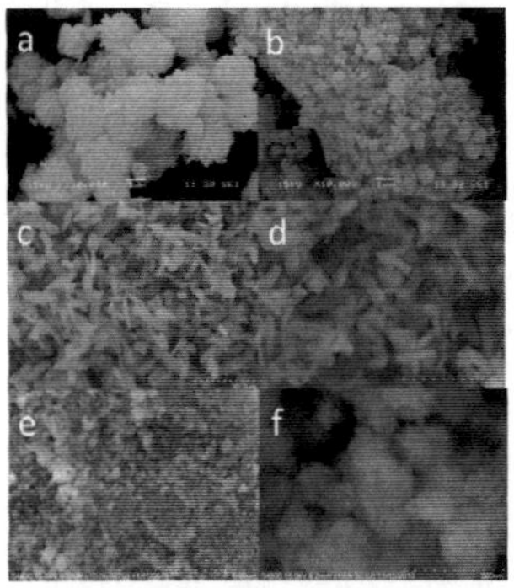

Fig.4 Diffusion of reactants and products inside zeolite crystal

THE GRAPHIC ON THIS PAGE WAS NOT AVAILABLE ON THE E-MEDIA VERSION OF THIS PAPER.

Fig.5 SEM image of TNU-9 samples applied in this work

530135 Catalytic Removal of Polybrominated Diphenyl Ethers (PBDEs) in Effluent Gas from Thermal Desorption Treated Soils

Abstract:

Over the past three decades, polybrominated diphenyl ethers (PBDEs) have been widely used as flame retardants in various customer products, such as electronics, electronic appliances, furniture, and textile [1, 2]. Generally, PBDEs are simply blended with the materials, and thus are likely to leak out during the production, usage and disposal of the products [3]. Due to the high hydrophobicity, persistence and bioaccumulation, PBDEs released from the products have caused serious environmental problems and threats to human health.

Up to now, several methods such as photochemical method, biological method, reduction by zero-valent iron and thermal desorption have been applied to degrade PBDEs [4-6]. Among them, thermal desorption was the preferred remediation method for PBDEs-contaminated soils due to its high removal efficiency and stable reaction process. However, the effluent-gas containing PBDEs were usually collected by activated carbon, which has several drawbacks, such as need of secondary treatment, complex operation and high cost. Therefore, it is important to develop an effective and feasible method for eliminating gaseous PBDEs rapidly and completely. Catalytic degradation was considered as a promising method for PBDEs removal due to its high treatment efficiency, low price and easy to operate.

In this study, a plate-type Î³-Al_2O_3/Al monolithic support was prepared through anodization technology [7, 8], and its catalytic performance for removal of gaseous PBDEs was studied. It was found that the structured support could not remove PBDEs effectively. However, Fe/Î³-Al_2O_3/Al catalyst was proved to have good activity for gaseous PBDEs elimination. As such, a series of Fe-based/Î³-Al_2O_3/Al catalysts with different iron loadings were further prepared by impregnation method, and the effects of Fe loading on the activity of catalysts were extensively investigated. The results show that the proper amount of iron loading could enhance the catalytic performance, and thus remove PBDEs significantly. In addition, based on the analysis of intermediate products, the degradation pathway and removal mechanism of PBDEs over Fe-based catalyst were speculated.

References

[1] L. Tan, S.Y. Lu, Z.Q. Fang *et al.*, Appl. Catal., B 200 (2017) 200-210.

[2] D.M. Orihel, T. Bisbicos, C.T.R. Darling *et al.*, Environ. Toxicol. Chem., 35 (2016) 573-583.

[3] Y.H. Peng, M.K. Chen, Y.H. Shih, J. Hazard. Mater., 260 (2013) 844-850.

[4] K.L. Chow, Y.B. Man, N.F.Y. Tam *et al.*, J. Hazard. Mater., 322 (2017) 263-269.

[5] J.H. He, K.R. Robrock, L. Alvarez-Cohen, Environ. Sci. Technol., 40 (2006) 4429-4434.

[6] F. Yang, Q.Q. Li, G.J. Su *et al.*, Chemosphere, 150 (2016) 445-452.

[7] F.Y. Fan, Q. Zhang, X. Wang *et al.*, Fuel, 186 (2016) 11-19.

[8] F.Y. Fan, Q. Zhang, H. Hou, Ind. Eng. Chem. Res., 57 (2018) 2426-2433.

533249 Fluoride Removal By Geopolymeric Adsorbent Synthesized from LD Slag

Abstract:

This research aims at investigating the utilization of steel plant LD (Linz-Donawitz) slag as a potential raw material for synthesis of LD slag based geopolymer (LDSGP) adsorbent to remove toxic fluoride ions from artificial wastewater. Steel-making LD slag, a solid waste of steel industries, is generated when the molten steel is separated from impurities, in LD (Linz-Donawitz) furnaces. It consists of mainly calcium oxide and iron oxide along with silica etc. Raw LD slag has been mechanically mixed with an alkaline activator (10M NaOH + sodium silicate (1:1 w/w)) at weight ratio of 2:1 and then cured for 3 days at 25^0 C to prepare LD slag based porous aluminosilicate geopolymeric adsorbent. The physical and chemical properties of new LD slag geopolymeric adsorbent has been characterized using BET, FTIR, XRD, SEM & TEM etc. which show an increase of BET surface area (from 4.85 to 30.84 m2/g) and formation of Ca_2SiO_4 and Ca_3SiO_5 in the aluminosilicate geopolymeric matrix. The micro-petal like morphology of LDSGP surface has been obtained. Batch adsorption experiments of fluoride ions onto LDSGP adsorbent has been carried out to investigate the effect of operating parameter i.e. different adsorbent dose, pH of the solution, initial fluoride concentration at different temperature. Isotherm study at equilibrium shows best fit to Freudlich model and the maximum experimental adsorption uptake has been achieved as 50.6 mg/g. Adsorption kinetic data fitted with pseudo-second-order model suggesting the nature of the adsorbent to be chemisorption. Thermodynamic study proves the spontaneous nature of the fluoride adsorption process onto LDSGP.

535070 Identification and Enzymatic Characterization of a Novel NADH Dependent Azoreductase, Encoded By *Azok* in *Klebsiella Pneumoniae*

Abstract:

Sulphonated azo dyes, known xenobiotic, are extensively used in textile, plastic, food, pharmaceutical and cosmetic products. These are highly carcinogenic and mutagenic to humans and their discharge into the surface water adversely affects the ecosystem. The microbial degradation of azo dyes involves the reductive cleavage of azo bond which is catalyzed by the group of NADH or NADPH dependent enzymes called azoreductases, few of which are flavin dependent. In this study, we have identified an azoreductase enzyme which is dependent on NADH for its catalytic activity. Azoreductase was encoded by *AzoK* from the previously isolated *Klebsiella pneumoniae* C1 strain. We have cloned *AzoK* gene from *Klebsiella pneumoniae* C1 strain and was heterologously expressed in *E. coli*. The azoreductase sequence encoded a functional protein of 201 amino acids with a molecular mass of approximately 23 kDa. The expressed azoreductase had 99% protein sequence similarity with the azoreductase of *Klebsiella pneumoniae subsp. pneumoniae* MGH 78578. The protein had only one amino acid change, serine replaced by glycine, at the 176th position. The recombinant azoreductase required NADH as electron donor for utilizing methyl orange as a substrate. The purification and characterization of recombinant enzyme for its optimal temperature, pH and thermostability is in progress. The cell fractionation method was used to determine the location of azoreductase within the periplasmic fraction of the cells and, thus, involved in the extracellular reduction pathway of azo dyes. Some amount was also present in the cytoplasmic fraction. The enzyme from recombinant strain could successfully decolorize the model azo dye, methyl orange within 12 h in an extracellular environment through a specific metabolic pathway. The decolorization efficiency of the enzyme was observed to increase in the presence of NADH and decolorized methyl orange within 4 h. Thus, the viability of the recombinant *AzoK* for the abatement of azo dyes from industrial wastewater was established. It was noted that *AzoK* represents the first azoreductase gene to be identified and cloned from *K. pneumoniae* and expressed in *E. coli*.

538194 Multi-Stage Hydrothermal Processing of Algae for Enhancing Biocrude Quality and Denitrogenation

Abstract:

Hydrothermal processing of algae into liquid hydrocarbon fuels has received considerable attention in the past few years due to the inherent ability of the process to convert wet biomass without energy intensive pretreatments such as drying. The present study investigates a two-step hydrothermal conversion of algae: a low temperature hydrothermal liquefaction (HTL) process for removal of N and O from algae for subsequent conversion by a second stage HTL at more severe conditions. The specific objectives under the above broad goals were to evaluate a first-stage (low temperature) HTL process at 175-250 ËšC and evaluate the recycling of nitrogen and phosphorous in the aqueous phase co-products (ACP) for growth of mixed consortia diatom and cyanobacteria algae. First-stage HTL experiments were conducted using a custom designed two-chamber reactor (TCR) system at different conditions of reaction temperature (175, 200, 225, and 250 ËšC) and time (1, 2.5, and 5 min). Treated solids and aqueous co-products (ACP) were analyzed by standard laboratory methods including mass yields, elemental C, H, N, and O contents, higher heating value (HHV), removal of N and O from the solids, total nitrogen, ammonia nitrogen, total phosphorous, total organic carbon, pH, and non-volatile residues (NVR) for the aqueous co-products. Statistical analyses were performed to evaluate solid yields in the first-stage HTL experiments. Solid yields were higher at less severe reaction conditions (lower temperature and shorter time) than at more severe conditions (higher temperature and longer time). The highest solid yield (~60%) was obtained at 170 ËšC and 1 min; the lowest solid yield (~35%) was obtained at 250 ËšC and 5 min HTL conditions. Recycling of ACP and further growth of algae was tested in batch cultivation of diatom-cyanobacteria mixed consortia algae and compared with growth using WC standard medium. Results showed that use of ACP medium, combined with phosphorous-depleted WC medium, provided significant growth.

531422 $Ca_xM_yO_z$ Solid Sorbents for CO_2Â Capture: An in Situ X-Ray Diffraction Study

Abstract:

Global climate change presents a constant challenge to humankind. The release of CO_2 to the environment, the main cause of climate change, is greatly exacerbated by industrial activity. One way to mitigate this problem is post-combustion CO_2 capture using solid sorbents. Calcium oxide-based materials, an economically favorable option due to its abundance and high adsorption capacity, suffers from significant long-term surface area loss during the carbonation/regeneration cycles. A simple solution is to introduce inert second phase by mechanical mixing. This will minimize the sintering of the calcium containing particles by limiting direct contact among CaO or $CaCO_3$ particles. In this research, different kind of metals, such as Fe, Co, Cu, are used to synthesize single-phase calcium metal oxide in form of $Ca_xM_yO_z$, then this solid precursor will be decomposed to CaO and the derivatives of metal oxides by reduction under hydrogen. The former is used as sorbent for CO_2, and the latter is a support to resist sintering and facilitate heat transfer. The temperature dependent carbonation/regeneration kinetics and CO_2 capture capacity change over cycling are investigated by collecting in situ X-ray diffraction pattern. Phase quantification of the CaO-based sorbents and their conversion to $CaCO_3$ will be carried out by Rietveld refinement.

532754 Effect of Aerosol on MEA Slip in Capturing Carbon Dioxide

Abstract:

Large-scale deployment of post-combustion CO2 chemical capture, absorption by chemical solvents have been demonstrated to be possible. One of the remaining concerns is the leakage of amines, which may result in the damage of human health and environment.

Literature revealed that amine slip increases in the presence of aerosol particles in the flue gas. However, systematic investigation of the mechanism is still absence. In this research, aerosol was generated by an generator TSI-3076. A particle size analyzer TSI model 3330 was used to analyze the particle size distribution and particle counts in the aerosol created. The FTIR instrument Gasmet DX-4000 was calibrated to measure MEA concentrations in the outlet gas stream.

If no aerosol was introduced into the inlet gas of Packed Bed, Aspen plus shows that the concentration of MEA in the slip stream is about 37 ppm. This value is out of the range of the Gasmet, so analysis shows that there is no MEA in the slip stream. It was found that the average size of the particle generated is about 0.3 mm and the particle count is about 10^5~10^6 #/cm^3. With the introduction of sulphuric acid aerosol, the concentration of MEA in the slip stream is about 1100~1400 ppm, whereas the particle will grow to about 1.2 mm and the particle counts was not changed much. If the sucrose aerosol was introduced into the inlet gas, the concentration of MEA would reduce to about 400~750 ppm. The result indicates that the property of nuclei can affect the MEA emission.

If the sulfuric acid aerosol was introduced into the Rotating Packed Bed, the concentration of MEA is about 10~200 ppm. It shows that RPB has lower MEA emission than normal absorber.

Acknowledgement: This work is financially supported by National Energy Program-Phase II: MOST 107-3113-E-007-002

532912 Kinetics and Mass Transfer Performance of CO_2 Absorption into DEEA/MAE Solution

Abstract:

Post-combustion carbon dioxide capture by amine scrubbing has been considered as one of the promising approaches for potential reduction of CO_2 emissions. However, amine based CO_2 absorption process encounters a considerable amount of energy requirement, which is significantly associated with the amine chemical structures and process configurations. Blended amine solutions with mixing carbamate formers (i.e. primary/secondary amines) as an activators and preferential bicarbonate producers (tertiary amines) possibly reduce the energy requirements for regeneration.

A tertiary amine, N,N-diethylethanolamine (DEEA), has received significant interest by various researchers and can be prepared from renewable and/or cheap resources. Blends of DEEA having characteristics of tertiary amine and absorption promoter is now intensively studied. For design and application of this CO_2 separation process the knowledge of vaporâ€"liquid equilibrium data, CO_2 cyclic capacities and mass transfer performance of the DEEA blends system is required. The objective of the present work described here is to investigate the interaction between DEEA and secondary amine 2-(methylamino)ethanol (MAE). Then the reaction kinetics using stopped-flow technique, the optimal ratio of DEEA and MAE using a developed solvent screening method, and the mass transfer performance in a hollow fiber membrane contactor were studied in the present work. Furthermore, both the models for reaction rate constants and overall gas phase mass transfer coefficient were proposed and discussed in this work.

The hybrid behavior of CO_2 absorption into blended DEEA/MAE was investigated by evaluating the experimental data of equilibrium solubility of CO_2, CO_2 cyclic capacity, reaction kinetics and mass transfer performance. The height of absorber column is determined by the overall volumetric gas-phase mass transfer coefficient (K_Ga_v), which will be affected by the CO_2 cyclic capacity. Furthermore, the CO_2 cyclic capacity also can influence the energy requirement of solvent regeneration in turn affects the CO_2 capture cost. The different length of reaction time in the screening experiments was regarded as the various heights of the absorber and stripper, and the CO_2 cyclic capacities of different lengths of reaction time were calculated.

Based on the investigations in the present work, it can be concluded that:

(i) The CO_2 absorption kinetics of newly formulated DEEA/MAE was determined using stopped-flow apparatus. In addition, it was found that the reaction rate constant values can be predicted through the proposed mechanism and model with an acceptable AARD. By comparing with the individual amines, the new amine blends were found to have much faster reaction kinetics than that of DEEA, indicating an interaction between secondary and tertiary amines.
(ii) An improved rapid absorbent screening method was applied to investigate the CO_2 absorption rate, regeneration rate and CO_2 cyclic capacity. Experimental results reports that the introduction of the rate activator can significantly enhance the absorption rate compared with 2M DEEA solution, and the addition of DEEA into the amine blend can obviously improve the regeneration performance of the MAE solution, resulting in the highest CO_2 cyclic capacity.

(iii) Furthermore, the mass transfer performance in terms of CO_2 flux in a hollow fiber membrane contactor were studied to verify the accuracy of the improved rapid absorbent screening method. The values of CO_2 flux of 1.0M DEEA+1.0M MAE solution show more than two times higher than that of DEEA solution, and increase with the increasing liquid velocity and liquid feed temperature. These valuable results indicate an interaction effect are happened during the CO_2 absorption into the amine blends, leading to the potential CO_2 capture absorbent (i.e. DEEA/MAE).

References

[1] Fu, K., Rongwong, W., Liang, Z., Na, Y., Idem, R., Tontiwachwuthikul, P., 2015. Experimental analyses of mass transfer and heat transfer of post-combustion CO_2 absorption using hybrid solvent MEAâ€"MeOH in an absorber. *Chem. Eng. J.* 260, 11-19.

[2] Rongwong, W., Cao, F., Liang, Z., Zhang, R., Idem, R., Tontiwachwuthikul, P., 2015. Investigation of the effects of operating parameters on the local mass transfer coefficient and membrane wetting in a membrane gas absorption process. *J. Membrane Sci.* 490, 236-246.

Vertical Gardening as Means for Sustainable Food Production in Food Insecure Urban Communities

Dr. Robert W. Peters, University of Alabama at Birmingham, Birmingham, AL, USA
Dr. Lee Moradi, University of Alabama at Birmingham, Birmingham, AL, USA
Julia Ashlyn Manzella, University of Alabama at Birmingham, Birmingham, AL, USA

Abstract

The University of Alabama at Birmingham undergraduate and graduate programs' continued involvement in the Birmingham-Jefferson County community have identified food insecure areas, known as food deserts, and associated health issues negatively impacting the community due to poor diets. Farm stands and community gardens around the Birmingham-metropolitan area have increased availability during the last several years to encourage healthier eating habits. To combat food insecurity, sustainable alternative growing techniques for urban agriculture need to be implemented and incorporated into local community gardens. Alternative garden techniques such as vertical gardens maximize yield in limited spaces. This research seeks to analyze various vertical growth methods (Tower Garden, Pyramid, and Vertical-Pallet), water quality (collected city and lake), and overall yield. Biomass, chlorophyll a and b, and amount and type of water required for irrigation will be assessed to determine the growth system providing the most productive system in terms of production. The collected waters used will also be analyzed for the impact of the water source on production. The goal of this study is to determine conditions and techniques for vertical gardening in urban areas located in the Southeastern United States with limited resources such as space, water, and soil.

The analysis will use tools, such as biomass, to identify overall health and yield of the spinach grown in the three vertical growth methods with different water sources. The resulting leaves will be dried and weighed for biomass analysis (dry versus wet weight). Aldridge Gardens located in Hoover, Alabama, has provided their land space and resources for this research. This data will be culminated to determine best growing practices for communities in climate zones like Birmingham, Alabama growing spinach and similar leafy greens. The goal is to define cost-effective means to grow leafy greens during the fall and spring planting seasons.

Introduction

As the world's population continues to rise, researchers have identified the need to create healthy, equitable, and sustainable communities, especially in terms of food security, in order to support the growing population and changing environment. Food security and ways to ensure food security for those living in areas with limited resources have been on the forefront of research throughout the world, especially the United States, "17.4 million U.S. households are food insecure" in 2014 (Food Insecurity, 2014). An initiative by Healthy People 2020 Food Insecurity (2014), has identified food insecurity as "the disruption of food intake or eating patterns because of lack of money and other resources." Food insecurity can be due to many factors ranging from financial to geographic, and those affected have a higher risk of negative health outcomes, such as cardiovascular disease and diabetes. ZIP Codes census tracts within Birmingham, Alabama, among other areas in Alabama have been identified as food insecure or food deserts. Food deserts, as defined by the United States Department of Agriculture (USDA), are both urban (1 mile) and rural (10 miles) areas that have limited access to affordable fresh and healthy foods by distance (Food Deserts, undated).

Birmingham, Alabama was analyzed for food desert conditions in the 2015 non-thesis report, *Use of GIS Spatial Analysis to Identify Food Deserts in Birmingham, Alabama*, by J. Ashlyn Manzella (2015). The report identified that twenty of the twenty-eight Birmingham-metropolitan ZIP Codes were low-income food deserts as defined by the USDA. The report identified 65% of the residents of these ZIP Codes affected as being African-American which corroborated previous research conducted by Jefferson County PLACE MATTERS (The Jefferson

County PLACE MATTERS Team, 2013) and the Mari Gallagher Research and Consulting Group (Mari Gallagher, 2010). The overall history and city planning of Birmingham has negatively impacted the residents of the city's food security. The research recommended community public and home gardens to provide access to fresh foods (Manzella, 2015). The United States Department of Agriculture National Resources Conservation Service (NRCS) published the "Community Garden Guide: Vegetable Garden Planning and Development" to assist communities in the development of community gardens and was used in the development of this project (United States Department of Agriculture: National Resources Conservation Service, 2009).

The United States Environmental Protection Agency (EPA) identified North Birmingham, within the city of Birmingham, as an area in need of environmental cleanup and remediation due to years of industrial activities negatively impacting the environment and, ultimately, placing the public's health at risk. In conjunction with the Alabama Department of Environmental Management (ADEM) and the Jefferson County Department of Health, the North Birmingham project sponsored by the EPA is addressing environmental issues and enforcing environmental regulations such as air, water, and environmental justice, etc. The soil cleanup process addressed residential homes, schools, and other properties where soil contaminants concentrations exceeded established caps. The City of Birmingham is collaborating with EPA to establish a watershed management plan to address water concerns (U.S. Agency, undated). Due to the stigma associated with the negative environmental impacts to the brownfield sites and soils in Birmingham, Alabama, this research chose to not use the site's soil. Rather, this research focuses on vertical gardening techniques using organic soil purchased from a local hardware store. The secondary research focus of this project, is to provide a low-cost method option to growing spinach and leafy greens instead of the need to purchase from a local store if that option is available. Therefore, the lowest cost organic soil mixture was chosen. It should also be noted that the city of Birmingham located in Jefferson County, consists of two soil areas, Appalachian Plateau and Limestone Valleys and Uplands as established by Charles C. Mitchell, Jr., Auburn University Professor of Agronomy and Soils (Mitchell, 2008). Appalachian Plateau soils are sandstone or shale, where Limestone Valleys and Uplands are predominately limestone (Mitchell, 2008). Therefore, these soils are not ideal for farming needs.

Numerous research studies have focused on the health, economic, and societal impacts of food insecurity. This research seeks to define ways to increase the availability of fresh fruits and vegetables via community gardens. Specifically, this research seeks to grow sustainable vertical gardens in climates and locations to similar to Birmingham, Alabama with limited resources, such as soil, space, and water. Three different water sources and three different vertical growth techniques will be utilized to grow three different types of plants.

Two different water sources, municipal city provided water and an on-site lake, for the project will be studied for impact to overall plant health and yield. The on-site lake provides a lower-cost water source as compared to the Birmingham Water Works Board provided water and Jefferson County sewer costs. The vertical growth techniques chosen were the Tower Gardens©, pallet gardens, and pyramid gardens, as seen in Figure 1. Tower Garden® can accommodate twenty plants in approximately three-square feet of horizontal space, the pallet gardens assembled can accommodate between eight to twelve plants in approximately 1.5 square feet of horizontal space, and the pyramid style garden selected can accommodate eighteen plants in approximately four-square feet of horizontal space. Kale, tomato, and yellow squash (squash) were originally selected and tested because each is a varying species known to grow in Birmingham's climatic conditions. Aldridge Gardens has successfully grown kale, yellow squash, and tomatoes in Tower Gardens® and raised-bed gardens per discussions with Mr. Rip Weaver, Executive Director at Aldridge Gardens (2017). However, due to initial failures and the need for replicability, spinach was chosen ultimately for fall and spring planting. The water sources, growth techniques, and plants are further described in this paper.

Objectives

This research seeks to define guidelines for producing high-nutrient valued plants using urban and vertical farming techniques in the metropolitan Birmingham, Alabama area to enhance the community's food security. It is suspected that if local water sources, such as roof runoff and pond/lake water, can be used without the addition pretreatment water chemicals, spinach can be grown sustainably in urban community-based garden using commercially available organic soils and fertilizers. This research also seeks to reduce soil, water, and energy requirements for growing healthy foods for consumption while increasing the viability of these techniques in meeting population-based nutritional needs. The two growth mechanisms requiring soil will be monitored for the volume of soil used and added throughout the project to address erosion. The soil will be monitored for pH as spinach has ideal pH ranges. The amount of additional water and rainfall data will be tabulated to monitor water demands. The amount of energy consumed to power the Tower Gardens® will also be addressed.

Methodology

The overall methodology is the analysis of the various growing techniques and water sources for each methodology's ability to produce spinach. The NRCS "Community Garden Guide" (2009) provided insight for location and orientation of garden techniques, however, the land space was previously allocated and could not be addressed to reflect the NRCS suggestions. Water quality analysis will be performed in accordance with *Standard Methods for the Examination of Water and Wastewater* (1998) as well as at home test methods. The method-ology is further discussed in the following sections.

Literature Review

For this study, community gardens health and environmental impacts, water sources and conservation, and community garden vegetable yields were explored. It was established that community gardens play an important role in health of both the community and environment. The research explored studies analyzing various races and socioeconomic groups within the population impacted by community gardens. Numerous studies have corroborated the positive physical and mental health impacts of community gardening on participants (Hartwig and Mason, 2016; Hite *et al.*, 2017; and Spees *et al.*, 2015). Studies have researched and proven the successful use of community gardens to reduce stormwater and grey water impacts on the water system, while also providing fresh produce to the community (Gittleman *et al.*, 2016; Jennings *et al.*, 2013; and Kulabako *et al.*, 2011). Current community garden research has analyzed crop yields from raised-bed community gardens to encourage community involvement and provide lessons learned to others. However, current data does not specifically exist for crop yields for vertical gardens (Algert *et al.*, 2014; and Farming Concrete, undated).

Current community and urban gardening research primarily focuses on the health impacts on humans associated with community gardens and not necessarily on the techniques and methods for growing sustainable community and urban gardens. There is limited research regarding vertical gardens and the resources necessary to develop and maintain sustainable community gardens. Major urban areas typically do not offer large plots of land for gardening because the available free space is allotted to public parts or future land development.

Materials and Methods

This research focuses on cost-effectively growing produce vertically for urban areas with limited access to horizontal land space and desirable soils. The other focus of this research is to establish the impact on the garden's production using city water versus lake water without pre-treatment chemicals; however, organic fertilizers will be used. All vertical growth mechanisms will be planted with the same plant from a single vendor in a constant orientation, further defined below. The goal is to identify the most productive and cost-effective vertical growth mechanism and water source.

Research will be conducted at Aldridge Gardens in Hoover, Alabama.

Vertical Growth

The following three vertical growth mechanisms were chosen for analysis: Tower Gardens®, pallet gardens, and raised-bed pyramid gardens (see Figure 1). There are a total of four mechanisms of each garden type. Each garden type will be evaluated for cost and yield.

Tower Gardens® are a vertical, aeroponic/hydroponic system that can support up to 20 plants in less than three square-feet and initial costs are around $550. The pallets with an assembly costs of less than $60 are upcycled and can support a range of 9 to 12 plants in roughly 1.5 square feet. The $120 pyramid gardens are a commercially purchased product that can support 15 to 18 plants in four square-feet of space. The pyramid gardens were commercially purchased but could be built and assembled at home with the correct tools for a lower price. Both the pallet and pyramid gardens both require soil, which is discussed below.

Figure 1: Tower Garden® (R), Pallet (C), and Pyramid (L)

Water Sources

This project seeks to determine the impact of various water sources on overall plant growth and production. Aldridge Gardens has provided access to city water (Birmingham Water Works Board) and an on-site lake water.

Aldridge Gardens offers a six-acre lake on-site for water retrieval. Aldridge Gardens' Executive Director, Mr. Rip Weaver (2017), explained that a unique aspect of Aldridge Gardens is its lack of off-site water contribution due to the "bowl" shaped topography of the thirty-acre property; approximately 85% of the water that falls on the property stays on the property, flowing into the six-acre lake. The lake has no outside or additional water source beyond rain runoff. A small amount of offsite water contribution comes from the adjoining property's playground and parking lot (7th Day Adventist Church). There is minimal water contribution that flows over any concrete surface, and none that flows directly into the lake without first going through a landscaped area. Additionally, Aldridge Gardens uses very little in the way of herbicides, fungicides or insecticides (Weaver, 2017).

Soil and Fertilizers

An organic garden soil was selected using a local commercial soil. The soil selected, Kellogg Garden Organics All Natural Garden Soil for Flowers and Vegetables, was the lowest priced per cubic foot. Pennington UltraGreen All Purpose Plant Food (10-10-10) was selected due to its 10% Nitrogen, 10% Phosphorus, and 10% Potassium mixture and its lower cost per cubic foot.

Plants

The plants originally chosen for this project represented fruits, leafy greens, and vegetables that are typically grown in this region. Tomatoes, kale, and squash were chosen due to their successful, historical growth in Tower Gardens® and raised-bed gardens at Aldridge Gardens (Weaver, 2017). Each garden type was planted with each plant in the same order. Squash was planted on the bottom, kale in the middle, and tomatoes on top due to sunlight and growth requirements discussed below. All seedlings were procured from the same farm for consistent species. However, throughout the growing season numerous problems were incurred and the project was halted. The numerous variables in terms of plant health, water conditions, and fertilization needed to be addressed. There also needed to be a replicability factor that was not available.

Therefore, the project needed to go in a different direction. The Alabama Cooperative Extension System through Alabama A&M and Auburn Universities' ANR-0063 "Planting Guide for Home Gardening in Alabama" (2013) was assessed for vegetables that had both a fall and spring harvesting guideline. Due to time constraints with the project, spinach was chosen due to planting dates in September and February to March (Musgrove *et al.*, 2013).

Spinach is a cool season crop that can grow in both full sun and partial shade that requires moist soil (Growing Guide: Spinach, 2006). Per Cornell University's Growing Guide (2006), the germination temperature ranges from 40°F to 75°F with failure to germinate at warmer temperatures and dryer soil. Spinach prefers a soil pH range of 6.4-6.8, and a high pH may be indicated by slow growth and yellowing of leaves (Sanders, 2001). However, the Cornell Univeristy's Growing Guide (2006) suggests spinach can be sustained in soils with a pH range of 6.0-7.5 (Old Farmer's Almanac, 2018). The recorded pH of the soils in the pallet and pyramid gardens has ranged from 6.8 to 7.2.

Data Collection

The data collection begins with the amount and type of each water source used for the various vertical growth mechanisms. Water was measured by the same one-liter measuring cup. The lake water will be collected in sterile glass bottles and then analyzed in accordance with *The Standard Methods for Examination of Water and Wastewater* (1998) under the supervision of the University of Alabama at Birmingham's Department of Civil, Construction, and Environmental Engineering. The lake water will be tested for alkalinity (Test 2320), conductance (Test 2510), hydrogen ion (Test 400-H^+), and turbidity (Test 213) in duplicate bimonthly.

Once harvesting begins, the harvested spinach will be measured for overall yield, wet weight, dry weight, and chlorophyll under the supervision of the University of Alabama at Birmingham's Department of Biology. This data will indicate which technique provides the best yield. Further, visual analysis will be provided during the course of the study as an indicator of plant health.

A minimum of three samples from each plant will harvested by cutting the stem at soil level and placed inside sealed containers prior to transport to the University of Alabama at Birmingham. A Denver Instrument SI-602 will used to measure both the wet and dry weight to 0.01 mg. A Yamata

DVS 600 Gravity Convection Oven will be employed to dry the samples by allowing the samples to dry at 45˚C ± 3˚C for a minimum of 24 hours. Additional one sample will undergo chlorophyll analysis using the Opti-Sciences OS30p Chlorophyll Fluorometer.

Results

At the time of this publication, no statistically significant data has been collected due to limitations of the study to be discussed in the next section.

Of all 200 spinach plants that were planted, only 26 plants survived until harvesting. Of those 26 plants, only 11 plants were healthy enough for harvesting. The results of the 11 plants are shown in Table 1.

Of the 11 plants, the average wet weight of a single leave was 0.58 mg and dry weight of 0.04 mg. The average percent moisture content was found to be 94.13%. The chlorophyll fluorescent measurement average was 0.752, and anything greater than 0.7 is indicative of a healthy plant.

Note: Water analysis has yet to be performed at this time due to lab space availability but will begin prior to the spring planting.

Limitations

There are numerous limitations to this study as there are numerous factors affecting spinach health. The major issue for this project was the unusually hot temperatures after the initial planting date as average temperatures were in the mid-80s ˚F. The site is also uncontrolled and is open to the public for use; however, this is more realistic to the actual community gardens. Further, not having water quality analysis available impaired the overall analysis testing.

Conclusions

The testing is on-going at this time. It is hoped more data will be collected that is statistically significant during spring planting.

Table 1: Plant Collected Data

Location		Total No. Leaves	Wet Weight (mg)	Dry Weight (mg)	% Moisture Content	Chlorophyll (F_v/F_m)
Pyramid	C1-Top Left	7	0.21	0.01	95.24	0.751
			0.41	0.02	95.12	
			0.22	0.01	95.45	
	C1-Top Right	5	0.23	0.01	95.65	0.749
			0.24	0.01	95.83	
			0.30	0.02	93.33	
	C1-Top Back	4	0.31	0.02	93.55	0.752
			0.21	0.01	95.24	
			0.19	0.01	94.74	
Tower Garden	C1-5C	8	0.34	0.02	94.12	0.762
			0.44	0.02	95.45	
			0.32	0.02	93.75	
	C2-4C	9	0.22	0.01	95.45	0.732
			0.19	0.01	94.74	
			0.42	0.03	92.86	
	L1-5B	17	0.46	0.02	95.65	0.758
			0.30	0.02	93.33	
			0.26	0.02	92.31	
	L1-1C	20	0.68	0.05	92.65	0.773
			0.96	0.02	97.92	
			0.67	0.05	92.54	
	L1-3C	19	0.96	0.06	93.75	0.720
			1.01	0.06	94.06	
			1.51	0.10	93.38	
	L1-4C	13	0.78	0.05	93.59	0.763
			0.51	0.04	92.16	
			1.06	0.07	93.40	
	L2-3C	11	0.40	0.03	92.50	0.766
			0.75	0.02	97.33	
			0.68	0.05	92.65	
	L2-5D	12	1.14	0.09	92.11	0.750
			1.55	0.11	92.90	
			1.24	0.08	93.55	

References

Agency, U.S., undated. *Cleanup Process in the North Birmingham Environmental Collaboration Project.* Retrieved July 4, 2018, from https://www.epa.gov/north-birmingham-project/cleanup-process-north-birmingham-environmental-collaboration-project

Algert, S.J., A. Baameur, and M.J. Renvall, 2014 (April 19). Vegetable Output and Cost Savings of Community Gardens in San Jose, California. *Journal of the Academy of Nutrition and Dietics, 114* (7): 1072–1076. doi:10.1016/j.jand.2014.02.030

Eaton, A.D., L.S. Clesceri, A.E. Greenberg, and M.A.H. Franson, 1998. *Standard Methods for the Examination of Water and Wastewater*, 20th Edition, American Public Health Association, Washington, D.C.

Farming Concrete, undated. Rtrieved July 19, 2018, from Farming Concrete, https://farmingconcrete.rog/

Food Deserts, undated. Retrieved June 15, 2018, from United States Department of Agriculture: Agricultre Marketing Service: https://apps.ams.usda.gov/fooddeserts/fooddeserts.aspx

Food Insecurity, 2014. Retrieved June 15, 2018, from Healthy People 2020: https://www.healthypeople.gov/2020/topics-objectives/topic/social-determinants-health/interventions-resources/food-insecurity

Growing Guide: Spinach, 2006. Retrieved August 3, 2018, from Cornell University Home Gardening: http://www.gardening.cornell.edu/homegardening/scene4c19.html

Manzella, J.A., 2015. *Use of GIS Spatial Analysis to Identify Food Deserts in Birmingham, Alabama.* Non-thesis, University of Alabama at Birmingham, Departent of Civil, Construction, and Environmental Engineering, Birmingham, Alabama.

Mari Gallagher, M.G., 2010 (August). *Examining the impact of food deserts & food imbalance of public health in Birmingham, AL*. Retrieved June 12, 2018, from Mari Gallagher Research and Consulting Group: http://www.marigallagher.com/projects/

Mitchell, C.J., 2008. *Soils of Alabama.* Auburn: Alabama Cooperative Extension System. Retrieved June 25, 2018, from http://www.aces.edu/pubs/docs/A/ANR-0340/index2.tmpl

Musgrove, M.B., J. Kemble, E. Bauske, D. Williams, D. Bond, and K. Smith, 2013. *Planting Guide for Home Gardening in Alabama.* Alabama Cooperative Extension System. Retrieved July 30, 2018, from http://www.aces.edu/pubs/docs/A/ANR-0063/ANR-0063.pdf

Old Farmer's Almanac, 2018. *Soil pH Levels for Plants*. (Yankee Publishing) Retrieved October 27, 2018, from The Old Farmer's Almanac: https://www.almanac.com/content/soil-ph-levels

Sanders, D., 2001 (January 1). *Spinach: Horticulture Information Leaflets*. Retrieved August 3, 2018, from NC State Extension Publications: https://content.ces.ncsu.edu/spinach

The Jefferson County PLACE MATTERS Team, 2013), *PLACE MATTERS for Health in Jefferson County, Alabama: The Status of Health Equity on the 50th Anniversary of the Civil Rights Movement in Birmingham, Alabama.* Uiversity of Alabama at Birmingham. Retrieved June 25, 2018, from http://jointcenter.org/docs/Jefferson%20Report.pdf

United States Department of Agriculture: National Resources Conservation Service, 2009 (November). *Community Garden Guide: Vegetable Garden Planning and Development.* Retrieved August 2018, from https://www.nrcs.usda.gov/Internet/FSE_PLANTMATERIALS/publications/mipmcot9407.pdf

Weaver, R., 2017 (October 19). Executive Director, Aldridge Gardens. *Personal Communication*. (J. A. Manzella, Interviewer), Birmingham, AL.

538619 Adsorption Modeling for CO_2 Capture in Water Stable MOFs

Abstract:

In comparison with the pre-combustion, and oxy-fuel combustion, post-combustion for capturing and or conversion of carbon dioxide is growing attention because it is most feasible as it just needs retrofitting of the existing power plants. In fact, CO2 capture via post combustion using MOFs is a promising and sustainable emerging technology because of high BET surface area of MOFs, low operating cost, easy regeneration of the adsorbate, and little conventional process modification (less capital cost). CO2 control resulting from conventional fossil fuel burning which is the main contributor to the global warming. But still a cost-effective continuous process for CO2 capturing and conversion, needs further optimization to be implanted on the commercial scale for continuous large-scale operation. The current study briefly discusses the recent advance in synthesis methods, metal-ion/organic linkers, structure modifiers or fillers, and adsorption isotherms for ZIF-300 and ZIF-302. Finally, the comparison of the fitting parameters (equilibrium adsorption constant, order of adsorption, and enthalpy of adsorption) with the already published literature is highlighted.

Modeling a Water Wash Sieve Tray for Aerosols Scavenging Using Computational Fluid Dynamics (ID# 527774)

*Siao-Han Huang[a], Jia-Lin Kang[b], Abhay S Zambare[a], David Shan-Hill Wong[a], Shi-Shang Jang[a]**

a Department of Chemical Engineering
National Tsing Hua University, Hsinchu 300 Taiwan

b Department of Chemical and Materials Engineering
Tamkang University, New Taipei City 251 Taiwan

* *Corresponding Author's E-mail:* ssjang@mx.nthu.edu.tw

Keywords: CFD; aerosol scavenging; sieve tray; multi-phase.

Abstract

The chemical absorption method of post-combustion CO_2 capture is widely used in the capture of carbon dioxide processes, which is flue gas post-process. Flue gas composition is complex and has fine size condensation nucleus. After the chemical absorption process, part of the solvent will be adsorbed on the nucleus surface to form aerosols and emission into the atmosphere with the flue gas. Because the size of the aerosol is extremely fine, it does not only cause environmental pollution but if inadvertently inhaled, it will harm the human body. Therefore, as CO_2 capture technology matures, aerosol becomes a challenge for the next generation. The common dust removal equipment in the factory includes cyclone dust collectors、bag filters、electrostatic precipitators and wet scrubbers, etc., and only wet scrubbers can remove solid dust and harmful gases simultaneously.

Using the most basic water-washed sieve plate tower to solve the problem of industrial aerosol emission will investigate in this paper. However, the experiment of fluid flow behavior in the sieve plates has the high time cost, and there are few studies well discussing this topic. The purpose of this study is to use the CFD (Computational Fluid Dynamics) to establish a model of a sieve plate tower to observe the movement and removal efficiency of the particles. The research was divided into two parts: (1) to build a two-dimensional sieve plate tower, and verify the results of the hydrodynamics behavior, and (2) to investigate the particle motion and removal efficiency in the sieve tray by using the particle tracker.

The simulated results of the upstream and downstream liquid velocities on the sieve tray are consistent with the empirical correlation. In addition, the trend of the hydrodynamics in the two-dimensional sieve tray model is correct, the error between the simulated results (including the clear liquid height, froth density, and froth height) and corresponding correlations were less than 22%.

For the particle removal simulation, two different turbulent models: standard k-ε and Realizable k-ε were used and compared. The simulated results show that both of the turbulent models can accurately simulate the average air velocity in perforations and froth density. And under the fixed particle size and liquid flow rate, the removal efficiency was improved when raising the air velocity. In summary, the establishment of this model can not only predicted the hydrodynamics and particle direction in the sieve tray but also serve as the reference for factories to deployed sieve plate tower.

Acknowledgment: This research was supported by Grant MOST 107-3113-E-007-002- from the Ministry of Science and Technology of the Republic of China.

BACKGROUND

With the industrial development, CO2 emissions increase day by day. The chemical absorption method of post-combustion CO2 capture is widely used in the capture of carbon dioxide processes, but it's not completely harmless. After the chemical absorption process, part of the solvent will be adsorbed on the nucleus surface to form aerosols and emission into the atmosphere with the flue gas. Because the size of the aerosol is extremely fine, the aerosol issue becomes a challenge for the next generation.

There are several types of particle pollution control device used in factories: gravity-settling chamber, cyclone dust collector, bag filter, electrostatic precipitator, and wet scrubber. An aerosol defined as a suspension of fine solid particles or liquid droplets in the air or another gas, and the diameter range is between 0.001~10 μm. Therefore, the wet scrubbers are the most suitable device for aerosol control which can simultaneously remove solid dust and harmful gas.

The sieve plate is widely used in industry as a gas-liquid contact equipment because of its simple construction, high plate efficiency, low price, and low kinetic loss. Using the most basic water-washed tower to solve the aerosol emission issue will be investigated in this paper. However, the experiment of fluid flow behavior in the sieve plates has the high time cost, and there are few studies well discussing this topic. In recent years, computational fluid dynamics (CFD) software has been continuously improved. The hydrodynamic simulation becomes important for industrial practice, it can not only grasp the comprehensive planning before building the factory but also reduce the wrong investment. So far, many studies have used CFD software to simulate the gas-liquid contact behavior of sieve tray towers. However, there are no simulation studies of the complicated aerosol removal behavior in the sieve tray towers. In order to overcome the limitations related to the experiment, the purpose of this study is to use the computational fluid dynamics model to establish a sieve plate tower to observe the movement and removal efficiency of the aerosols in the tower.

METHODOLOGY

Simulation Software

The FLUENT computational fluid dynamics simulation software from ANSYS, USA uses in this research. It can analyze include laminar flow, turbulent flow, compressible flow, incompressible flow, particle diffusion flow, multiphase flow, chemical reaction flow field, steady-state or transient flow, and so on.

Numerical Method

Fluent uses the finite volume method. The geometry is to divide into a series of control volumes, then do the integral calculation. In the model of this paper, the mass conservation equation, momentum conservation equation, multiphase, turbulent equations are used. The numerical simulations are based on three-fluid in Euler-Lagrange approach, the dense discrete phase model was used. The turbulent viscosities used the standard k -epsilon and realizable k-epsilon. The continuity equation and momentum equation of fth phase are given by equation 1 and 2, respectively.

$$\frac{\partial}{\partial t}(\alpha_f \rho_f) + \nabla \bullet (\alpha_f \rho_f \vec{v}_f) = 0 \tag{1}$$

$$\frac{\partial}{\partial t}(\alpha_f \rho_f \vec{v}_f) + \nabla \cdot (\alpha_f \rho_f \vec{v}_f \vec{v}_f) = -\alpha_f \nabla P + \nabla \cdot \tau_f + \alpha_f \rho_f \vec{g} + \sum_{q=1}^{n\ phases}\left(K_{qf}(\vec{v}_q - \vec{v}_f) + \dot{m}_{qf}\vec{v}_{qf} - \dot{m}_{fq}\vec{v}_{fq}\right) \tag{2}$$

where, α, ρ, $\vec{v}$, P, $K_{qf}(\vec{v}_q - \vec{v}_f)$ represent the volume fraction, density, velocity vector, static pressure ,and interphase momentum transfer term.

Interphase Momentum Transfer Term

The interphase momentum transfer term is basically interphase drag force per unit volume. With the gas as the secondary phase, the interphase momentum transfer term shows below:

$$K_{GL}(\vec{v}_G - \vec{v}_L) = \frac{3}{4}\frac{\alpha_G}{d_G}\rho_L C_D |\vec{v}_G - \vec{v}_L|(\vec{v}_G - \vec{v}_L) \quad (3)$$

where α_G, d_G, C_D represents the gas volume fraction, Sauter-mean diameter and drag coefficient. The following equation (4) is the interphase momentum transfer by Krishna et al. (1999) [1]. The equation was used to describe the rise of a swarm of large bubbles in the churn turbulent regime.

$$C_D = \frac{\rho_L - \rho_G}{\rho_L} g d_G \frac{1}{V_{slip}^2} \quad (4)$$

where, V_{slip} represent the slip velocity, which can estimate from the superficial velocity V_s and the average froth density $r_G^{average}$ (eq. 5), and Bennett et al. correlation [2] are used to estimate the average froth density, the equations (6) (7) shown below :

$$V_{slip} = \frac{V_s}{r_G^{average}} \quad (5)$$

$$r_G^{average} = 1 - r_L^{average} \quad (6)$$

$$r_L^{average} = exp\left[-12.55\left(V_s\sqrt{\frac{\rho_G}{\rho_L - \rho_G}}\right)^{0.91}\right] \quad (7)$$

From the given equations, the interphase momentum transfer term becomes independent of bubble diameter. This obviates the need for its input that could have been difficult.

Turbulent Equation

The Standard k-ε model is a semi-empirical based on the experimental phenomenon, it was proposed by Launder and Spalding [3]. Robustness, economy, and reasonable accuracy for a wide range of turbulent flows explain its popularity in industrial flow and heat transfer simulations.

The realizable k-ε model is more accurately predicts the spreading rate of both planar and round jets. It is also likely to provide superior performance for flows involving rotation, boundary layers under strong adverse pressure gradients, separation, and recirculation. The difference between the realizable k-ε model and the standard k-ε model is that C_μ is no longer constant. For more details refer to reference [4].

Dense Discrete Phase Model

In the standard formulation of the Lagrangian multiphase model, the assumption is that the volume fraction of the discrete phase is sufficiently low: it is not taken into account when assembling the continuous phase equations. The dense discrete phase model provided by Popoff .B (2007) [5] is used to overcome this limitation, the mass and momentum conservation equations of fth phase show below :

$$\frac{\partial}{\partial t}(\alpha_f \rho_f) + \nabla \cdot (\alpha_f \rho_f \vec{v}_f) = \sum_{q=1}^{n\,phases}(\dot{m}_{qf} - \dot{m}_{fq}) \quad (8)$$

$$\frac{\partial}{\partial t}(\alpha_f \rho_f \vec{v}_f) + \nabla \cdot (\alpha_f \rho_f \vec{v}_f \vec{v}_f) = -\alpha_f \nabla P + \nabla \cdot [\alpha_f \mu_f (\nabla \vec{v}_f + \vec{v}_f^T)] + \alpha_f \rho_f \vec{g} + F_{DPM} \quad (9)$$

where F_{DPM} represents the force between phases. However, the momentum equation does not solve an individual particle, the solver will trace a swarm of particles as a parcel. Every parcel has the same particle characteristic (such as velocity, position, mass). The number of particles per parcel can be obtained by the following equation (10) :

$$n_p = \frac{\dot{m}_{parcel}\, \Delta t}{m_p} \quad (10)$$

where, $\Delta t, \dot{m}_{parcel}, m_p$ represent the transient time step size, an individual parcel mass flow rate, and an individual particle mass (calculated by particle density and diameter).

Particle Motion Theory

Fluent software predicts the trajectory of a discrete phase particle by integrating the force balance on the particle. This force balance equation the particle inertia with the forces acting on the particle can be written as equation (11) :

$$\frac{d\vec{u}_p}{dt} = F_D(\vec{u}_f - \vec{u}_p) + \frac{\vec{g}(\rho_p - \rho_f)}{\rho_p} + \vec{F}_{additional} \quad (11)$$

where $\vec{F}_{additional}$ and $F_D(\vec{u}_f - \vec{u}_p)$ represent additional forces and drag force per particle mass (force/per particle mass). The additional forces included virtual mass force and pressure gradient force in this study, are shown below :

$$\vec{F}_{virtual\ mass} = \frac{1}{2}\frac{\rho_f}{\rho_p}\frac{d}{dt}(\vec{u}_f - \vec{u}_p) \quad (12)$$

$$\vec{F}_{pressure\ gradient} = \left(\frac{\rho_f}{\rho_p}\right)\vec{u}_p \nabla \vec{u}_f \quad (13)$$

The following equations (14) is the stokes-cunningham drag law [6], which suitable for particles smaller than submicron size.

$$F_D = \frac{18\,\mu}{\rho_p\, d_p^2\, C_c} \quad (14)$$

where, μ, ρ_f, ρ_p, d_p represent fluid viscosity, fluid density, particle density ,and particle diameter. C_c represents Cunningham correction to Stokes drag law.

Sieve Plate Geometry

In this section, the model was established to confirm whether the user-defined drag force is correct under the two-dimensional flow field. The geometry (Figure 1.) was based on the experimental work of Solari & Bell (1986) [7], the specification of the sieve tray shows in table 1.

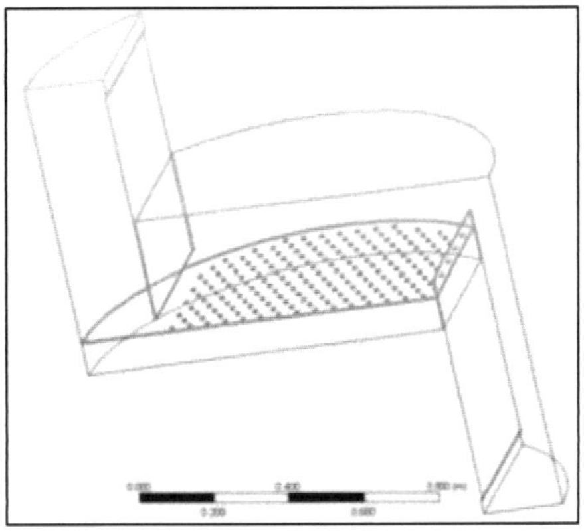

Figure 1. The 3-D sieve tray geometry

Table 1. Tray Specification

tray diameter	1.213m	thickness	0.005 m
tray spacing	0.61m	downcomer area	13% (total area)
holes diameter	0.0127m	hole area	5% (over bubbling area)
weir height	0.05m	pitch, triangular	0.05 m
downcomer clearance	0.038m	weir length	0.925 m

To reduce the mesh quantity of 3-D geometry, a 2-D sieve tray geometry base on the Solari & Bell experimental work was built in this paper. Figure 2. shows the experimental probes positions of Solari & Bell (1986). The plane between x=0.209 and x=0.438 is called upstream, and the plane between x=0.438 and x=0.667 is called downstream.

The plane z=0.33 (shown in Fig 3) was chosen to build the 2-D sieve tray geometry (shown in Fig 4), and compared the Solari & Bell experimental values and simulation results.

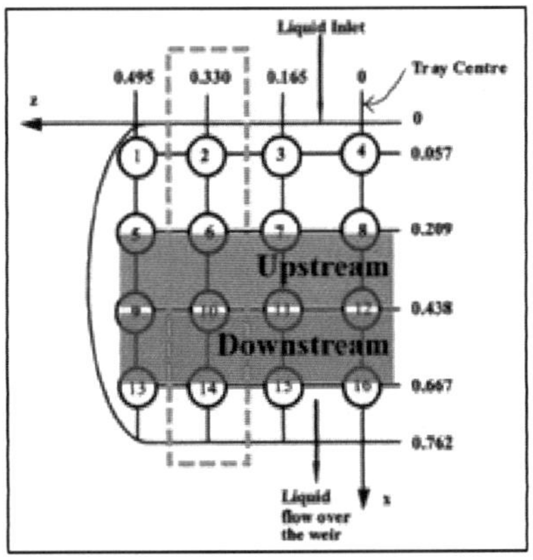

Fig 2. Experimental probes positions of Solari & Bell (1986)

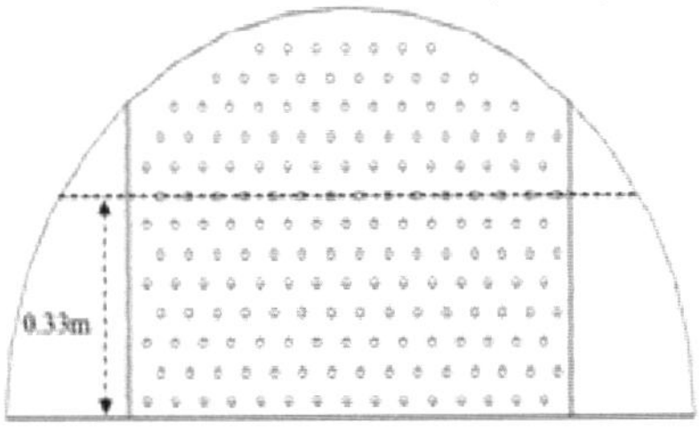

Fig 3. The view from the sieve tray top.

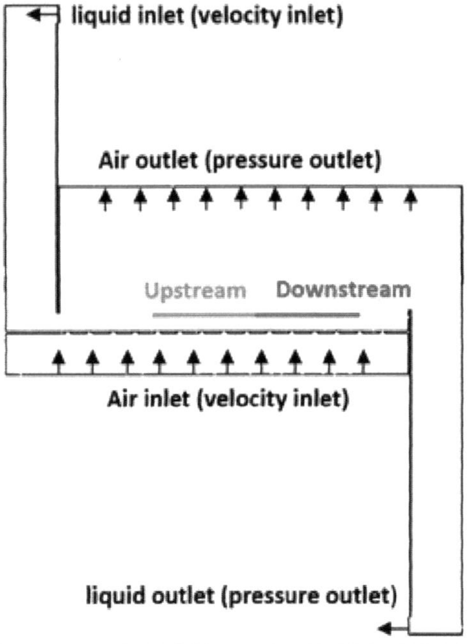

Fig 4. The 2-D flow geometry and boundary conditions.

Taheri & Calvert Experimental Geometry

The model was built to compare the flow behavior and removal efficiency of the particles with the experimental data of Taheri & Calvert (1968) [8]. The geometry specification is shown in below Table 2, and Fig 5 is the boundary conditions. The inlet boundaries are set as velocity inlet and the outlet boundaries are set as outflow. The particles were injected after the flow field reached steady (after 20s flow time), the discrete phase boundary is set as wall-jet when impact the walls, escape when reaching the outlet zone, and reflect when reaching the liquid inlet zone.

Table 2 The sieve plate specification

Taheri & Calvert rectangular sieve plate	
Cross-section area	9.1912 in2
Perforation Diameter	1/8 in
Free area percentage	6.8
Thickness of Plate	1/8 in

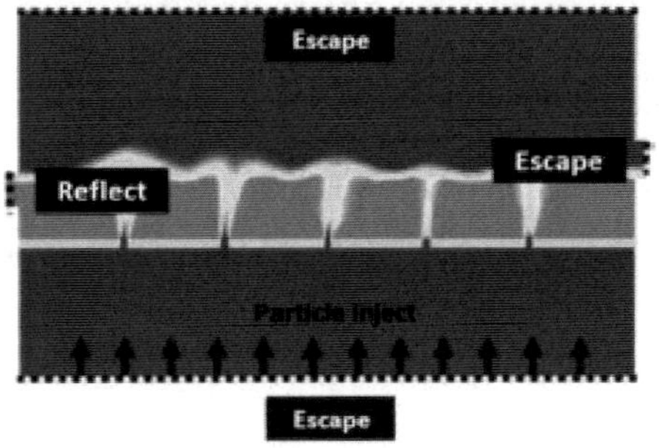

Fig 5 Taheri & Calvert sieve plate geometry and the discrete phase boundary

RESULTS AND DISCUSSIONS

The hydrodynamics verification of sieve plate

The literature used to verify the user-defined drag force and hydrodynamics in the study are shown below: Bennet correlation(1983) [2], Solari & Bell experimental data [7], the CFD model developed by Getye Gesit et al (2003) [9], the CFD model developed by Rahbar Rahimi et al (2011) [10]. The simulation results of clear liquid height, froth height, and the average x-velocity of liquid will be compared with the literature.

Table 3 Results of x-velocity of liquid ($Q_L = 0.0178m^3/s$ $F_s = 0.462$)

	upstream	downstream
Solari & Bell [7]	0.18	0.2
CFD Gesit et al [9]	0.29	0.29
CFD Rahbar [10]	0.2	0.19
CFD Results	0.1953	0.205

Table 4 Results of clear liquid height ($Q_L = 0.0178m^3/s$ $F_s = 0.462$)

clear liquid height	
Solari & Bell (Exp) [7]	0.064
CFD Rahbar [10]	0.074
CFD Gesit et al [9]	0.078
Bennet correlation [2]	0.071
CFD Results	0.0835

Table 5 Results of froth height ($Q_L = 0.0178m^3/s$ $F_s = 0.462$)

froth height	
CFD Rahbar [10]	0.105
CFD Gesit et al [9]	0.126
Bennet correlation [2]	0.09
CFD Results	0.0121

Table 6 Results of froth density ($Q_L = 0.0178m^3/s$ $F_s = 0.462$)

Average liquid volume fraction in froth	
CFD Rahbar [10]	0.77
CFD Gesit et al [9]	0.7
Bennet correlation [2]	0.77
CFD Results	0.69

Table 3 shows the results of average x-velocity of liquid, and its close with the literature results. Because of the geometry completeness, table 4, 5, 6 show that the results from Rahbar [10] are more close to Bennett correlation [2]. The 2-D geometry is used in this study, although it showed the correctness of the hydrodynamics, it also leads to some error of results.

The particle removal efficiency analysis

Taheri & Calvert et al [8] used a large amount of experimental data to derive an empirical correlation for the particle removal efficiency of the sieve plate, the equation)15) shows below:

$$\eta = 1 - e^{-40\,(r_G^{ave})^2\,K} \tag{15}$$

where, η, r_G^{ave} represent scavenging efficiency and froth density, and K is the inertial parameter which can be represented by the equation (16):

$$K = \frac{V_h\,\rho_p\,D_p^2}{9\,\mu_G\,D_h} \tag{16}$$

where, V_h, ρ_p, D_p, μ_G, D_h are representing the average air velocity in holes, particle density, particle diameter, the viscosity of gas phase and perforation diameter。

The Taheri & Calvert correlation shows the particle size and froth density are the main factor affects the scavenging efficiency and the froth density affected directly by air velocity. In this study, two different turbulence models (Standard k-ε and Realizable k-ε viscosity model) were used to simulate the scavenging efficiency of particles under the different particle injected concentration.

Table 7 and Table 8 are shown the particle initial condition and the different injected concentration. The results of froth density and average air velocity in holes were obtained by Ansys monitor, fig 7 shows the monitored regions. The obtention of particle scavenging efficiency results was divided into two ways:

(1) To substitute the Ansys monitored results (V_h and r_G^{ave}) into the Taheri & Calvert correlation, and calculated the particle removal efficiency.

(2) To use the Ansys particle tracks to monitor the number of escape particles from outlet zones, and calculated the scavenging efficiency from the formula below :

$$\eta_{\,particle\ tracker} = \frac{Particle_{\,liquid\ outlet}}{Particle_{\,liquid\ outlet} + Particle_{\,gas\ outlet}} \times 100\% \tag{17}$$

where $Particle_{\,liquid\ outlet}$ and $Particle_{\,gas\ outlet}$ are representing the number of particle escape from liquid outlet zone and gas outlet zone.

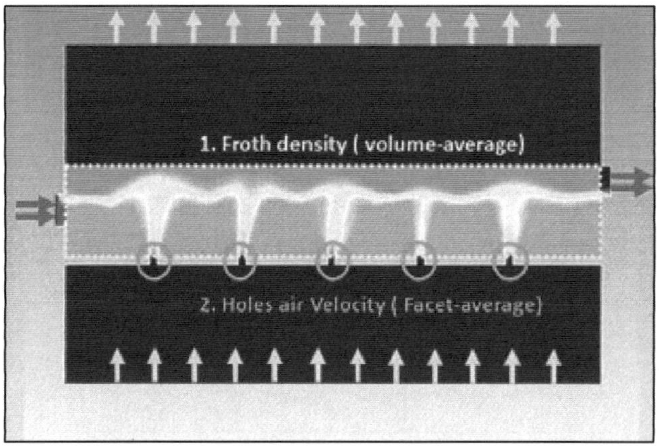

Fig 7 Monitored region for V_h and r_G^{ave}.

Table 7 The particle inject condition

Particle inject condition	
Particle diameter	2.28 μm
Particle density	998.2 kg/m3
Injection time	20s - 40s
Inject velocity	The same as air velocity

Table 8 Injected concentration

No	Inject time step size	Inject type	Inject parcel no. (per inject time)	total mass flow (kg/s)
1.	0.01 s	surface	84	5×10^{-13}
2.	0.05 s	surface	84	5×10^{-13}
3.	0.01 s	surface	84	5×10^{-09}
4.	0.005 s	group	1471	7.7×10^{-06}

Results of average air velocity in holes

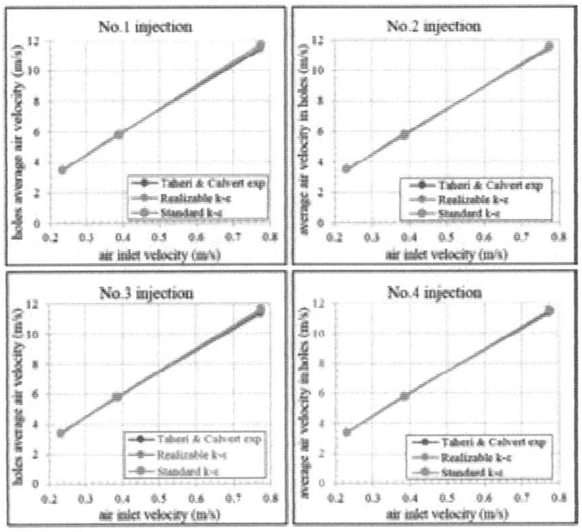

Fig 8 Air inlet velocity v.s average air velocity in holes

Results of froth density

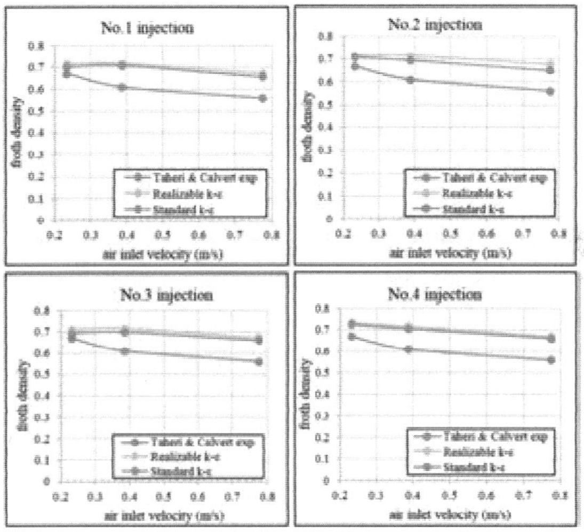

Fig 9 Air inlet velocity v.s froth density

Fig 8 shows the different particle inject concentration does not affect the average air velocity in holes, and the results are consistent with the experimental data. Fig 9 shows the higher the air inlet velocity, the smaller the froth density. The trend is consistent with experimental, but the simulation results overestimated the value of froth density, and the Realizable k-ε turbulence model will obtain the higher froth density than Standard k-ε.

The particle scavenging efficiency

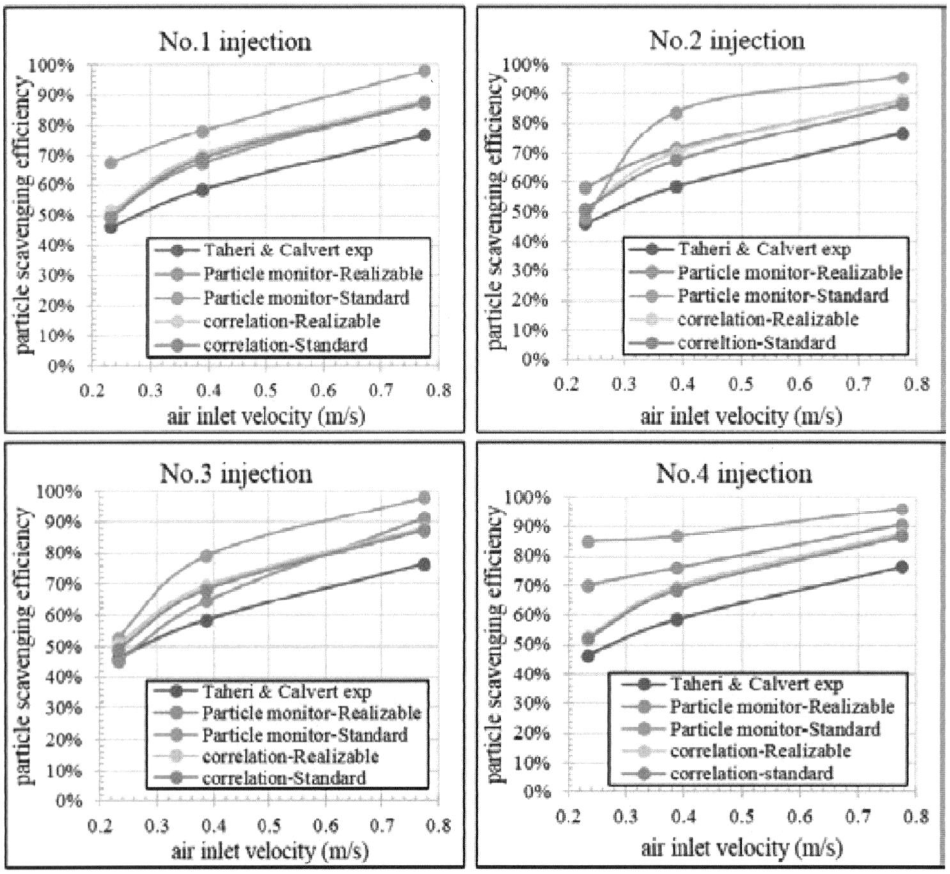

Fig 10 Air inlet velocity v.s particle scavenging efficiency

Fig 10. show that the results of Realizable k-ε turbulence model are much more consistent with the Taheri & Calvert experimental trend than Standard k-ε model. NO.1,2,3 injection shows the particle tracks results of the realizable k-ε model (the orange line) fits the Taheri & Calvert correlation very well, and the error will not exceed 15%.

The Taheri & Calvert correlation does not mention the effects of particle inject concentration. No.4 injection shows the simulation result has higher removal efficiency than the correlation and experimental data, because of a large number of injecting particles will lead its coalescence, and the simulated removal efficiency will be raised.

CONCLUSIONS

To use the 2-D geometry in sieve tray simulation can reduce the time cost and the error between the simulated results and corresponding correlations were less than 22%. In the scavenging efficiency simulation part, the particle removal efficiency can be accurately simulated when the injected concentration is low (no interaction between particles). The Taheri & Calvert correlation does not mention the effect of particle injected concentration, however, the model can simulate the interaction between particles which affect the removal efficiency.

REFERENCES

[1] Krishna, R., Urseanu, M.I., van Baten, J.M. and Ellenberger, J. (1999). "Rise velocity of a swarm of large gas bubbles in liquids." Chem. Eng. Sci., Vol. 54, PP. 171-183.

[2] Bennett, D.L., Rakesh Agrawal and Cook, P.J. (1983). "New pressure drop correlation for sieve tray distillation columns." AIChE J., Vol. 29, PP. 434-442.

[3] B. E. Launder and D. B. Spalding. Lectures in Mathematical Models of Turbulence. Academic Press, London, England, 1972.

[4] T.-H. Shih, W. W. Liou, A. Shabbir, Z. Yang, and J. Zhu. A New k- ϵ Eddy-Viscosity Model for High Reynolds Number Turbulent Flows - Model Development and Validation. Computers Fluids, 24(3):227-238, 1995.

[5] Krishna, R., Van Baten, J., Ellenberger, J., Higler, A. and Taylor, R. (1999). "CFD Simulations of Sieve Tray Hydrodynamics. " Chemical Engineering Research and Design, 77(7), pp.639-646.

[6] H. Ounis, G. Ahmadi, and J. B. McLaughlin. Brownian Diffusion of Submicrometer Particles in the Viscous Sublayer. Journal of Colloid and Interface Science, 143(1):266-277, 1991.

[7] Solari, R.B. and Bell, R.L., (1986). "Fluid flow patterns and velocity distribution on commercial-scale sieve trays. " AIChE J., Vol. 32, PP. 640-649.

[8] Taheri, M. and Calvert, S. (1968). Removal of Small Particles from Air by Foam in a Sieve-Plate Column. Journal of the Air Pollution Control Association, 18(4), pp.240-245.

[9] Gesit, G., Nandakumar, K. and Chuang, K. (2003). CFD modeling of flow patterns and hydraulics of commercial-scale sieve trays. AIChE Journal, 49(4), pp.910-924.

[10] Rahimi, R., Ameri, A., Setoodeh, N. (2011). Effect of Inlet Downcomer on the Hydrodynamic Parameters of Sieve Trays Using CFD Analysis. JChPE, 45(1), pp.27-38.

Assessment of Carbon Bio-Fixation By Mixed Indigenous Microalgae

Almomani, F.[1]*, Judd, S[2]., Shurair, M[1]., Bhosale, R. [1], Majeda Khraisheh[(1)], Kumar, A., Khaled Al- Jaml

Department of Chemical Engineering, Qatar University, P.O box 2713, Doha, Qatar

1) Gas Processing Center, , Qatar University, P.O box 2713, Doha, Qatar

2) falmomani@qu.edu.qa

0. Abstract

This study assesses the potential use of mixed Indigenous Microalgae (MIMA) for carbon capture from flue gases in a photo-bioreactor. The effect of inlet CO_2 concentration on the carbon bio-fixation rate was investigated. The PBR wasoperated under ambient conditions of solar irradiation and temperature for 3-months. Under the conditions studied, MIMA showed a significant carbon bio-fixation rates in order of 0.34 $g_C.L^{-1}.d^{-1}$ suggesting that Microalgae with fast growth rate and naturally grown in the water system can be a sustainable and effective technology for CO_2 uptake.

Keywords: growth rate, biomass production, Carbon dioxide capturing

1. Introduction:

There has been increasing focus on the use of microalgae culture technology (MCT) for CO_2 bio-fixation from flue gases[1-3] and as an advanced wastewater treatment [4-8]) The dual action of microalgae in MCT is considered as an economic and feasible technology [9, 10] that offers a potentially low-energy requirements in a single process. The objective, here in, is to use the biology of algae which considered the most economical alternative [11-13] if compared with chemical and physical unit operation. Thus, CO_2 can be removed from gas stream, wastewater can be treated for possible reuse and a valuable biofuel can be generated [14-17]. However, a well-established

*Corresponding author: Fares Almomani, Associate professor, College of Engineering, Department of Chemical Engineering, Qatar University, P. O. Box 2713, Doha, Qatar. Tel: +974 4403 4140 Fax: +974 4403 4131. E-mail: falmomani@qu.edu.qa

performance of such technologies still in development phase and any improvement in process performance will be add up value for the progress of this technology development.

Wastewater generated by human use residential, commercial and industrial activities have always been a permanent problem that required adequate solution. Conventional and advanced wastewater unit operations are effective in treating organic matter, solids and bacteria exist in water[18]. However, there is special attention focus on the removal of non-biodegradable organic matter and nutrients [19, 20]. Nutrients in water can cause a lot of problems to human and ecosystem. Non-biodegradable organic matters also pose a serious problem on the environment. In most of the developed technologies the removal efficiencies of these two contaminates are limited and required advanced treatment processes.

Different technologies were proposed for nutrient removal and CO_2 Capturing [21-25]. However, the cost, efficiency and scale up requirements limits their large scale operation. Algae, on other hand, can be used effectively as advanced wastewater treatment process if adequate concentration of CO_2 and sunlight is a variable for the growth. Different studies have proved the efficiency of algae in treating different pollutant from wastewater. With is in mind, algae can be applied as a successful technology that offers a promising solution for wastewater treatment , CO_2 capturing and clean biofuel production. Algae in this scenario is advantageous over any other treatment (chemical or physical) in cost and simple unit operation. In addition, this process is safe does not requires chemicals, special equipment or trained personal.

Although the rule of algae treating wastewater and capture of CO_2 is well-known, a very small attention was given to optimize and scale up this technology. Moreover, most of trails performed on this technology deals with the ability of single strain to achieve promising outcome or conducted on synthetic wastewater and artificial light source, ignoring the effect of water matrix and ambient conditions on the final performance. Also, the performance of mixed culture compared with single strain is yet not determined. Thus, the current study aims at evaluating MCT for CO_2 bio-fixation at pilot scale under ambient conditions in hot climate region. The work was conducted over a period of 3 months, during which the CO_2 bio-fixation capacity by mixed indigenous microalgae, *MIMA* was studied and evaluated. The effect of operational conditions on CO2 capturing were assisted

2. MATERIALS AND METHODS

2.1 Algal culture

A mixed indigenous microalgae (*MIMA*) culture collected from a secondary basin of Doha south wastewater treatment plant was used in the study. Stock solutions of microalgae were grown at room temperature under continuous fluorescent light providing an irradiance of 180 $\mu E.m^{-2}.s^{-1}$, and mixed by aeration with filtered air. The growth media used in this study is presented in our previous work[10]

2,2 Pilot plant photo-bioreactor design

Experiments were carried out in 250 L photo-bioreactor consist of eight high-grade polyethyl terephthalate columns connected in series. The tubes were fitted with CO_2 gas injection ports and sampling points. The algal biomass was circulated through the PBR using a variable speed pump. The plant was installed at the Qatar University campus in Doha (25.2854° N, 51.5310° E).

Experimental procedure

Experiments were conducted over a period of 3 months (May 2016 – July 2017) under ambient conditions of light and temperature ($T_{Abm, avg.}$~30°C, $T_{PBR, avg}$~ 29.9, $Li_{, avg.}$~ 9.05 $E.m^{-2}$). Secondary effluent wastewater were used in trails.

Experimental Procedure

Experiments were conducted by circulating the wastewater through the reactor for one hour before adding the algal inoculation media. Cultures were allowed to grow for 2 days at. The PBR was initiated when algae solution achieved at an OD_{690nm} of 0.024±0.005 AU, at which point the wastewater was injected with CO_2-enriched air into the reactor at concentrations of 2.5-35% v/v and a normalized flow rate of 0.2 vvm (volume of gas per working volume per minute). Samples were collected for determination of biomass productivity and CO_2 bio-fixation rate.

2.4 Analysis

Carbon Dioxide Bio-Fixation Rate

Carbon dioxide fixation rate (R_{CO_2}, in $g_C.L^{-1}.d^{-1}$) was determined from the ratio of microalgae carbon content to average biomass productivities according to Equation (3).

$$R_{CO_2} = C_C.P_{bio}.\frac{M_{CO2}}{M_C} \quad (4)$$

where C_C is the carbon content of microalgae biomass (in % w/w), P_{bio} is the average biomass productivities, M_{CO2} and M_C are the molecular weight of CO_2 and carbon($g.mol^{-1}$), respectively.The microalgae biomass carbon content of was determined using a Flash EA1112 CHNS analyzer (Thermo Finnigan CE Instruments, Italy) equipped with a gas chromatography column and a thermal conductivity detector.

RESULTS AND DISCUSSION

Algae Growth

The growth of *MIMA* in wastewater is presented in Figure 1. Growth curve follows a typical growth regimes curve consisting of lag phase, exponential growth phase and stationary phase. A very short lag time of 0.5 day was observed, exponential growth phase lasted for 4 day and after that algal cells entered the stationary phase. The specific growth rate (μ) of *MIMA* was determined to be 1.205 day^{-1}. The μ values suggest that algae can grow fast in this wastewater within 5 days under ambient conditions of temperature in the range 29 to 33 ◦C and total light intensity in the range 8.64 to 10.71 $E.m^{-2}$. Temperature and light intensity were the limiting factor for algae growth in the PBR, increasing the light intensities, photosynthesis increases until a point is reached where the growth rate is the maximum attainable (saturation point) [26]. The μ values reported in this study are higher than the values reported in literature. Almomani and Ormeci [6] reported a μ value for *C. vulgaris, N. Oleoabundans* and *MIMA* of 0.52, 0.26 and 0.21 day^{-1}, respectively. Singh et al., [17] also reported lower μ values for *Chlorella AU1 sp.*, *Scenedesmus AU1 sp* and *S. platensis* of 0.30, 0.28 and 0.26, respectively,.

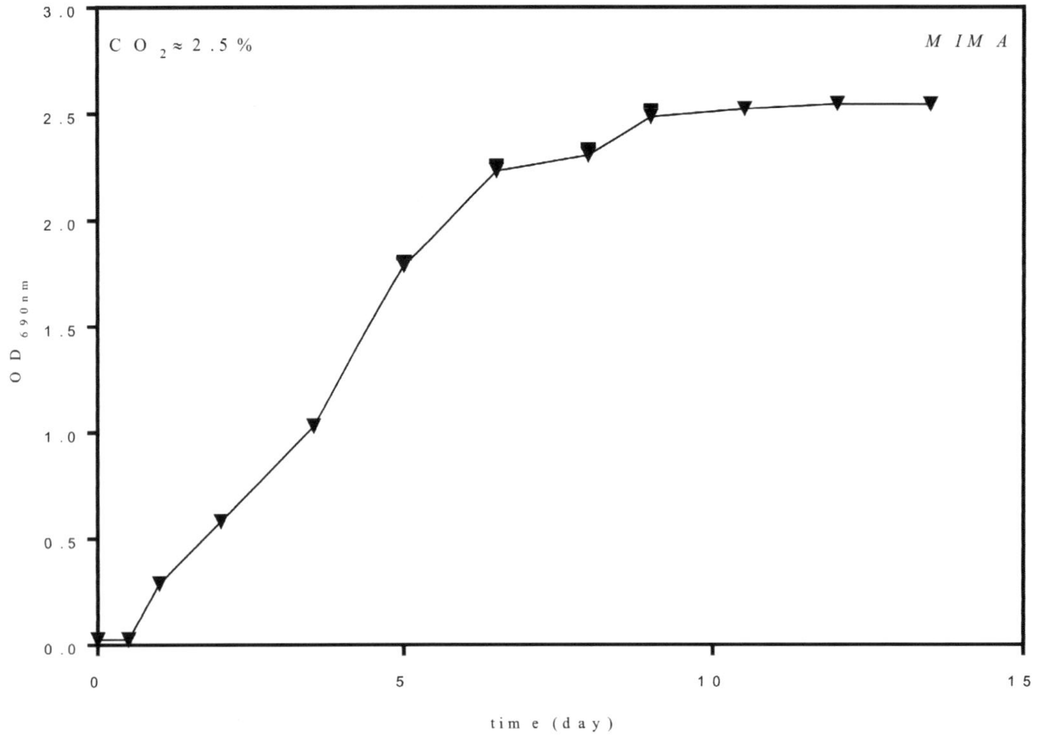

Figure 1: The change in optical density of *MIMA* as a function of cultivation time

The growth rate of algae was used to calculate dry cell weight (DCW; $g.L^{-1}.day^{-1}$), biomass productivity (BP; $g.L^{-1}.day^{-1}$) (Table 1). The BP values ranged from 0.028 to 0.235 $g.L^{-1}.d^{-1}$ for *MIMA, with* the lowest values achieved in cultures fed with high concentrations of CO_2.
The dry weight cell (DCW) achieved in the last day of culturing ranged from 0 0.232 to 0.950 $g_{dw.}$ $L^{-1}.d_{-1}$(for *MIMA*). The reported values are in the same order of magnitude for the values reported for *Chlorella sp.* cultivated with CO_2 concentrations ranging from 0.03 to 15% (v/v) [27].

Table 2: Growth rate and biomass productivity (BP) of *MIMA* at different periods and different CO_2 dosage

μ (Day^{-1})	BP

	(g.L-1.day-1)
0.288	0.235
0.302	0.283
0.347	0.376
0.336	0.293
0.147	0.041
0.110	0.028

Effect of CO_2 dosage on CO_2 bio-fixation rate

The carbon content of algae strain was analyzed using elemental CHN analyzer. The carbon content for MIMA ranged from 44.20 ± 1.33 to 56.20 ± 3.13 (%w/w. The results are in the same order of magnitude with the values reported in literature [28, 29].

Figure 2 shows the R_{CO_2} for the experiments carried out under varied CO_2 concentrations. For *MIMA* the values for R_{CO_2} ranged from 0.096 to1.074 $g_C\ L^{-1}\ d^{-1}$ with a maximum value occurs at 10% CO_2. The values reported in this study are slightly higher than the values reported in literatures for other algal strains. The high R_{CO2} in this study may be aimed to the natural growth condition used in the pilot plan. [28] reported R_{CO2} values for *S. obliquus* and *C. pyrenoidosa* in the range of 0.105 ±0.006 to 0.288±0.004 $g_C\ L^{-1}\ d^{-1}$.

The effect of increasing CO_2 concentrations on R_{CO2} was similar to the one observed for μ and BP. For both strains, the lowest R_{CO2} values were observed to occur for cultures cultivated at high temperature periods and high concentration of CO_2. Increasing the concentration of CO_2 in the gas stream from 2.5% to 10% and 15% led to an increase in R_{CO2} values for *SP.PL* and *MIMA*, respectively, followed by a decrease determined for cultures grown with higher CO_2 concentration

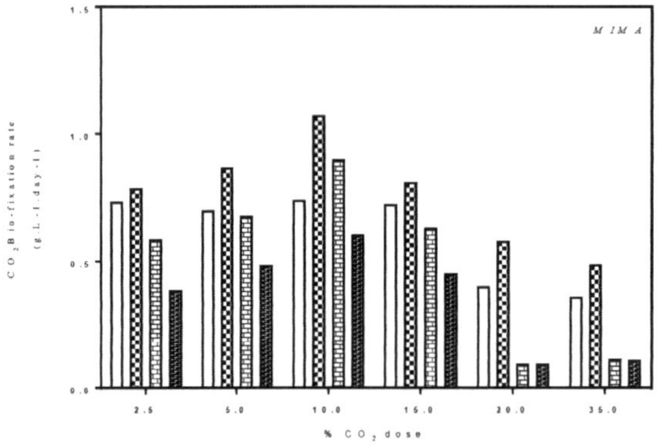

Figure 4: Carbon bio-fixation rate under varied CO_2 concentrations for *MIMA*.

Conclusion

Mixed indigenous microalgae (*MIMA*) showed a very effective tendency toward carbon dioxide capturing. This simple and cost-effective treatment option was successfully employed for 3 month under hot climate condition to remove nutrient from wastewater and capture CO2 from flue gases. The results of growth rate, biomass productivity, CO2 bio-fixation rate are promising suggesting that this technology and be a suitable alternative to chemical process when space and solar lighr are available. *MIMA* grow well in wastewater in the presence of CO_2 with a maximum CO_2 uptake of 35 v/v%.

References

[1] S.A. Razzak, M.M. Hossain, R.A. Lucky, A.S. Bassi, H. de Lasa, Integrated CO2 capture, wastewater treatment and biofuel production by microalgae culturingâ€"A review, Renewable and Sustainable Energy Reviews, 27 (2013) 622-653.

[2] M. Adamczyk, J. Lasek, A. SkawiÅ„ska, CO(2) Biofixation and Growth Kinetics of Chlorella vulgaris and Nannochloropsis gaditana, Applied Biochemistry and Biotechnology, 179 (2016) 1248-1261.

[3] W. Zhou, J. Wang, P. Chen, C. Ji, Q. Kang, B. Lu, K. Li, J. Liu, R. Ruan, Bio-mitigation of carbon dioxide using microalgal systems: Advances and perspectives, Renewable and Sustainable Energy Reviews, 76 (2017) 1163-1175.

[4] D.L. Sutherland, C. Howard-Williams, M.H. Turnbull, P.A. Broady, R.J. Craggs, Seasonal variation in light utilisation, biomass production and nutrient removal by wastewater microalgae in a full-scale high-rate algal pond, Journal of Applied Phycology, 26 (2014) 1317-1329.

[5] D.L. Sutherland, C.H. Williams, M.H. Turnbull, P.A. Broady, Craggs R. J., Frequency of CO2 supply affects wastewater microalgal photosynthesis, productivity and nutrient removal efficiency in mesocosms: implications for full-scale high rate algal ponds, , J. Appl. Phycol. , 27 (2015) 1901–1911.

[6] F.A. AlMomani, B. Ormeci, Performance Of Chlorella Vulgaris, Neochloris Oleoabundans, and mixed indigenous microalgae for treatment of primary effluent, secondary effluent and centrate, Ecological Engineering, 95 (2016) 280-289.

[7] H. Znad, A.M.D. Al Ketife, S. Judd, F. AlMomani, H.B. Vuthaluru, Bioremediation and nutrient removal from wastewater by Chlorella vulgaris, Ecological Engineering, 110 (2018) 1-7.

[8] W. Zhou, Y. Li, Y. Gao, Z. H2., Nutrients removal and recovery from saline wastewater by Spirulina platensis. B, ioresource Technology, 245 (2017) 10-17.

[9] S.J. Judd, F.A.O. Al Momani, H. Znad, A.M.D. Al Ketife, The cost benefit of algal technology for combined CO2 mitigation and nutrient abatement, Renewable and Sustainable Energy Reviews, 71 (2017) 379-387.

[10] M. Shurair, F. Almomani, J. Simon, R. Bhosale, A. Kumar, U. Gosh, M. Khreisheh, Advanced wastewater treatment using microalgae: effect of temperature on removal of nutrients and organic carbon, 7th International Conference on Environment and Industrial Innovation 67(2017)012032 (2017).

[11] S. Contreras, M. Rodríguez, F. Al Momani, C. Sans, S. Esplugas, Contribution of the ozonation pre-treatment to the biodegradation of aqueous solutions of 2,4-dichlorophenol, Water Research, 37 (2003) 3164-3171.

[12] F. Al Momani, O. Gonzalez, C. Sans, S. Esplugas, Combining photo-Fenton process with biological sequencing batch reactor for 2,4-dichlorophenol degradation, Water Science and Technology, 2004, pp. 293-298.

[13] F.A. Almomani, R. Delatolla, B. Örmeci, Field study of moving bed biofilm reactor technology for post-treatment of wastewater lagoon effluent at 1°C, Environmental Technology (United Kingdom), 35 (2014) 1596-1604.

[14] Sánchez-Fernández J.F., González-López C.V., Acién F.G., Fernández-Sevilla J.M., Molina E., Utilization of Anabaena sp. in CO2 removal processes: modelling of biomass, exopolysaccharides productivities and CO2 fixation rate. , Appl. Microbiol. Biotechnol, 94 (2012).
[15] J. Zhang, O. Nwani, Y. Tan, D.W. Agar, Carbon dioxide absorption into biphasic amine solvent with solvent loss reduction, Chemical Engineering Research and Design, 89 (2011) 1190-1196.
[16] I. de Godos, J.L. Mendoza, F.G. AciÃ©n, E. Molina, C.J. Banks, S. Heaven, F. Rogalla, Evaluation of carbon dioxide mass transfer in raceway reactors for microalgae culture using flue gases, Bioresource Technology, 153 (2014) 307-314.
[17] S.K. Singh, A. Rahman, K. Dixit, A. Nath, S. Sundaram, Evaluation of promising algal strains for sustainable exploitation coupled with CO2 fixation, Environmental Technology, 37 (2016) 613-622.
[18] F.A. Al Momani, B. Örmeci, Measurement of polyacrylamide polymers in water and wastewater using an in-line UV-vis spectrophotometer, Journal of Environmental Chemical Engineering, 2 (2014) 765-772.
[19] N.B. Abderrazik, F. Al Momani, C. Sans, S. Esplugas, Combined advanced oxidation with biological treatment, Afinidad, 59 (2002) 141-146.
[20] N.B. Abderrazik, F. Al Momani, M. Rodríguez, A. Azmani, C. Sans, S. Esplugas, Biodegradability improvement by photo Fenton reaction, Afinidad, 59 (2002) 391-398.
[21] A. Allagui, S. Sarfraz, S. Ntais, F. Al Momani, E.A. Baranova, Electrochemical behavior of ammonia on Ni98Pd2 nano-structured catalyst, International Journal of Hydrogen Energy, 39 (2014) 41-48.
[22] A. Allagui, N. Brazeau, H. Alawadhi, F. Al-momani, E.A. Baranova, Cathodic contact glow discharge electrolysis for the degradation of liquid ammonia solutions, Plasma Processes and Polymers, 12 (2015) 25-31.
[23] A. Allagui, S. Sarfraz, B. Middleton, F. Almomani, E.A. Baranova, Ammonia electrooxidation on NiPd nanoparticles in alkaline media: Effect of pH and concentration, ECS Transactions, 2012, pp. 1897-1906.
[24] R.R. Bhosale, A. Kumar, F. AlMomani, Kinetics of reactive absorption of CO2 using aqueous blend of potassium carbonate, ethylaminoethanol, and N-methyl-2-Pyrollidone (APCEN solvent), Journal of the Taiwan Institute of Chemical Engineers, 89 (2018) 191-197.

[25] R.R. Bhosale, A. Kumar, F. Almomani, M. Khraisheh, I. Alxneit, Catalytic Reduction of CO2 into Solar Fuels via Ferrite Based Thermochemical Redox Reactions, MRS Advances, 2017, pp. 3389-3395.

[26] A. Richmond, Physiological principles and modes of cultivation in mass production of photoautotrophic microalgae. In: Cohen Z(ed) Chemicals from microalgae., 1999.

[27] A.B. Fulke, S.N. Mudliar, R. Yadav, A. Shekh, N. Srinivasan, R. Ramanan, K. Krishnamurthi, S.S. Devi, T. Chakrabarti, Bio-mitigation of CO2, calcite formation and simultaneous biodiesel precursors production using Chlorella sp, Bioresource Technology, 101 (2010) 8473-8476.

[28] D. Tang, W. Han, P. Li, X. Miao, J. Zhong, CO2 biofixation and fatty acid composition of Scenedesmus obliquus and Chlorella pyrenoidosa in response to different CO2 levels, Bioresource Technology, 102 (2011) 3071-3076.

[29] A.L. Gonaalves, C.M. Rodrigues, J.C.M. Pires, M. Simaes, The effect of increasing CO2 concentrations on its capture, biomass production and wastewater bioremediation by microalgae and cyanobacteria, Algal Research, 14 (2016) 127-136.

531394 Fluid Flow and Nutrient Retention in Biochar Amended Soils

Abstract:

Scope and Significance: Over the past several decades, food production has kept pace with human population growth only because of the successful development of new high-yielding crop varieties grown with the help of fertilizers. Thus, vast amounts of energy are consumed every year to produce the approximately 200 Mt of ammonia needed to make the nitrogen (N) fertilizers used by farmers around the globe. Unfortunately, more than half of the N fertilizer applied to a field is not available for plant growth due to losses caused by surface runoff, leaching into surface and ground water, or volatilization. These losses cause serious water pollution problems ranging from the hypoxic zone in the Gulf of Mexico to the contamination of wells and groundwater with nitrates. In addition, fertilizer losses trigger the release into the atmosphere of large amounts of nitrous oxide (a major greenhouse gas) and NO_x.

The amendment of soils with biochar has been heralded as a sustainable method for preventing pollution problems caused by fertilizer losses and, at the same time, increasing crop yields. However, the available experimental data are highly variable. While some field studies report significant beneficial effects with up to 60% higher yield after biochar addition, other studies report that biochar amendments either have no effect or even result in lower agricultural productivity [1-2]. A major source for the variability of experimental data is the fact that biochar is not a single entity. Dozens of biomass feedstocks are used to produce biochars in different types of reactors under widely varying temperature programs and pyrolysis atmospheres. Thus, biochars have a very broad spectrum of chemical and physical properties. To further complicate things, soil properties, climate conditions (like rainfall and temperature), plant requirements and many other parameters vary from application to application. Clearly, much better understanding of the complex dynamic interactions among the various components of the biochar/soil/climate system is needed before we can predict the environmental performance of a specific biochar.

Our Earlier Work in this Area: To begin bridging this knowledge gap, our recent research we focused on understanding the mechanisms that control the ability of biochars to improve crop nutrient availability and mitigate fertilizer leaching. First, we developed and tested a mathematical model that describes nutrient transport in soils amended with biochar[3]. Simulation results showed that biochar is effective in increasing nutrient availability for plants and minimizing fertilizer losses through leaching, but only if the biochar/soil ratio and some key biochar properties (like its adsorption capacity and affinity to the nutrient) are carefully matched to the soil properties (water velocity, soil type) and the amount of rainfall or irrigation.

A subsequent publication focused on the adsorption/desorption dynamics within biochar particles [4]. While it confirmed our original hypothesis that a biochar's ability to retain nutrients is influenced by the differential in adsorption and desorption rates, the second study revealed a more complicated dynamic behavior that originally envisioned. The ability of a biochar to adsorb and then slowly release the fertilizer is determined by a complex interplay of external mass transfer, intraparticle diffusion (occurring through either pore or surface diffusion), and adsorption dynamics. Biochar particle size and porosity modulate the interactions between adsorption/desorption and intraparticle or external mass transfer. And, fast surface diffusion improves the ability of a biochar to act as a slow release medium at high water interstitial velocities, pointing out that biochars with fast intraparticle transport rates are good candidates for high-permeability fields or high rainfall areas. These findings explain why biochars with similar properties can have widely different impacts on crop yields, as reported in the literature.

However, both these publications considered the response of biochar-amended soils to a single pulse of nutrient that simulates a single application. Continuous flow of water before and after the end of the fertilizer pulse was assumed.

Fluid Flow in Biochar-Amended Soils: This study presents the development and testing of a mathematical model that provides a much more realistic description of fluid flow in structured soils amended with biochars. We assume a triple-porosity structured soil that consists of uniform soil aggregates and biochar particles packed in a simple cubic structure. This structured soil has three pore systems: (i) the interstitial pores formed between the soil aggregates and the biochar particles; (ii) the less permeable intrapores of soil aggregate pores; and (iii) the intraparticle pores of the biochar. Water in all pore systems is assumed to be mobile. We should note here that biochars are very porous materials with total porosities as high as 0.8 and cavities connected to the exterior with micron-size mouths [5].

The solution of the flow equation of water in soils requires the water retention curve (WRC) that describes the relationship between the volumetric water content, or water saturation, and the capillary pressure. We used the van Genuchten and lognormal Kosugi models to fit the available experimental data for the WRC of (a) sandy soils and (b) soils amended with biochars having different particle size distributions. While the bimodal van Genuchten model provided very good approximations for the WRC of soils, the trimodal Kosugi model did a better job fitting the experimental data for biochar-amended soils. Because of the semi-empirical nature of the van Genuchten model parameters, bimodal and trimodal Kosugi models were used to fit the WRC of all porous media considered in this study.

Richards' equation was used to describe two-phase flow in our triple-porosity medium with water as the invading fluid. Water transfer between the interstitial pores and the pores of soil aggregate or biochar particles was modeled with first-order rate equations. The pressures of both fluids in the three pore systems were assumed to be equal at each spatial position (local equilibrium). We also assumed constant densities and viscosities, and negligible gravity effects. With these assumptions, we obtained a system of coupled partial differential equations for the air and water saturations that was solved together with the appropriate boundary and initial conditions using the IMPES finite difference method. A full parametric study was conducted to select the time step and grid sizes that satisfy the stability criterion and minimize mass balance errors. In all cases, the PDEs were integrated for long enough times to describe the wetting and drying dynamics, as well as the temporal evolution of water velocities in the topsoil. As is well known, only the top approximately 30 cm of soil are critical for nutrient transport and crop uptake.

Simulation results with the WRC obtained experimentally for (a) sandy soils and (b) soils amended with 2 wt% biochar show that the addition of biochar increases only slightly either the time required to fully saturate the topsoil (about 12 hours in both cases) or the water holding capacity of the amended soil. Small effects on the topsoil saturation times were also observed with biochar/soil ratios as high as 5 wt% and 0.5-0.8 total biochar porosities [5], the typical range of these parameter values for field applications. Clearly, the beneficial effect of biochar comes primarily from its ability to adsorb and slowly release nutrients (fertilizer).

Having computed the temporal evolution of water velocities in the top soil during wetting and drying events, we revisited our earlier model to study the adsorption/desorption dynamics of biochar-amended soils with realistic velocity profiles. We will present simulation results describing how the way in which the fertilizer is applied (flow rate, nutrient concentration etc.) affects the initial spatial distribution of nutrient in the topsoil and how this distribution changes over time in response to irrigation or rain events and dry periods. These simulations confirm our earlier results regarding the benefits of biochar amendments and provide guidelines for rational management of fertilizer use.

References:

[1] Spokas, K. A.; Cantrell, K. B.; Novak, J. M.; et al. "Biochar: A Synthesis of Its Agronomic Impact beyond Carbon Sequestration."*J. Environmental Quality***41**(2012) 973-989.

[2] Crane-Droesch, A; S. Abiven; S. Jeffery; M.S. Torn. "Heterogeneous global crop yield response to biochar: a meta-regression analysis." *Environmental Research Letters***8**(2013) 044049.

[3] Sun, S; C. E. Brewer; C. A. Masiello; K. Zygourakis. "Nutrient Transport in Soils Amended with Biochar: A Transient Model with Two Stationary Phases and Intraparticle Diffusion" *Industrial and Engineering Chemistry Research*, **54**(2015) 4123–4135.

[4] Zygourakis, K. "Biochar soil amendments for increased crop yields: How to design a "designer" biochar," *AIChE J.*, **63**(2017), 5435-5437.

[5] Brewer, C.E.; V.J. Chuang; C.A. Masiello; H. Gonnermann; X. Gao; B. Dugan; L. E. Driver; P. Panzacchi; K. Zygourakis; C.A. Davies, "Beyond Biochar Chemistry: Measuring Density Towards an Understanding of Porosity and Environmental Interactions," *Biomass and Bioenergy***66**(2014) 176-185.

537505 Sorption Characteristics of Nitrogen and Phosphorus Onto Biochar from Aqueous Solution

Abstract:

Nitrogen (N) and phosphorus (P) pollution in water has created a serious concern worldwide due to its adverse effects on the environment. N and P concentrations have increased in water due to the discharge from point and non-point sources. High N concentration causes various health problems such as methemoglobinemia, or blue baby syndrome. On the other hand, P is one of the limiting nutrients to control eutrophication in natural ecosystems. Eutrophication causes excessive algal growth, known as algal bloom, which has various detrimental effects such as increase of water treatment cost, decrease of oxygen content and recreational use of water. Different technologies are available to remove N and P from water. Among these, the adsorption process has proven to be an effective and low-cost technique for the removal of these pollutants. Nowadays, biochar as an adsorbent has received significant attention because of its distinctive characteristics. Therefore, this study focuses on the removal of N and P from aqueous solution by batch adsorption experiments using biochar derived from wood chips (spruce, pine, and fir). Sorption characteristics of N and P onto biochar were evaluated as a function of biochar dosages, initial concentrations of N and P, and pH using synthetic water. At the end of each batch test, samples were collected and analyzed to determine the residual NH_4^+-N and P concentrations using a HACH DR 2800 Spectrophotometer. Results demonstrated that N removal increased to 80% with the increase of biochar dosage at 100 g/L whereas N removal decreased with the increase of initial N concentration. The highest N removal was achieved in a pH range from 4 to 6. In case of P removal, the percentage of P removal increased consistently with the increase of biochar dosages except at the end of the test. The highest P removal was 60% at a biochar dosage of 80 g/L. P removal deteriorated with the increase of initial P concentration. The maximum P removal was obtained at a pH value of 2. The sorption isotherm followed the Langmuir model for both N and P sorption onto biochar. The sorption kinetics of N and P onto biochar followed a second order kinetic model. This study indicated that biochar has the potential to remove both N and P from aqueous solution. Further experiments are in progress to investigate the removal mechanism of N and P onto biochar using Fourier Transform Infrared (FTIR) Spectroscopy.

525230 Lead Removal from Water Using Insoluble Bacterial Carboxymethyl Cellulose

Abstract:

INTRODUCTION

Among the most common water pollutants, heavy metals are highly disturbing due to their non-biodegradability and their high toxicity. For instance, lead (Pb) is a potent neurotoxin which may cause critical diseases in human beings, including cancer. Thus, many researches invest their efforts in the development of systems that allow the elimination of lead from water to safe levels[1]. In this context, many cellulose-containing materials and cellulose derivatives have been reported as useful for the adsorption of heavy metals[2]. Both, the natural abundance of cellulose, and the possibility of obtaining it not only from pulps derived from wood but also from a number of alternative abundant and underutilized cheaper cellulose sources, make this polysaccharide very attractive as raw material. On the other hand, the separation processes of cellulose from plant sources require the use of reagents that can have a negative environmental impact[3]. Conversely, bacterial cellulose (BC) has the same molecular formula as plant cellulose, but it has the advantage of being produced free of lignin, hemicelluloses and pectin, thus avoiding the need for chemical treatments devoted to the removal of these compounds prior to cellulose derivatization.

Taking into account the above discussion and considering that Pb (II) exists as a positive ion, the use of insoluble carboxymethyl cellulose (CMC) as cation exchanger could be a good alternative for lead removal from water. Water solubility of CMC is determined by the degree of substitution[4] (DS). Previously, some of us reported a systematic study of reaction variables to tailor the DS of carboxymethylated BC (CMBC)[3]. Based on our previous experience, in this work, insoluble carboxymethyl cellulose was synthetized and lead adsorption studies were carried out.

EXPERIMENTAL

BC and CMBC were produced as we previously described[3,5]. The DS of CMBC was determined by conductometric titrations as it was described elsewhere[4]. *FT-IR* spectra of the samples were acquired on an IR Affinity-1 Shimadzu Fourier Transform Infrared Spectrophotometer in absorbance mode.

In order to study the water solubility of CMBC, a film was made by drying the sample between glasses at 110°C for 2 h. The film obtained was weighed before been placed in a beaker with 50 mL of distilled water under static conditions for 24 h. Within this interval, the film was periodically removed from water, dried at 110°C during 1 h and finally weighed.

Lead concentration was determined on an air-acetylene flame type atomic absorption spectrometer (FAAS) (Model iCE 3000, Thermo Scientific).

Adsorption Experiments

Adsorption kinetics

Native BC or CMBC (50 mg) was suspended in 100 mL of lead ion solution (100 $mg.L^{-1}$) at pH 7 and the mixture was stirred at 400 rpm at room temperature (25°C). Samples were taken at various time intervals and lead concentration was determined by FAAS. As was expected, native BC shows a maximum lead adsorption capacity almost ten times lower than CMBC.

Effect of solutions pH values

The effect of pH on the adsorption of lead ion was studied in a pH range of 2.5-7.0 at 25°C. 25 mL of an aqueous lead ion solution (100 $mg.L^{-1}$) was shaken with 25 mg of CMBC at 400 rpm for 2 h. The pH was adjusted using HCl (0.01 M) or NaOH (0.01 M). After filtration, lead concentration was determined by FAAS.

Adsorption isotherms

The adsorption isotherm experiments were conducted with initial lead ion concentration in the range of 50-150 mg.L^{-1}. 25 mg of CMBC was suspended in 25 mL of lead solutions with different concentration at pH 7 and 25°C. The mixture was stirred at 400 rpm for 2 h. After filtration, lead concentration was determined by FAAS.

Column experiment: Breakthrough curve

Column experiments were carried out using an acrylic column (I.D. 6 mm, L. 22 mm) filled with CMBC (50 mg) and connected to a peristaltic pump by tygon tubes. Aqueous lead ion solution (7 mg.L^{-1}) at pH 7 was sent into the column at a flow rate of 2.5 mL.min^{-1}. Effluent samples were taken every 10 mL and the concentration of lead ion in all samples were determined by FAAS.

RESULTS AND DISCUSSION

Characterization of the CMBC

The DS of CMBC determined by conductometric titration showed to be 0.17, and the insolubility of the sample was confirmed by no significant change in sample's weight after being 24 h in water. Comparison of *FT-IR* spectra of BC and CMBC samples also provided evidence of etherification (Fig. 1). In the CMBC spectrum, apart from the typical bands of cellulose[6], the band characteristic of the stretching vibration of the carboxylate groups at 1607 cm^{-1} was observed, confirming that the carboxymethylation had occurred.

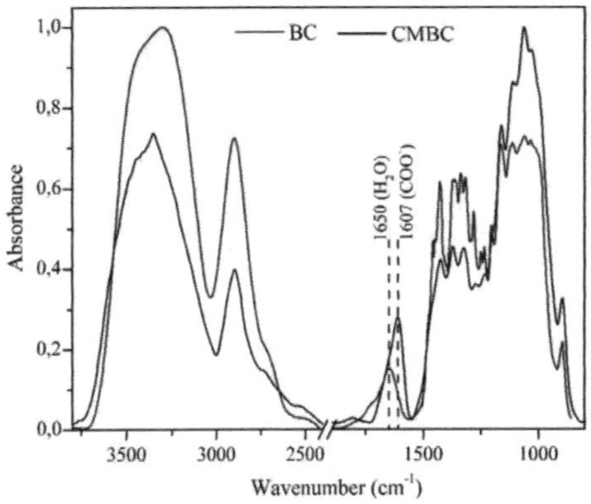

Fig.1. *FT-IR* spectra of BC and CMBC.

Adsorption kinetics studies

The uptake of lead by CMBC increased rapidly in the first 20 min and then the sorption slowed down significantly (Fig. 2).

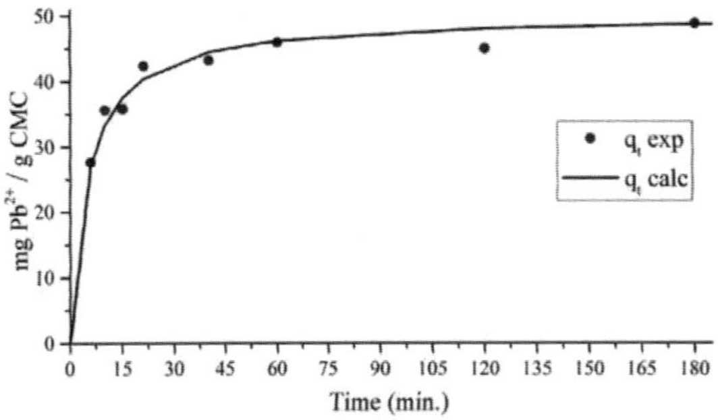

Fig.2. Amount adsorbed of Pb^{2+} on CMBC at different adsorption time (initial metal ion 100 mg.L^{-1}, pH 7).

To study the controlling mechanism of the adsorption process, experimental data were adjusted to pseudo-first-order and pseudo-second-order models (eq. 1 and 2, respectively).

eq.1 $$\ln(q_e - q_t) = -k_1 t + \ln q_e$$

eq.2 $$\frac{t}{q_t} = \frac{t}{q_e} + \frac{1}{k_2 q_e^2}$$

Results undoubtedly showed that the pseudo-second-order kinetic equation more appropriately described the lead adsorption on CMBC (Table 1). Therefore, as it was reported elsewhere, the adsorption mechanism is kinetically controlled mainly by chemical interactions[7,8]. Moreover, the excellent concordance between the experimental data and those calculated by means of the pseudo-second-order kinetic equation, is evidenced in Figure 2.

Pseudo-first-order model			Pseudo-second-order model		
k_1 (min^{-1})	q_e ($mg\ g^{-1}$)	R^2	k_2 ($g\ mg^{-1}\ min^{-1}$)	q_e	R^2
0.033	19.6	0.8962	0.004	50.0	0.9998

Table 1. Comparison of the adsorption kinetic models

Effect of pH

Figure 3 shows the strong pH dependence of lead adsorption on CMBC. The critical role of the pH on the adsorption capacity of the material was in concordance with that expected considering that the pka values of the carboxylic acids is around 5.

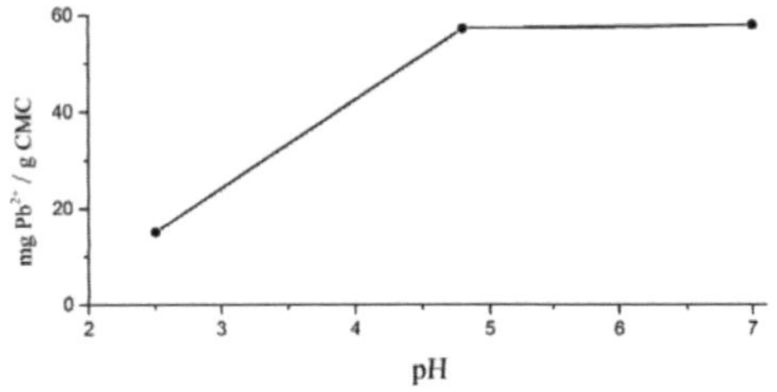

Fig.3. Effect of pH on removal of lead: adsorbent dose, 50 mg/25 mL; Pb^{2+} concentration, 100 $mg.L^{-1}$

Adsorption isotherms

Experimental isotherm data were fitted to different models of sorption isotherms: Langmuir, Freunlich, Dubinin-Radushkevich and Frumkin. As can be seen in Table 2, the Langmuir equation gives the best fit of our data. The experimental data of Pb^{2+} sorption on CMBC as well as the isotherm calculated by the Langmuir model are shown in Figure 4.

Model	Equation	Parameters		
Langmuir	$\frac{C_e}{q_e} = \frac{1}{bQ_0} + \frac{C_e}{Qo}$	Q_0	59.54	$mg\ g^{-1}$
		b	0.5146	$L\ mg^{-1}$
	$R_L = \frac{1}{1 + bC_0}$	R^2	0.9995	
Freundlich	$\log q_e = \log k_F + \frac{1}{n} \log C_e$	K_f	4.95	
		n	11.02	
		R^2	0.9503	
Dubinin–Radushkevich	$\ln q_e = \ln q_{D-R} - k_{D-R}\, \varepsilon^2$	q_{D-R}	55.6	$mg\ g^{-1}$
	$\varepsilon = RT \ln\left(1 + \frac{1}{C_e}\right)$	k_{D-R}	5.06 10^{-7}	$mol^2\ J^{-2}$
		E	994.4	$J\ mol^{-1}$
	$E = \frac{1}{\sqrt{2k_{D-R}}}$	R^2	0.7033	
Frumkin	$\ln\left[\left(\frac{\theta}{1-\theta}\right)\frac{1}{C_e}\right] = \ln k + 2a\theta$	a	11.76	
		k	2.22 10^{-7}	
	$\theta = \frac{q_e}{q_{max}}$	R^2	0.969	

Table 2. Adsorption isotherm model parameters.

On the other hand, the separation factor R_L indicates the type of the isotherm to be either unfavorable ($R_L>1$), linear ($R_L=1$), favorable ($0<R_L<1$) or irreversible ($R_L=0$). In our case, the values of R_L factor were always between 0 and 1 for any value of initial concentration (C_0), indicating a favorable adsorption of lead onto CMBC[9,10].

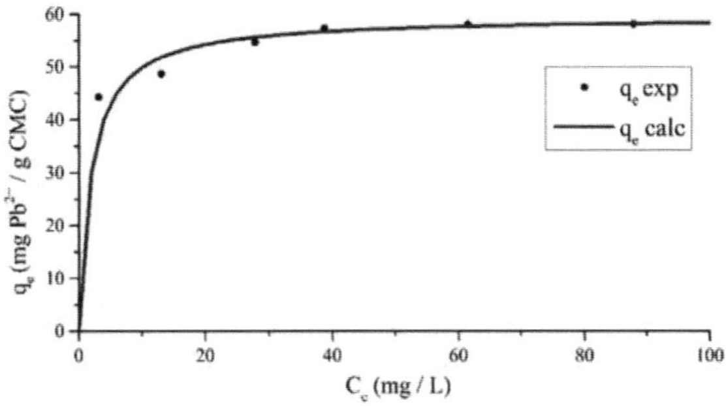

Fig.4. Adsorption isotherm of lead on CMBC (pH 7, 25 °C)

Column experiment: Breakthrough curve

In order to obtain the characteristic parameters for fixed-bed systems, experiments using a column packed with CMBC under continuous flow were conducted. The performance of this experiment can be described in terms of time vs. effluent concentration (breakthrough curve) (Fig. 5).

Several simple mathematical models have been developed to predict the dynamic behavior of the column. Among them, Bohart-Adams, Clark and modified dose-response models are recommended because of the usefulness of the designed parameters that they provide[11]. Thus, those models were used here to estimate and analyze the column adsorption performance. In order to evaluate the goodness-of-fit of the equation to the experimental data, the correlation coefficient (R^2) as well as the residual root mean square error (RMSE) and the chi-square test (χ^2) are presented.

As can be seen, the Bohart-Adams model gives the best fit to the experimental data (Table 3).

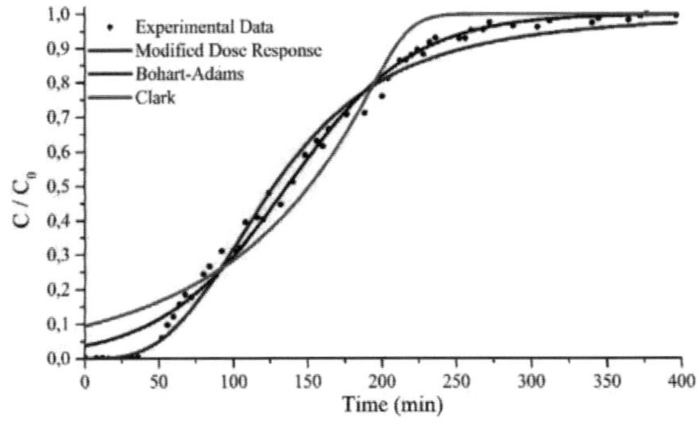

Fig.5. Breakthrough curve.

Model	Equation	Parameters		
Modified Dose Response	$\frac{C}{C_0} = 1 - \frac{1}{1 + \left(\frac{C_0 Q t}{q_0 x}\right)^a}$	q_0	39.2	mg g^{-1}
		a	3.15	
		R^2	0.990	
		RMSE	0.071	
		χ^2	0.00137	
Bohart-Adams	$\frac{C}{C_0} = \frac{e^{k_{BA} C_0 t}}{e^{k_{BA} N_0 \frac{Z}{u}} - 1 + e^{k_{BA} C_0 t}}$	k_{BA}	0.00365	L min^{-1} mg^{-1}
		N_0	2891.4	mg L^{-1}
		$q_0 = N_0 / \rho_L$	33.3	mg g^{-1}
		R^2	0.992	
		RMSE	0.056	
		χ^2	0.00108	
Clark	$\frac{C}{C_0} = \left(\frac{1}{Ae^{-rt} + 1}\right)^{\frac{1}{n-1}}$	A	1.74 10^{10}	
		r	0.11	min^{-1}
		R^2	0.962	
		RMSE	0.263	
		χ^2	0.00506	

Table 3. Breakthrough curve model parameters.

CONCLUSIONS

In conclusion, water insoluble CMBC was synthesized and its adsorption capacity of lead ions was investigated. CMBC showed good adsorption performance in the pH range of 4.5-7.0. The adsorption was a fast process following the pseudo-second-order kinetic model, indicating the importance of chemical interactions in the process. The experimental adsorption isotherm data and the breakthrough curve were well fitted with the Langmuir and Bohart-Adams models respectively.

REFERENCES

(1) Hooper, C.; Shi, X. *Journal of environmental health* **2012**, *75*, 14.

(2) O'Connell, D. W.; Birkinshaw, C.; O'Dwyer, T. F. *Bioresource technology* **2008**, *99*, 6709.

(3) Casaburi, A.; Montoya Rojo, Ú.; Cerrutti, P.; Vázquez, A.; Foresti, M. L. *Food Hydrocolloids* **2018**, *75*, 147.

(4) Eyler, R.; Klug, E.; Diephuis, F. *Analytical chemistry* **1947**, *19*, 24.

(5) Cerrutti, P.; Roldán, P.; García, R. M.; Galvagno, M. A.; Vázquez, A.; Foresti, M. L. *Journal of Applied Polymer Science* **2016**, *133*.

(6) Ashori, A.; Babaee, M.; Jonoobi, M.; Hamzeh, Y. *Carbohydrate Polymers* **2014**, *102*, 369.

(7) Ho, Y. S.; McKay, G. *Process Biochemistry* **1999**, *34*, 451.

(8) Ren, H.; Gao, Z.; Wu, D.; Jiang, J.; Sun, Y.; Luo, C. *Carbohydrate Polymers* **2016**, *137*, 402.

(9) Dolphen, R.; Sakkayawong, N.; Thiravetyan, P.; Nakbanpote, W. *Journal of Hazardous Materials* **2007**, *145*, 250.

(10) Singh, J.; Mishra, N. S.; Uma; Banerjee, S.; Sharma, Y. C. *BioResources* **2011**, *6*, 2732.

(11) Lee, C.-G.; Kim, J.-H.; Kang, J.-K.; Kim, S.-B.; Park, S.-J.; Lee, S.-H.; Choi, J.-W. *Desalination and Water Treatment* **2015**, *55*, 1795.

528350 Design of an Electrochemical Cell for Desalination of Seawater

Abstract:

With rapidly increasing water-stressed regions and global fresh water demand rising at an alarming rate, there is a need to develop an efficient technology to supply fresh water. About 70% of earth being seawater, one of the obvious solutions to meet the freshwater crisis is desalination of seawater. One of the biggest challenges here is that seawater is heavily saline and must be desalinated to about 99% in order derive any application from it such as irrigation or drinking. Present technologies used commercially for seawater desalination are Reverse Osmosis (RO) are Multi-Stage Flash Distillation (MSF). However, there is a trade-off between energy efficiency and salt rejection efficiency in these methods. In the present work, we propose a stack-type electrochemical cell employing bipolar membranes to desalinate seawater. The cell stack of a bipolar membrane, anion-exchange membrane, cation-exchange membrane, and bipolar membrane divide seawater into a brine stream and a clean water stream by electrodialysis. The feasibility of this design was analyzed using COMSOL Multiphysics. The simulation results show that a stacked arrangement of such cell can successfully reach a desalination efficiency of 99% at near neutral pH. To operate such a system renewably, the power of operation may be extracted from a solar cell device. A detailed balance (Shockley-Queisser limit) analysis was performed to determine the current-voltage characteristics of multijunction light absorbers. An operating point (current density and voltage) was obtained for ideal single, double, and triple junction light absorbers for the electrochemical desalination of seawater. Making an integration with solar cell will make this design portable and would be easily accessible in the water-stressed regions.

528805 An Investigation of Flow Obstructions to Minimize Media Loss in Simultaneous Air/Water Backwash Operations in Gray Water Filtration Systems

Abstract:

Simultaneous air/water backwash in water filtration systems has been shown to be more effective in cleaning filtration media than water-only backwash or sequential water and air backwash. However, simultaneous air/water backwash operations can result in unacceptable media loss. This work is an investigation of troughs as flow obstructions in gray water filtration tanks during backwash operations. These flow obstructions can generate quiescent zones near the liquid surface and allow for the settling of media particles thus decreasing media loss. Media loss during simultaneous air/water backwashing in a pilot scale (D= 42 in) vertical system was investigated as a function of air/water flow rates. Flow velocity profiles near the trough were measured to obtain velocities in two directions and a measure of turbulence in the system. Industrial scale water and air flow rates of 3 GPM/sq.ft. and 5 GPM/sq.ft. and 0.8 scfm/sq.ft. and 1.5 scfm/sq.ft. respectively were used. Commercial anthracite and Greensand PlusTM were used as filtration media. The rationale for the trough design used will be discussed in the context of the complexity of these systems. These results will be compared to laboratory scale measurements and the effect of system and air/water flow rate scales will be discussed.

529536 Appilcation of Polysaccharide Derivatives As Novel Draw Solutes in Forward Osmosis for Protein Concentration

Abstract:

In this study, a series of anionic polyelectrolytes, sodium carboxymethyl dextrans with different molecular weights (CM-dextran-1000/5000/20000), are facilely synthesized via a one-step reaction and employed as novel draw solutes in FO process. Various characterizations (FTIR, NMR, conductivity titration and cytotoxicity test) are conducted to confirm the chemical structure, substitution degree and toxicity of the synthesized CM-dextrans. Effects of the concentration, molecular weight and substitution degree on the physicochemical properties (solubility, relative viscosity, conductivity and osmotic pressure) of CM-dextrans as well as their corresponding FO performance are investigated systematically. Compared with 1 M NaCl draw solution, 40 wt% CM-dextran-1000 draw solution shows a higher FO water flux (24.9 LMH) and a much lower reverse solute flux (0.97 gMH) in PRO mode. The recyclability of the diluted CM-dextran draw solutions is evaluated by an ultrafiltration process. Furthermore, the developed CM-dextrans with low reverse solute diffusion and nontoxicity can be potentially employed as promising draw solutes for the protein concentration via FO.

531919 Analysis and Control of Al Concentration in Groundwater Based on Mathematical Modeling and Laboratory Tests

Abstract:

Aluminium (Al) is present in all types of natural water sources and additionally comes to drinking and industrial waters from alum used as a coagulant in water treatment technologies (Sposito, 1996). The recommended upper boundary limit of Al concentration in drinking water is 0.20 $mg \cdot L^{-1}$ in accordance with the U.S. Environmental Protection Agency (USEPA, 2012). The problem of elevated Al level has to be considered as a growing number of epidemiological studies linked the high Al content of drinking water as being proportional to the degree of neurological disease incidence (Wang et al., 2016). Current investigation has an industrial dimension: the existing problem of elevated Al concentrations in the groundwater of the Khibiny alkaline massif (Kola Peninsula) — area with a large-scale industrial production of apatite (Konukhin et al., 2012). A high Al level in water of this aquifer is described as a serious health concern when used for household water supply. According to the available data set of groundwater quality monitoring, during the past 10 – 12 years there is a general increase of Al concentrations in groundwater of Khibiny water aquifer. The current concentrations exceed the guideline up to nine times. The main objectives of current investigation are the quantitative analysis of the long-term results of field data monitoring program by using chemometric methods and a study of reliable adsorbents to remove Al effectively under the water aquifer-specific natural conditions.

The time series (1999 – 2012) of 12 physico-chemical variables monitored from ten groundwater wells were examined in both time and frequency domains by using multiple and multivariate regression methods as well as spectral analysis based on fast Fourier transform (FFT) algorithm. Computed pairwise correlation coefficients matrix revealed that pH values, concentrations of sulfate SO_4^{2-}, nitrate NO_3^-, chloride Cl^-, and total dissolved solids TDS correlate with Al concentrations at a statistically significant level (α-level=0.05). According to the coefficient of determination R^2 the multiple regression model considering the result of multicollinearity check explained up to 54% of Al temporal variation linked to pH values, concentrations of Cl^- and NO_3^- in groundwater. The sequential chain of contribution to the R^2 percentage is as follows: pH > NO_3^- > Cl^-. The cluster analysis (CA) identified three different groups of similarity based on correlation between variables within the dataset. The factor analysis/principal component analysis (FA/PCA) revealed three factors defined as linear combination of the original variables, capturing 58.7% of the total variance of the data set and allowed to group the selected variables according to common features. The first factor accounting for the most of total variance (34%) of the data set includes Al, pH, TDS, NO_3^-, Cl^-, SO_4^{3-}. Considering the results of multiple, multivariate regression analyses and multicollinearity check the Fourier coefficients, power spectral density (PSD) and cumulative spectral power (CSP) were calculated for Al, pH, NO_3^-, and Cl^- time series to identify and analyze the strength of their variations as a function of frequency.

Calculations revealed the spectrums of these compounds include three frequency bands corresponding approximately to 5 – 7, 13 – 17 and 20 – 34 months periods. The fluctuations within these bands contribute mostly to the total temporal variation of Al, Cl^-, NO_3^- concentrations and pH values.

The Khibiny alkaline massif groundwater most of the time does not meet established drinking water standards at least for two variables: the level of pH and concentration of Al. These variables are highly correlated, have similar variability patterns and belong to one cluster as the results of mathematical modeling revealed. The laboratory study has been undertaken to analyze adsorbents that can be reliable and cost-effective when applying to the water treatment process in the investigated area for Al removal under alkaline conditions – that is one of the major features of Khibiny massif groundwater. The initial Al concentration for all experiments was controlled at approximately 2 $mg \cdot L^{-1}$– the concentration is close to the maximum levels of Al observed in the groundwater based on the monitoring data set. Titanium dioxide Degussa P25 (TiO_2), a crushed vermiculite concrete modified by iron oxyhydroxide (VC-FeOOH) (Martemianov et al., 2017), montmorillonite K10 and silica gel (SiO_2) were four adsorbents tested in a concentration range 0.5 – 10 $g \cdot L^{-1}$.

The results of laboratory tests revealed that K10 and SiO_2 are favourable for the adsorption of Al in the acidic pH range, however no significant adsorption was found at pH > 8. An equilibration time of 2 hours for K10 and 10 hours for SiO_2 resulted in the removal of 99% and 57% of Al at pH 4 respectively. The alkaline conditions were further investigated for VC-FeOOH and TiO_2. The maximum adsorption of Al up to 90 – 91% onto VC-FeOOH took place within 4 hours under the constant alkaline conditions (pH 9). Laboratory testing showed that VC-FeOOH can act as a coagulant that accelerates the Al removal rate.VC-FeOOH loading ≤ 2 $g \cdot L^{-1}$ resulted in up to 95% removal due to Al precipitation within 30 min at final pH 8.1. Laboratory experiments with TiO_2 showed Al removal up to 75 – 78% at pH 9. The Al concentration decreases in three to five times within initial 7.5 min of experiments where TiO_2 loading was ≥ 1 $g \cdot L^{-1}$ and reached an equilibrium in 30 min. The mathematical description of equilibrium data was fitted most satisfactorily with Langmuir isotherm model.

The current investigation is an initial and necessary step to develop and implement a reliable method to reach a target level of Al removal in the groundwater of Khibiny alkaline massif. The application of chemometric methods allowed better understanding the major features of Al concentration variability as well as associations between Al and other compounds examined at both time and frequency domains. The evidence of direct anthropogenic influence on the elevated level of Al in groundwater was not releveled. The results obtained rather support that high Al concentration is caused by aluminum leaching from minerals. At big depths (more than 100 m) there are reduction conditions formed promoting the leaching in accordance with the recent geochemical analysis conducted in the area (Mazukhina et al., 2012). The implementation of adsorption process to the current technology at the Khibiny local water treatment facility can be an effective way to decrease Al concentrations below the guideline. Moreover, it is well known that TiO_2 can act as catalyst (Chen & Ray, 2001), therefore, the main direction of future works is to study how photocatalysis can improve Al removal. Although the current investigation has been done for the particular metal and groundwater source, similar analysis can be successfully applied to any other groundwater sources and could be extended to other toxic metals found in the groundwater.

537425 Modeling the Ionic Transport in an Electrodialysis Cell: Investigating the Impacts of Non-Ideal Solution Behavior in the Cell

Abstract:

Electrodialysis (ED) is an electro-membrane desalination technique in which ions are separated from a saline solution via selective transport through ion exchange membranes (IEMs) in the existence of an imposed electric field. Modeling the transport phenomena is a key to identify the optimum operating conditions to reduce energy consumption of the ED process. We use the fundamental applied equations for electrochemical systems including the Navier-Stokes, continuity, molar mass balance, and the Nernst-Planck to develop a 2-dimensional ion transport model for desalination of aqueous NaCl solution in an ED cell, considering the co-ion transport through IEMs. This study addresses the impact of the non-ideal behavior of electrolytes on ionic transport in both spacer-free and spacer-filled ED systems. The non-ideality of the electrolyte solution inside channels and IEMs is modeled by applying the polyelectrolyte NRTL activity coefficient model. The flow pattern in spacer-filled channels is determined considering the laminar flow assumption and submerged spacer configuration. The COMSOL Multiphysics software is used to numerically solve the set of developed equations employing the finite element method. The comparison of the ideal and the non-ideal solution models results demonstrates a slight deviation among them for concentrations up to 0.5 mol/kg NaCl for both spacer-free and spacer-filled ED systems. However, the eddy promotion in spacer-filled channels reduces the thickness of diffusion boundary layers, thus, the limiting current is reached at higher imposed potential compared to that of the spacer-free cell.

537633 Modeling the Ion Transport and Adsorption in a Capacitive Deionization Cell

Abstract:

Capacitive Deionization (CDI) is a promising technology for removing salts from brackish water by applying an electric field to drive ionic species toward oppositely charged electrodes. Thus ions are stored within the electrical double layer (EDL) formed near the surface of the electrode during charging process and pushed back to the bulk solution during discharging process, respectively. In this study, the CDI process is analyzed from both equilibrium state and transient transport state. A one-dimensional overlapped EDL adsorption model with changing stern layer thickness versus concentration was developed and the corresponding isotherm fit the experimental data well. A two-dimensional transient transport model incorporated with the adsorption model was built to study the diffusion, advection and electromigration of the ionic species in channel and electrode macropores and captured the ion adsorption dynamics based on changing inlet concentration, flow rate as well as voltage in the CDI system. Whatâ€™s more, the improvement in performance associated the use of spacers to aid mixing was explored with the model. Various operational regimes and the resulting power requirements were also evaluated to optimize the system.

537721 Using Methanotroph-Microalgae Coculture for Wastewater Treatment

Abstract:

Currently, the amount of excess fixed nitrogen in the world is causing increasingly negative consequences for our ecosystems and the public health, including worsening of the greenhouse effect, reducing the protection ozone layer, adding to smog, contributing to acid rain, and contaminating drinking water[1,2]. Consequently, effective treatment of wastewater (rich in nutrient nitrogen) plays a key role in managing the nitrogen cycle, and improving our ecosystems and the public health.

In the water resource recovery facilities (WRRFs), activated sludge is the most commonly applied approach to reduce biochemical oxygen demand (BOD) and remove nutrient (N and P). Anaerobic digestion (AD) is the most widely employed method to reduce the amount of sludge from both primary and secondary treatment, and to produce biogas, a renewable source of energy and feedstock to produce high-density fuels and chemicals. Besides, AD can mitigates pathogens and reduce odor. Because of these advantages, AD has been widely adopted by WRRFs in the US. It is estimated that 48% of total wastewater flows in the US is treated by AD, with at least 1238 WRRFs that currently process solids through AD. However, due to impurities (e.g., CO_2, H_2S, etc.) that are difficult to remove economically, the value of biogas is low. As a result, among 1238 WRRFs that has AD installed, about 80% (~975) of them flares the produced biogas or burn it for heating and cooking, while only 20% (~263) of them use biogas for power generation or driving machinery. The AD effluent is fed back to primary treatment to recover the nutrient.

In this work, we aim to develop a new wastewater treatment platform where we use coculture of methanotroph-microalgae to augment the AD treatment. In the proposed framework, the methanotroph-microalgae coculture not only recovers energy and carbon from biogas without requiring external oxygen supply, but also recover nutrient from AD effluent. In addition, the interdependent yet compartmentalized configuration of the coculture offers flexibility and more options for metabolic engineering for producing high-density fuels and commodity chemicals from various wastes. For any biotechnology to become potential industrial applications, the stability and robustness of the biocatalyst has to be established, in particular with respect to various contaminations contained in wastewater. Such robustness will determine the level of required pretreatment of wastewater, which could have a significant impact on the economic feasibility of the new platform.

In our previous work, we have assembled several synthetic methanotroph-photoautotroph cocultures that exhibit stable growth under various substrate delivery and illumination regimes[3]. In this work, using *Methylosinus trichosporium - Chlorella sorokiniana* as the model coculture system, we examine its potential in the combined biogas conversion and nutrient recovery from wastewater. Specifically, we first adapted both strains on diluted AD effluent using clarifier water (the water to be discharged from a WRRF), with all water samples provided from Columbus Water Works, and examined the performance of the coculture-based treatment. Our results show that cultivation of the coculture on AD effluent pretreated by filtering can support better growth than on AD effluent pretreated by a combination of filtering and autoclaving. In addition, we have shown that the coculture can be supported on AD effluent after allowing the sludge to settle.

Our results show that methanotroph-microalgae coculture offer a very promising route for simultaneous biogas conversion and nutrient recovery from AD effluent, while potentially producing value added products from wastes.

Reference

(1) Driscoll, C.; Whitall, D.; Aber, J.; Boyer, E.; Castro, M.; Cronan, C.; Goodale, C.; Groffman, P.; Hopkinson, C.; Lambert, K. Nitrogen Pollution: Sources and Consequences in the US Northeast. Environ. Sci. Policy Sustain. Dev. 2003, 45, 8.

(2) Galloway, J. N.; Dentener, F. J.; Capone, D. G.; Boyer, E. W.; Howarth, R. W.; Seitzinger, S. P.; Asner, G. P.; Cleveland, C. C.; Green, P. A.; Holland, E. A. Nitrogen Cycles: Past, Present, and Future. Biogeochemistry 2004, 70, 153.

(3) Roberts, N.; Hilliard, M.; Bahr, K.; He, Q. P.; Wang, J. Coculture of Methanotrophs and Microalgae â€“ a Flexible Platform for Biological CH4\/CO2 Co-Utilization, 2017 AIChE Annual Conference, Oct. 28 â€“ Nov. 3, 2017, Minneapolis, MN. This work won 2017 AIChE Sessionâ€™s Best Paper Award.

525695 Application of Biosurfactant Surfactin for Efficient Oil Separation from Waste Crude Oil Via Two-Step Process

Abstract:

Waste crude oil, a W/O type of crude oil emulsion which is inevitable produced during petroleum production, transportation and refining processes, needed efficient oil separation before discharge due to environmental risk and economic reasons, while it was still a challenge. This study developed a two-step process using surfactin as novel demulsifier for efficient oil separation. Firstly, surfactin was added to adsorb onto oil-water interface to co-stabilize emulsion droplet, and then the addition of acid caused surfactin precipitate and should be resulted in a sufficient oil separation. Such process is mainly based on the pH-switch emulsification-demulsification activity of surfactin as a result of dissolution or precipitation nature of surfactin under alkaline or acid condition. As found, surfactin at 200 mg/L can readily co-stabilize emulsion droplet with the native emulsifiers in emulsion, and then acid addition at 1% of 1M H_2SO_4 can achieve a quite well oil separation (over 95 %) during 10 min at 50 ^oC. The validity of this process was further confirmed on the waste crude oil, over 95 % of oil was separated within 30 min at 70 ^oC. This process have of great feasibility in industrial application for oil separation from crude oil emulsion.

521830 Impact of Seasonal Salinity Variations in Estuarine Systems: Thermodynamic Feasibility Analysis of Pressure Retarded Osmosis (PRO) and Reverse Osmosis (RO) Combinations

Abstract:

In a recent effort by the authors, the potential to extract energy from the salinity gradients present during the mixing of river and sea water was investigated for locations where river water properties vary significantly with the seasons. It was found that estuarine systems coupled with heavy rainfall pose considerable challenges if the technology of pressure retarded osmosis (PRO) is to be used to generate energy that can be fed back to the grid. The feasibility of such a process is heavily dependent on the salinity of the river water, which as a result of precipitation, changes drastically over a 12-month period for which the analysis was carried out. The highest mixing energies were 0.77 kWh/m^3, which is close to the thermodynamic limit, while the lowest obtained was 0.02 kWh/m^3. The benefit of energy generation through PRO- that the only by product is brackish water- is outweighed by the fact that it is not feasible in such locations. In order to make such a technology feasible, the membrane configurations must change with the change in salinity. Although energetically feasible, such configurations are not commercially viable yet, as membranes with different power densities will be required during different seasons so as to maintain a constant rated output throughout the year. This is a significant finding, since it demonstrates that for a difference of 10 bars in osmotic pressure differential, the membrane permeability has to be increased almost 3 folds. In estuarine systems, seawater intrusion is also a reason why salinities vary. Here, salinities vary with the distance from the point of confluence with the sea. A representative analysis has graphically shown that the pumping energy requirements exceed the energy generated through the PRO process. Temperature effects were also accounted for by using literature data from 3 decades ago, comparing with the latest data and analyzing the difference. The overall variation in temperature for the 12 months is approximately 4-5.C, for both the periods. However, an interesting observation is that the average temperatures throughout both the 12-month periods under study, varies by 1.2.C, with the average temperature being higher in 2014-15 period, than in 1981.The results are indicative of change in hydrological regimes of rivers, due to temperature change, which leads to increased mixing energies. Thus, having carried out the above studies, we conclude that using PRO process to generate energy is not a viable option in locations with seasonally varying river water properties. Alternate strategies must be adopted so as to make it energetically feasible, and economically viable. In attempts to address this problem, and eventually arrive at a solution, a hybrid system consisting of PRO and reverse osmosis (RO) system has been discussed.

The result of the above study was that greater the salinity gradient, greater the mixing energies emanated. However, for locations where the salinity and other properties change seasonally, the PRO unit alone was proving to be infeasible for an all year round operation as a result of varying salinities. To address the problems raised, the authors envisioned a configuration consisting of a RO desalination unit that pretreats the sea water entering the PRO, making it more saline for the draw solution, while the treated low salinity water from the RO unit can be used as the feed solution for the PRO unit. Energetically, this proved to be infeasible- using the thermodynamic limits, set by Gibbs energy formulations. However, such a system leads to membrane fouling that further prove its commercial infeasibility and ineffectiveness at tackling a major issue that the authors had sought to solve. It was thus found that if the PRO unit is thought of as a supplement to desalination RO unit, rather than the focus of the plant, then there could be multiple benefits, as opposed to having a hybrid PRO-RO system- which is not energetically feasible.

To address the problems raised in the previous study, the authors have worked on a different configuration of PRO process- in conjunction with an RO unit, such that the primary focus shall be on the RO unit that desalinates brackish water so as to provide potable water for use by households. A novelty of the configuration being studied is that OPEX of RO desalination process can be reduced by supplementing the energy from the PRO unit. Here, the PRO unit is being studied as an accessory to the desalination plant, rather than an energy generating unit by itself. Another major advantage of such a supplement to the normal RO desalination plant is that high salinity brine discharge to the water bodies can be avoided. Previously, high saline discharge would lead to degradation of the marine ecosystem in the region. Thus, energy requirements can be reduced and balance in the marine ecosystem is maintained as a result of this configuration change in the standard RO desalination unit. Reduced energy requirements benefit the consumer of the potable water as now the cost of producing same volume of potable water requires lesser energy than before.

The authors have attempted to tackle issues of energy and water- the two top challenges for the recent millennium and discover a solution which address both. They firmly believe that this is impactful, and can be further investigated before commercialization.

533324 Detection of Metallic Ions in Solution Using Optical Emission Spectroscopy of Plasma Driven By Bipolar Pulsed Power Sources

Abstract:

Detection of metallic ions in solution using optical emission spectroscopy in plasmas offers advantages including simultaneous detection of multiple elements, minimization of cross-element interference, and rapid detection. In this work, plasma ignited in metallic-salt-containing-solution using bipolar pulsed DC power is studied. The goal is to understand how the modulation of the bipolar pulsed power influences the plasma optical emission of metallic elements. A power source that provides asymmetric bipolar voltage pulses of a duration between 10 Î¼s to 100 ms with an amplitude up to Â±1 kV is used. Plasma is ignited inside the gas-phase formed in the solution. The gas phase composes of water vapor, oxygen, and/or hydrogen through joule heating and/or electrolysis. The plasma behavior and therefore the optical emission are greatly influenced by the width, polarity, and amplitude of the bipolar pulses. Voltage and current probe are used to conduct electrical analysis; optical emission spectrometer and photo multiplier tube are used to perform time-averaged optical emission spectroscopy and time-resolved emission intensity, respectively.

While using repetitive 7500 Î¼s-wide low voltage pre-pulses followed by 100 Î¼s-wide 600 V pulses to ignite plasma in solution consisting of 0.1 M HNO_3, 500 ppm Pb, and 250 ppm Na, we observe a low peak current with the application of the pre-pulse. The intensities and ratios of H, Pb, and Na emission lines are varied with different polarity and amplitude of the pre-pulse. The above observations show the modulation of the voltage pulses greatly influences the plasma behavior and therefore its optical emission. This work provides an effective way to tailor the plasma behavior and its optical emission, which could potentially be used as a strategy to obtain more reproducible and sensitive detection of metallic ion in water for analytical purposes.

523650 Iron(III)-Based Metal Organic Frameworks As Heterogeneous Fenton-like Catalysts for Organic Pollutants Degradation

Abstract:

Iron(III)-based metal organic frameworks as heterogeneous Fenton-like catalysts for organic pollutants degradation

Xie Quan

Key Laboratory of Industrial Ecology and Environmental Engineering (Ministry of Education), School of Environmental Science and Technology, Dalian University of Technology, Dalian, 116024, China.

Heterogeneous Fenton-like reaction is a promising technology for degrading refractory organic pollutants. Nevertheless, the catalytic performance of heterogeneous catalysts is low and needs to be improved. Iron-based metal organic frameworks have emerged as appealing materials in the field of catalysis because of the high surface area and porous structure, which endow them with sufficiently exposed active sites and favorable mass transfer of reactants. However, their catalytic efficiency for Fenton-like reaction and the relationship between the catalytic efficiency and building units (iron cluster and organic ligands), which is important to clarify the catalytic mechanism, remains unclear. Here, Fe-based MOFs (MIL-88B-Fe, MIL-53-Fe and MIL-101-Fe) with different structures of iron cluster were employed as heterogeneous Fenton-like catalysts. The influence of structures of iron cluster on the catalytic activity was investigated. Furthermore, varying MIL-88B(Fe)-X (X=$-NH_2$, $-CH_3$, -H, -Br and $-NO_2$) with substituent of different electrophilicity on organic ligands were synthesized and the relationship between the catalytic performance and organic ligands was investigated. It was found that MIL-88B-Fe with the highest amount of coordinatively unsaturated irons among three Fe-based MOFs (MIL-88B-Fe, MIL-53-Fe and MIL-101-Fe) presented the best catalytic performance, which was about 1-3 orders of magnitude higher than those for the three conventional catalysts (Fe_2O_3, Î±-FeOOH and Fe_3O_4). Moreover, the catalytic activity of MIL-88B-Fe could be improved by increasing the electrophilicity of substituent ($-NH_2 < -CH_3 < -H < -Br < -NO_2$) on organic ligands, which was attributed to the facilitated reduction of Fe(III) to Fe(II) resulted from the reduced electron density of Fe(III) by substituent with high electrophilicity. This work could provide new insights into the rational design and development of highly effective Fenton-like catalysts.

531236 Novel Materials and System Architecture for Membrane Based Water Treatment Technology

Abstract:

Surging population and climate change will push society, ever more urgently, to harvest lower quality or impaired water supplies (e.g., sea water or brackish groundwater) for drinking water. For example, desalination of sea water has become a promising technique for providing additional sources of fresh water along coastal communities. However, the widespread adoption of desalination technologies (Reverse Osmosis and Electrodialysis) is currently limited due to its energy intensive nature and/or limited membrane lifetime and performance, especially during high operation rates.

This work focuses on developing a novel, cost-effective, electrodialysis (ED) unit. Specifically, we have replaced planar electrodes in traditional electrodialysis units with 3-D electrodes. We hypothesize that the 3-D electrode bed configuration provides large surface area per unit volume, thereby allowing high currents to be drawn at current densities short of the diffusion-limited regime. Further, iR losses due to mass transfer limitation will be minimal.

533254 Developing a Prototype: "Portable" Solution for "Potable" Water

Abstract:

The development of reverse osmosis happened in 1960. Along the years this technique has been widely accepted all over world for varied applications ranging from seawater desalination to municipal water treatment. Forward osmosis (FO) is one of the foremost technologies for low energy consuming solutions for water purification. Water treatment is emerging as one of the most popular fields of engineering. Providing safe drinking water has become a challenge for various government agencies and municipal corporations across the world. While the nature of the problem remains same, the specifics related to treatment changes according to source and geographical locations. In this study we have carried out a detailed analysis on selection, design, development and pricing for a prototype of potable water system for purifying water in emergency situations. The portable and light purification system is envisaged to be driven by FO. This pouch will help serve as an emergency water filtration device. The current effort employs a model to understand the interplay of permeability and area on the rate of purification of water from any impure source / brackish water. The draw solution for the FO pouch is considered to be a combination of salt and sugar such that dilution of the same would result in an oral rehydration solution (ORS) which is a boon for dehydrated patients. Experiments have been carried out in designing, developing and checking the feasibility of pouch under standard conditions. However, the effort takes an extra step to actually estimate the cost and pricing of designing such a prototype.

While the mathematical model yields the best membrane (compositions taken from literature) combination in terms of permeability and area, the experiments shows the feasibility of prototype and the pricing takes into account the viability of such a solution to be made available as a retail item. The expected outcome of this work is to design a pouch for collecting purified water from brackish water source in stipulated amount of time depending on various conditions. The product is envisaged to be a market competitor for packaged drinking water and ORS combination (costing around $0.5 combined) and thus, to be feasible has to be priced around the same range with greater margins in order to have a better distribution. Thus, a proper business plan and production of the same has been formulated in order to be a feasible solution for unprecedented calamities and emergency situations.

536561 Technoeconomic Optimization of Emerging Technologies for Regulatory Analysis: NH_4HCO_3 Forward Osmosis for Power Plant Wastewater Treatment

Abstract:

The regulatory design process requires the use of tools that can rapidly assess the cost and performance of potential compliance technologies. Performing technoeconomic assessments of emerging treatment options for integrated, sustainable technologies is particularly challenging as models of full-scale performance do not yet exist (for emerging technologies) or highly nonlinear and with greater uncertainty (for integrated processes). In this talk, we demonstrate the use of gray-box optimization to inform process assessment during the development of regulations. We do this with the Effluent Limitation Guidelines for flue gas desulfurization wastewater discharges as a case study. We build a set of computationally lean meta-models of ammonia-bicarbonate forward osmosis (NH_4HCO_3 FO) â€" an emerging process that can be integrated with the power plant steam cycle or waste heat. Using these meta-models, we perform technoeconomic optimization of NH_4HCO_3 FO and crystallization to assess the capital and operational costs. We compare these NH_4HCO_3 FO costs to the costs for mechanical vapor recompression and crystallization, the EPA identified best available technology for achieving zero liquid discharge. We estimate that the NH_4HCO_3 FO treatment train would reduce zero liquid discharge costs by \$0.30-\$1.20/m^3 or \$20-\$70 million nationwide annually. This work highlights the opportunities and challenges associated with deploying gray-box optimization techniques for the rapid assessment of emerging technologies in policy analysis.

520341 Modeling and Experimental Approach Towards of Photoelectrocatalytic Bacterial Inactivation of *E.coli* Using Vertically Aligned ZnO/CuI for Water Treatment

Abstract:

Microbial contamination is one of the major issue in water treatment, causing tons of water-borne diseases. Therefore, an excellent type-2 staggered band aligned composite ZnO/CuI was grown/ immobilized on FTO conducting substrate for microbial inactivation by photoelectrocatalysis. Vertically aligned ZnO nanorods were modified in other morphologies via hydrothermal approach. Thereafter, incorporation of CuI was achieved on the optimized morphology using SILAR (successive ionic layer adsorption and reaction) method. The synergistic effect of the composite was observed for *E. coli* inactivation with respect to ZnO and CuI. Various characterizations such as X-ray diffraction, diffused reflectance spectroscopy, photoluminescence, ultraviolet photoelectron microscopy, X-ray photoelectron microscopy and scanning electron microscopy were carried out for phase purity, band gaps of the semiconductor, charge carrier lifetime and defects in the semiconductor, band edge positions, oxidation states of the elements, morphology of the synthesised photocatalyst, respectively. The mechanism of the charge transfer dynamics was derived based on the scavenger studies and band edge determination. Significant enhancement in the absorption wavelength and suppressed recombination gives rise to the augmented charge separation and better photoactivity of the composite. The experimental data is plotted with the help of the model derived for photoelecrocatalytic inactivation of bacteria. Bacterial inactivation mechanism was confirmed by various methods such as K^+ ion leaching, thiobarbituric acid assay for detection of lipid per oxidation. This present work opens the doors to explore this composite for wide area of application in the field of nanotechnology.

534024 Optimization and Green Synthesis (*Delonix regia* mediated) of Zero Valent Iron Nanoparticles

Abstract:

The present study focused to investigate the green synthesis of zero valent iron nanoparticles (ZVI-NPs) from *D. regia*. In order to produce maximum conversion (%) of Fe^{3+} to ZVI-NPs and minimum hydrodynamic size of ZVI-NPs, one variable at a time (OVAT) and response surface methodology (RSM) were adopted for process optimization. Confirmation of biosynthesis of ZVI-NPs was primarily done using UV-Visible spectrophotometer. The crystalline phase of nanoparticles was detected using X-ray diffraction (XRD) analysis, morphology and particles size of ZVI-NPs were analyzed by transmission electron microscopy (TEM) and scanning electron microscopy (SEM) images were recorded further for analyzing surface topography using field emission gun-scanning electron microscopy (FEG-SEM). Based on outputs of OVAT and RSM, maximum conversion (%) and hydrodynamic size of colloidal ZVI-NPs was experimentally confirmed using ICP-MS and DLS analysis. Results showed that out of the five parameters tested (reaction time, salt concentration, volume ratio of leaf extract to salt, pH and temperature), factors like reaction time, salt concentration, volume ratio of leaf extract to salt (Fig. 1) contributed significantly except pH and temperature. The maximum conversion (%) obtained 31% by OVAT and 72% by RSM. Hydrodynamic size obtained 72 nm by OVAT to 40 nm by RSM. Further, ANOVA for interaction of parameters and its coefficients is studied for the conversion (%) and hydrodynamic size of ZVI-NPs. The adequacy of fitted model was checked by F-test.

References:

Aslan, N., 2008. Application of response surface methodology and central composite rotatable design for modeling and optimization of a multi-gravity separator for chromite concentration. Powder Technol. 185, 80â€“86.

Dauthal, P., Mukhopadhyay, M., 2013. Biosynthesis of palladium nanoparticles using *Delonix regia* leaf extract and its catalytic activity for nitro-aromatics hydrogenation. Ind. Eng. Chem. Res. 52, 18131â€“18139.

Dauthal, P., Mukhopadhyay, M., 2016. AuPd bimetallic nanoparticles: Single step biofabrication, structural characterization and catalytic activity. J. Ind. Eng. Chem. 35, 45â€“53.

520525 Optimization Study for Wet Air Oxidation of Ethylene Plant

Abstract:

Optimization study for wet air oxidation of Ethylene plant.

Shehzada Khurram P.E Staff Scientist Olefins & Aromatics Department SABIC Saudi Basic Industries Corporation Saudi Arabia.

Abstract:

The objective of this study is to optimize the spent caustic unit in order to increase the run length. The scope of this study is to explore the opportunities and find out the short coming of the process, which cause the short length of the unit. In addition to this, suggest solutions to overcome the problem to increase overall plants efficiency, and decrease environmental impact as part of sustainability. The drive of this paper is to provide insight into the design of a wet air oxidation System, as it is applied on a full scale basis for spent caustic treatment. There are numerous design parameters, including temperature and pressure. The variations in spent caustic characteristics and treatment objectives must be carefully considered.

Red oil formation is not new for ethylene producers. For numerous years ethylene producers have been over warned with red oil associated fouling problems in oxidizers. To deal with issues normal practice is to treat the Caustic tower. In normal process there is currently no treatment in Caustic tower. Caustic towers at ethylene plants remove acid gases, hydrogen sulfide (H2S) and carbon dioxide (CO2) from ethylene gas. The spent caustic from these towers contains mercaptans and sulfides, which are reactive and odorous. Accordingly, the spent caustic requires special handling and treatment before being discharged to a conventional wastewater treatment plant. Often, the spent caustic is most commonly mated in an oxidation reactor. Also known as wet air oxidation (WAO). This process converts the sulfides into oxidation products such as sulfate ions.

541533 Microbially-Mediated Moisture Retention in Emulated Soil Micromodels

Abstract:

The function of microbial communities is inextricable from the complex physical, chemical and biological microstructure of the soil itself. For example, pore-scale water content controls the hydraulic connectivity of microbial communities and the flux of aqueous and gaseous substrates. Soil bacteria regulate local moisture within porous structure through the secretion of extracellular polymeric substances (EPS). EPS promotes water retention as hydrogel in several ways: (i) by increasing the abundance of intra-aggregate micropores via soil aggregation, (ii) by swelling during wet soil conditions and remaining hydrated during dry soil conditions, and (iii) by altering soil surface wettability. The complex microbial process of soil moisture regulation can be understood by controlling the micro-structured environment systematically. Here we describe pore-scale changes in moisture distribution within emulated soil micromodels featuring systematically varied physical microstructures and surface water repellencies, loaded with well-defined pore solutions.Air infiltration into initially water-saturated soil microenvironments was recorded and quantified using microscopy. Experimental results showed that microstructure geometry was found to effect drying rate over a range of pore saturation values as well as control the spatial distribution of liquid and gas phases with unsaturated soil micromodels. EPS was cultured from stationary-phase *Sinorhizobium meliloti*bacteria, and was diluted to 0.25x, or concentrated via lyophilisation to 5x. The pristine EPS solution took twice as long to dry compared with pure water in identical micromodels. The 0.25x EPS solution took about four times slower than deionized water, and the 1x EPS solution took about eight times slower than deionized water. The influence of a relatively small amount of EPS is magnified by the porous structure to limit evaporation at pore throats. These results illustrate how natural microbial processes work within a complex soil microstructure to control the spatial distribution of water and limit the loss of soil moisture. We provide a systematical method to evaluate drying process within realistic soil microenvironments; we anticipate our approach may facilitate development of novel agriculture biotechnology to enable farmers to produce more crops using less water.

Green Synthesis of Nano Iron Carbide: Preparation, characterization and Application for Aqueous Phosphate Removal

Rabie S. Farag[1], Maha M. Elshfai[2], Ahmed S. Mahmoud[2], Mohamed K. Mostafa[3,4], and Robert W. Peters[5]

[1]Faculty of Science, Al-Azhar University

[2]Sanitary and Environmental Institute (SEI), Housing and Building National Research Center (HBRC), Giza, Egypt

[3]Faculty of Engineering and Technology, Badr University in Cairo

[4]Environmental Engineering Program, Zewail City of Science and Technology, Giza, Egypt

[5]Department of Civil, Construction and Environmental Engineering, University of Alabama at Birmingham, Birmingham, AL

Abstract

Nanoparticles especially iron nanoparticles has proved to be effective in wastewater treatment due to large surface area and reactivity. The main purpose of this study is to prepare and characterize low cost iron nanoparticles that produced by reduction of iron cations using extracted phenolic compounds from black tea to produce carbide compounds also to use iron carbide nanoparticles for phosphate removal from aqueous solution. The iron carbide was characterized by using XRD, SEM and EDAX. The effect of iron carbide on phosphate removal was studied at different contact time, absorbent dose, pH, stirring rate, and initial phosphate concentration. The results indicated that iron carbide is effective in the removal of aqueous phosphate, where a removal efficiency of 82% was achieved for 5 mg/L phosphate solution after 30 min of contact time using dose 0.6 g/L at pH 3 and stirring rate 250 rpm. The removal efficiency of phosphate solutions ranged from 66 to 92% for initial phosphate concentration of 10 and 1 mg/L, respectively, at the same environmental conditions. The equilibrium isotherm of phosphate was determined using different models and the results showed that Hill model with the lowest Chi error 0.0637 was the best fitting model. Modeling results showed that artificial neural network (ANN) with a structure of 5 - 2 – 1 and a sum of squares error of 0.33 and 0.623 for testing and training, respectively, is reliable in describing the adsorption of phosphate onto the iron carbide nanoparticles.

Introduction and Background

Phosphorus (P) presents in water in dissolved and particulate forms and the main forms of phosphorus in aqueous solutions are organic phosphates, orthophosphates and polyphosphates (Mezenner and Bensmaili, 2009). Excessive phosphorus in water is the main reason for growing unwanted organisms, such as algae. Phosphates may also present in some wastes such as, plastic wastes, food wastes, hazardous wastes, marine litter, and construction and demolition wastes (Naidu et al., 2012, Philander, 2008). High phosphate concentrations in water may lead to some diseases, such as osteoporosis, kidney damage, and hyperphosphatemia. Eutrophication of water bodies may also cause fish death in aquatic life. Eutrophication not only affects the aquatic life, but also affects the economic progress of communities that depend mainly on aquatic food (Cleary et al., 2009, Kanamarlapudi et al., 2015, Kröger and Law, 2005, LEBOUKH, 2014, OECD, 2011). Although, excessive phosphates have negative effect on aquatic life and water quality, it is important for agriculture to be used as a fertilizer (Cordell et al., 2009).

Nanocatalysts technology, such as zero-valent metal have been widely used in wastewater treatment due to their higher reactivity and surface area which can degrade or adsorb different wastewater contaminants (El Saliby et al., 2008). The treatment processes using nanotechnologies depends mainly on the particle size. Several studies have been conducted to study the removal of wastewater contaminant using magnetic nanosorbents, such as nZVI. Contaminants that normally found in domestic wastewater includes inorganic cations and anions, as well as organic compounds. Iron (oxy) hydroxides, formed during the reaction between nano iron and the water, react with phosphates ions forming stable inner-sphere complex (Tanboonchuy et al., 2012). The nano iron and bimetallic iron compouds have emerged as effective redox media for reducing a variety of phosphate removal as a tertiary treatment step. Coagulants are recently used for the phosphorus removal at the municipal wastewater treatment plants and they are effective and efficient, however it is not expected that coagulants are able to handle with pharmaceuticals, EDs and hormons in the future (Marek HOLBA, 2012). Nano iron can also be used in wastewater treatment for nitrogen removal through chemical reduction of nitrate and chemical phosphate precipitation.

This work attempts to examine the removal efficiency of negative phosphate ions in aqueous solution using green synthesis iron carbide. The relation between phosphate removal efficiency and iron carbide dosage, pH, contact time, stirring rate and concentration were studied. This study also aims to prepare low cost and high stability nano iron with particle size less than 90 nm.

Materials and Methods

Preparation of nano Iron carbide

Synthesis of green synthesis nano iron carbide using drop by drop method using black tea; About 25 g from black tea was mixed in one liter of deionized water for two hours at 300˚C. After cooling the extracted tea solution was filtrated using Whatman filter paper number 1. About 100 mL of extra pure ethanol solution was added to the infiltrated teat and mixing for 15 min then filtrate again into extracted tea solution. About 2.1624 g FeCl3.6H2O was dissolved in a 4/1 (v/v) ethanol/water mixture (96 mL ethanol + 24 mL deionized water). The extracted tea solution was poured in a burette and added at a rate of one drop per two second into ferric chloride solution and stirring at magnetic stirrer at 150 rpm. After complete reaction between phenolic compounds in black tea and $FeCl_3.6H_2O$ the black precipitate appeared. The black precipitate was collected by centrifuge at 5000 rpm. After washing and drying the precipitate using ethanol and dryer with fixed temperature 105 ^{0}C the black precipitate was ready to use for any environmental application.

Characterization of nano Iron carbide

X-Ray Diffraction patterns were recorded at a radiation wavelength (Cu K alpha = 1.5418 A°). The X-Ray current and voltage values were 40 kV and 40 mA, respectively. The diffraction angle (2θ) ranged from 5° to 70° at a step size of 0.0167°(Xi et al., 2010). The prepared sample was examined using EDAX and scanned using scanning electron microscope (SEM) to discover size and morphology of the prepared nano iron carbide particles.

Batch adsorption studies

The adsorption of phosphate ions onto encapsulated iron carbide was studied by batch technique. The phosphate removal efficiency was studied at initial phosphate concentration 1 - 10 mg/L, pH 3 - 11, adsorbent dose 0.1 - 1 g/L, stirring rate 50 - 500 rpm, and contact time 5 - 90 min. A known weight of adsorbent dosage 0.6 g was equilibrated with 100 ml of the aqueous phosphate solution of known concentration (5 mg/L) in 250 mL of Erlenmeyer flasks and shacked at 250 rpm at a room temperature of 25 ± 2 °C for a known period of time from 30 min. After equilibration, the suspension of the adsorbent was separated from solution by filtration using Whatman No. 41 filter paper and the concentration of phosphate remaining in solution was measured according to ASTM 2005 stander method for water and wastewater stannous chloride method (Federation and Association, 2005). The percentage of phosphate compounds removal was calculated using Eq. 1. The amount of phosphate sorbed by weight of nano iron carbide was calculated using Eq. 2.

$$\text{Sorption [\%]} = [C_o - C_e/C_o] \times 100 \qquad \text{Eq. 1}$$

where C_o is the initial concentration of phosphate solutions, (mg/L) and C_e is the equilibrium concentration of phosphate solutions (mg/l)

$$q_e\ [\text{mg/g}] = [[C_o - C_e]V]/\ m \qquad \text{Eq. 2}$$

where qe is the equilibrium adsorption capacity (mg /g), V is the volume of aqueous solution (L), and m is the dry weight of the adsorbent (g)

Isotherm study

Different isotherm studies were conducted for monolayer and multilayer adsorption process using nonlinear equations including Freundlich, Langmuir, Redlich–Peterson, Hill, Sips, Khan, Toth, Koble–Corrigan, Jovanovich, and Hossein, as shown in Table 1.

Table 1. Nonlinear equations of isotherm models

Isotherm model	nonlinear equations		Constants	References
Langmuir	$Q_e = \frac{q_m K_a C_e}{1 + K_a C_e}$	K_a	Langmuir constant (L/mg)	(Redlich and Peterson, 1959)
		q_m	maximum monolayer adsorption capacity (mg/g)	
Freundlich	$Q_e = K_F C_e^{1/n}$	n	Freundlich constant related to adsorption intensity (dimensionless)	(Redlich and Peterson, 1959)
		K_F	Freundlich constant related to adsorption capacity ((mg/g) (mg/L)-1/n)	
Redlich–Peterson	$Q_e = \frac{K_R C_e}{1 + a_R C_e^{b_R}}$	K_R	Redlich–Peterson isotherm constant (L/g)	(Redlich and Peterson, 1959)
		a_R	Redlich–Peterson isotherm constant (1/mg)	
		b_R	Redlich–Peterson isotherm exponent	
Hill	$Q_e = \frac{Q_H C_e^{n_H}}{K_D + C_e^{n_H}}$	Q_H	Hill isotherm maximum uptake saturation (mg/L)	(Hill, 1910)
		K_D	Hill constant	
		n_H	Hill cooperativity coefficient of the binding interaction	
Sips	$Q_e = \frac{Q_s (K_s C_e)^{B_e}}{1 + (K_s C_e)^{B_e}}$	K_s	Sips isotherm model constant (L/g)	(Koble and Corrigan, 1952)
		B_s	Sips isotherm model exponent	
Khan	$Q_e = \frac{Q_K B_K C_e}{(1 + B_K C_e)^{A_K}}$	Q_k	theoretical isotherm saturation capacity (mg/g)	(Khan et al., 1997)
		B_k	Khan isotherm model constant	
		A_k	Khan isotherm model exponent	
Toth	$Q_e = \frac{K_t C_e}{(a_t + C_e)^{\frac{1}{t}}}$	K_t	Toth isotherm constant (mg/g)	(Toth, 1971)
		a_t	Toth isotherm constant (L/mg)	
		T	Toth isotherm constant	
Koble–Corrigan	$Q_e = \frac{A C_e^D}{1 + B C_e^D}$	A	Koble–Corrigan isotherm constant (Lnmg1−n/g)	(Koble and Corrigan, 1952)
		B	Koble–Corrigan isotherm constant (L/mg)n	
		D	adsorption intensity	
Jovanovich	$Q_e = q_{max}(1 - e^{K_J C_e})$	qmax	(mg/g) is the maximum uptake	(Jaroniec, 1975)
		KJ	Javanovic isotherm constant (L/g)	

Error functions calculations

Different error parameters were calculated in order to identify the most suitable isotherm and kinetic models that describe the treatment process. Due to the ingrained bias that resulting from linear form of isotherm models, the nonlinear form was good alternative parameter which provide an accurate technique to determine isotherm and kinetic parameters from the original equation forms (Malek and Farooq, 1996, Khan et al., 1996). To evaluate the better fit of nonlinear models the error function is required (Ho et al., 2002). The five error functions were used for all models and the ideal model was selected from the summation of these errors. The error parameter equations show in Table 2 (Marquardt, 1963).

Table 2. shows errors mathematical equation.

Error name	Equations
The Sum of the Squares of the Errors (ERRSQ)	$\sum_{i=1}^{p}(q_{e,meas} - q_{e,calc})_i^2$
A Composite Fractional Error Function (HYBRD)	$\sum_{i=1}^{p}\left\|\frac{(q_{e,meas} - q_{e,calc})^2}{q_{e,meas}}\right\|_i$
A Derivative of Marquardt's Percent Standard Deviation (MPSD)	$\sum_{i=1}^{p}\left(\frac{q_{e,meas} - q_{e,calc}}{q_{e,meas}}\right)_i^2$
The Average Relative Error (ARE)	$\sum_{i=1}^{p}\left\|\frac{q_{e,meas} - q_{e,calc}}{q_{e,meas}}\right\|_i$
The Sum of the Absolute Errors (EABS)	$\sum_{i=1}^{p}\left\|q_{e,meas} - q_{e,calc}\right\|_i$
Chi error	$\sum_{i=1}^{p}\frac{(q_{e,calc} - q_{e,meas})^2}{q_{e,calc}}$

Neural network structure

An artificial neural network was developed to predict phosphate removal efficiency, which consists of input, hidden, and output layers. The neural network structure was expressed as 5 – 2 – 1 after using multilayer perceptron (MLP) and 5 – 10 – 1 after using Radial basis function (RBF). The data from the five independent variables are transferred to the input layer. All the available data are normally divided into standard values for training, validation, and testing procedures. The network type is "Multilayer perceptron backpropagation. The multilayer perceptron uses a feedforward architecture and can have multiple hidden layers. It is one of the most commonly used neural network architectures.

Response surface methodology

The RSM plots show simultaneous confidence band for the fitted response surface. A chart was plotted that displaying a contour of RSM for phosphate removal efficiency against the independent variables concentration, pH, time, adsorbent dose, and stirring rate. A pure-linear regression model was employed to describe the RSM plots by fitting the experimental data.

$$Y = \beta_0 + \beta_1 x_1 + \beta_2 x_2 + \beta_3 x_3 + \beta_4 x_4 + \beta_5 x_5 \qquad \text{Eq. 3}$$

where, Y is the predicted response of different wastewater contaminants removal efficiency (%); $x_1$is pH (3 – 11); x_2 is adsorbent dose (0.1 – 1 g); $x_3$is stirring rate (50 – 500 rpm); x_4 is contact time (5 – 90 min); x_5 is concentration (1 – 10 mg/L); β0 is the model intercept; β_1, β_2, β_3, β_4, β_5 and β_6 are the linear coefficients of x_1, x_2, x_3, x_4, x_5 and x_6, respectively.

Results and Discussion

Characterization of nano Iron carbide

X-ray diffraction analysis (XRD)

The XRD pattern as shown in Figure 1 shows Many beaks of iron carbide was observed at different angles indicated that the prepared sample in crystalline structure and the main of powder sample free of oxides and impurities. By analyzing the positions and the shape of the diffraction peaks, Scherrer formula can be used to calculate the average grain size of crystallites (Equation 4):

$$D = \frac{k\lambda}{\beta \cos \theta} \quad \text{Eq. 4}$$

where D is the crystallite size (nm), β is the full wave at half maximum (FWHM), θ is diffraction angle, and k is Scherrer constant (0.9), and λ is x-ray wavelength.

The calculated particle size ranged from 60 to 80 nm which is in a good agreement with the result obtained from the SEM analysis. The XRD analysis also proved that there is no formation of oxides and the nano iron carbide was predominantly present in the sample.

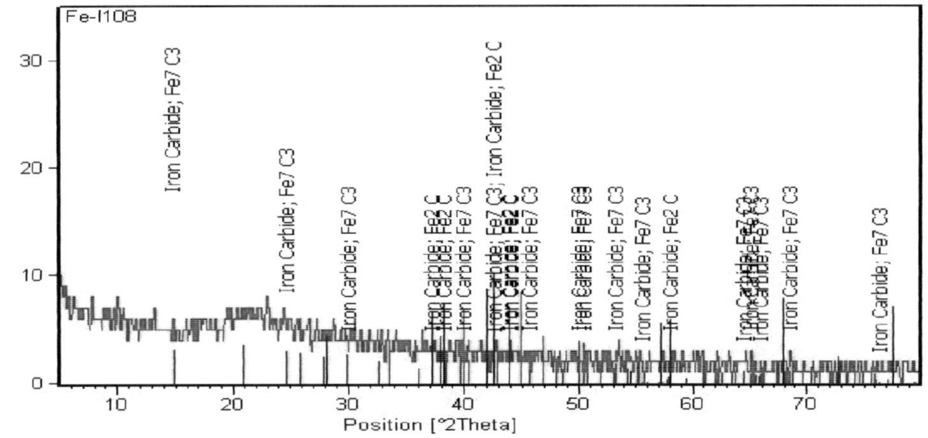

Figure 1. XRD of prepared nano Iron carbide.

Scanning Electron microscope (SEM) and EDAX

Figure 2 shows the SEM image of the prepared nano Iron carbide. The nano Iron carbide also formed an irregular surface structure with an average size of 70 nm. Many pores were observed which allows better mass transfer and diffusion of phosphate ions to the inner iron nanoparticles. Figure 3 shows EDAX analysis of coated nano Iron carbide with gold indicating formation of pure nano iron particles with few outer oxide layer. Many beaks were observed of Fe, C, O, P and Cl indicating the formation of Iron Carbide and presence of some natural impurities from extracted tea adsorbed by iron carbide crystals.

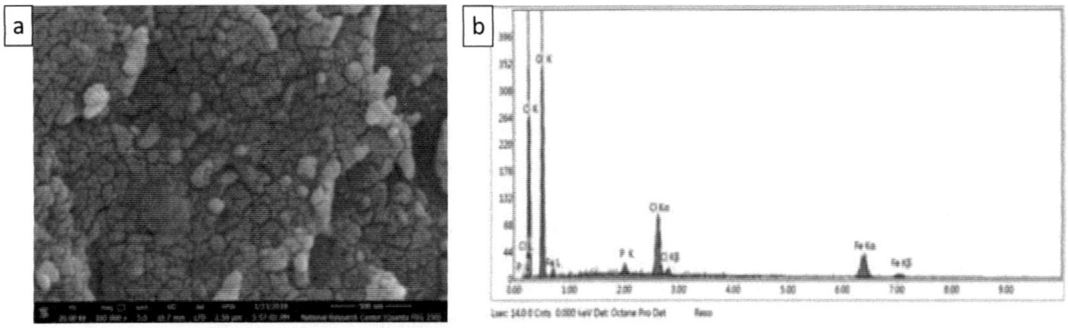

Figure 2. a) SEM prepared nano Iron carbide sample, b) EDAX of prepared nano Iron carbide sample.

Effect of operating parameters

Effect of pH

The effect of pH was studied at Acidic, Neutral and Alkaline pH of 3 - 11. The effect of pH on Phosphate removal efficiency was studied at contact time: 30min, temperature: 25 ± 3 °C, dose: 0.6 g, initial concentration: 5 mg/L, and stirring rate: 250 rpm. A decrease in phosphate removal efficiency from 82% to 80, 76, 75, 72 % was associated with an increase in pH from 3 to 4, 5, 6, 7, respectively. After pH 7 the removal efficiency was increased from 72 to 74, 76, 78, and 81% at pH 8, 9, 10, and 11, respectively as shown Fig 3a. From the obtained results the most efficient media for phosphate removal was at pH 3 due to creating a positive charge at the surface of the iron carde which attract the phosphate negative ions found in the solution.

Effect of contact time

The effect of contact time was studied at 5, 10, 15, 20, 25, 30, 45, 60, 75 and 90 min. The effect of contact time on phosphate removal efficiency was studied using dosage: 0.6 g, temperature: 25 ± 3 °C, pH: 3 ± 0.2, initial concentration: 5 mg/L, and stirring rate: 250 rpm. The results showed that an increase in phosphate removal efficiency from 63 to 67, 72, 76, 79, 82, 83 % was associated with an increase in contact time from 5 to 10, 15, 20, 25, 30, and 45 min, respectively. After 45 min the removal efficiencies was 82, 84 and 83 at 70, 75 and 90 min, respectively as shown Fig 3b. At the beginning of adsorption, a large number of abundant sites on the nano iron carbide surface were available to adsorb greater amounts of the phosphate ions. The minimum effective time was recorded as 30 min. The effective time was selected as 30 min due to the small variation of the removal efficiencies between times 30 and 45, 60, 75, 90 min.

Effect of adsorbent dose

The effect of dose was studied at 0.1, 0.2, 0.3, 0.4, 0.5, 0.6, 0.7, 0.8, 0.9 and 1 g/L. The effect of Iron Carbide dosage on Phosphate removal efficiency was studied at contact time: 30 min, temperature: 25 ± 3 °C, pH: 3 ± 0.2, initial concentration: 5 mg/L, and stirring rate: 250 rpm. The results showed that an increase in phosphate removal efficiency from 69 to 72, 74, 77, 80, 82, 84, 86, 88, and 91% was associated with an increase in adsorbent doses from 0.1 to 0.2, 0.3, 0.4, 0.5, 0.6, 0.7, 0.8, 0.9, and 1 g/L, respectively as shown Fig 3c. A large number of vacant sites on nano Iron Carbide surface were available as a result of increasing the adsorbent dosage, which adsorbed large amounts of phosphate ions.

Effect of stirring rate

The effect of stirring rate was studied at 50, 100, 150, 200, 250, 300, 350, 400, 450 and 500 rpm. The effect of stirring rate on Phosphate removal efficiency was studied at contact time: 30 min, temperature: 25 ± 3 °C, dose: 0.6 g, initial concentration: 5 mg/L, and pH: 3 ± 0.2. From the obtained results the minimum effective stirring rate for phosphate removal was 250 rpm as shown Fig 3d.

Effect of the initial concentration

The effect of initial phosphate concentration was studied at 1, 2, 3, 4, 5, 6, 7, 8, 9 and 10 mg/L. The effect of initial concentration on removal efficiency was studied at contact time: 30 min, temperature: 25 ± 3 °C, dose: 0.6 g, stirring rate: 250 rpm, and pH: 3 ± 0.2. A decrease in phosphate removal efficiency from 92 % to 89, 85, 84, 82, 78, 74, 70, 67, and 66% was associated with an increase in phosphate concentration from 1 mg/L to 2, 3, 4, 5, 6, 7, 8, 9, and 10 mg/L, respectively as shown Fig 3e. At low concentrations, the ratio of abundant sites to the total phosphate in the solution was high, and thus, all phosphate ions were bound to the iron carbide surface.

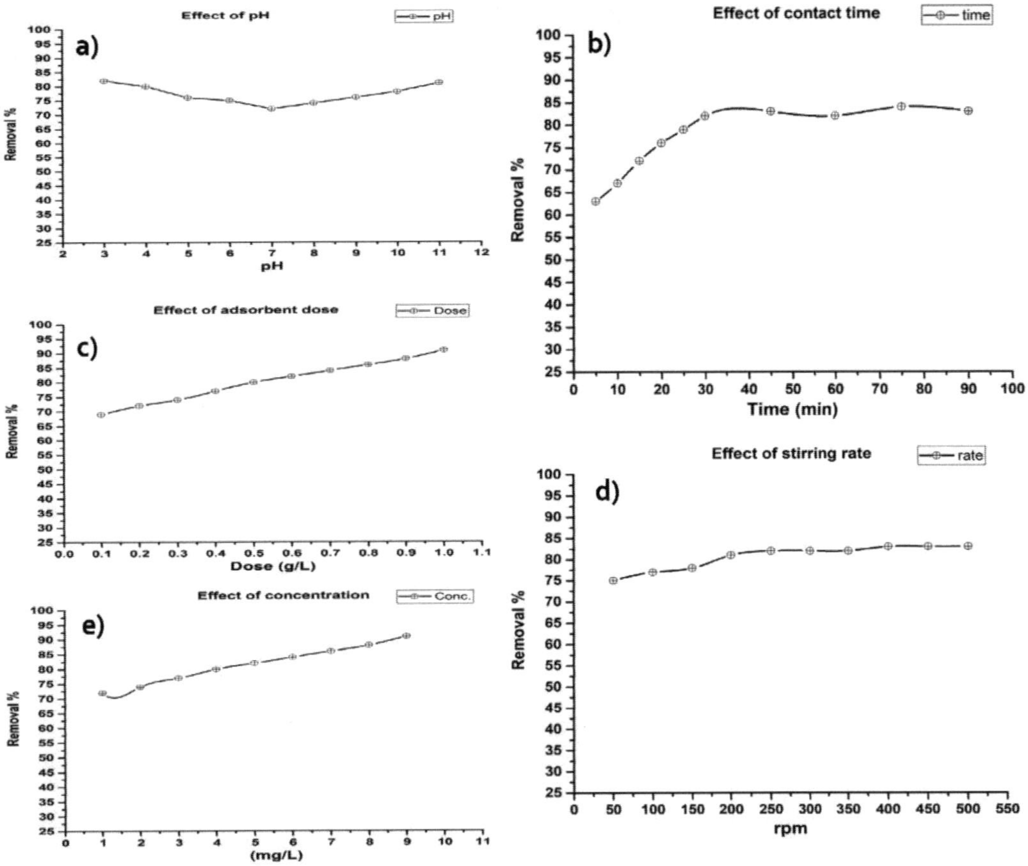

Figure 3. Effect of operating parameter on phosphate removal using nano iron carbide (a) pH, (b) contact time, (c) iron carbide dosage, (d) stirring rate, and (e) initial concentration

Adsorption study

Figure 4 and Table 3 describe nonlinear relations between different adsorption isotherm models. The resulted data from adsorption study using nonlinear models indicating that Hill is the most suitable model that describe the adsorption process with lowest summation of error of 1.694, as shown Table 3. This result indicated that the removal of phosphate by nano iron carbide means binding of different types into homogeneous substrates (Hill, 1910). This isotherm model assumes that one of the binding sites forming into macromolecule may affect the binding sites in the same molecule, so the adsorption process is a cooperative phenomenon between adsorbent and adsorbate (Ringot et al., 2007).

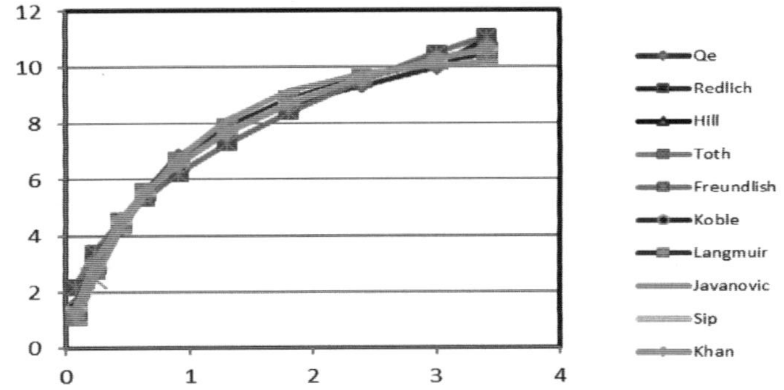

Figure 4. Isotherm studies for nonlinear models.

Table 3. shows the results of nonlinear adsorption models

	Redlich-Peterson		Hill		Sips		Khan		Toth	
Constants	Kr	22.98	QH	15.62	Qs	15.51	Qk	6.808	Kt	3.614
	Br	1.32	nH	0.804	Ks	.0759	Bk	2.846	at	0.854
	G	0.512	KD	1.267	Bs	0.814	Ak	0.768	t	3.614
ERRORS										
Chi error		0.085		0.064		0.064		0.092		0.221
ERRSQ		0.477		0.468		0.466		0.469		0.723
HYBRD		0.029		0.029		0.030		0.023		0.045
MPSD		0.020		0.005		0.007		0.021		0.077
ARE		0.220		0.136		0.154		0.218		0.370
EABS		1.112		0.992		1.026		1.039		1.447
Error sum		1.944		**1.694**		1.748		1.862		2.883

	Koble –Corrigan		Javanovic		Freundlich		Langmuir	
Constants	A	12.366	Qm	10.36	Kf	6.488	Q_0	13.068
	B	0.795	Kj	1.179	n	2.288	b	1.170
	D	0.813						
ERRORS								
Chi error		0.064		0.605		0.413		0.221
ERRSQ		0.466		1.772		1.804		0.723
HYBRD		0.030		0.105		0.036		0.045
MPSD		0.007		0.161		0.166		0.077
ARE		0.153		0.518		0.478		0.370
EABS		1.026		1.909		1.357		1.447
Error sum		1.745		5.069		4.254		2.883

Artificial neural network

Fig 5a shows ANN model using Multi-Layer Perceptron (MLP) statistical algorithms with structure of 5 - 2 – 1 with sum of squares error 0.33 and 0.623 for testing and training, respectively with relative error 0.044 and 0.043, respectively. Fig 5b 2 shows that the model was effective in simulating the performance of iron carbide for phosphate removal because the relation between predictive value and residual was between (-4 and +6%). Fig 5c shows the importance of each individual operating parameter in the removal efficiency showed the most effective parameter was concentration and the least one was pH.

Fig 6a shows ANN model using Radial Basis function (RBF) statistical algorithms with structure of 5 - 10 – 1 with sum of squares error 0.996 and 0.560 for testing and training, respectively with relative error 0.062 and 0.127, respectively. Fig 6b shows that the model was effective in simulating the performance of iron carbide for phosphate removal because the relation between predictive value and residual was between (-5 and+5%). Fig 6c shows the importance of each individual operating parameter in the removal efficiency showed the most effective parameter was adsorbent dose and the least one was pH.

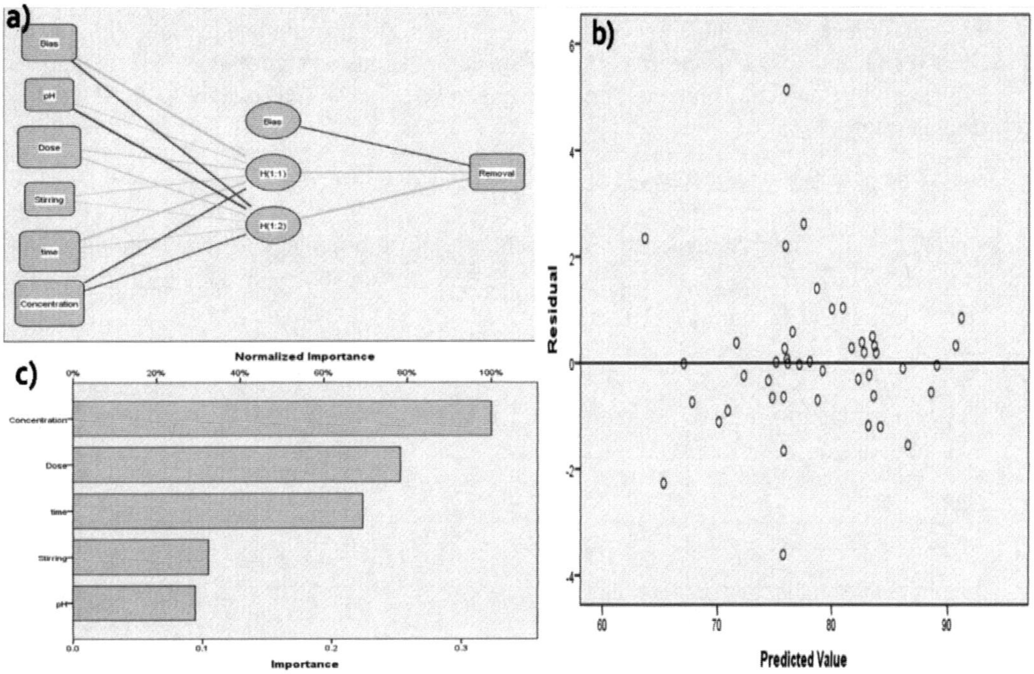

Figure 5. ANN- MLP results: a) Regression plot , b) Training, and c) Best validation performances

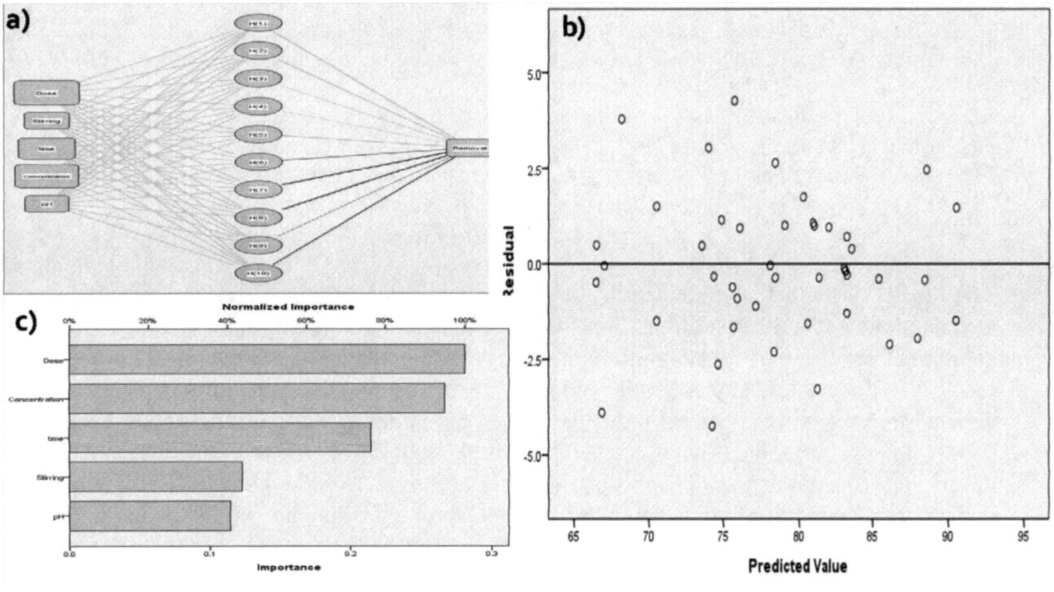

Figure 6. ANN- RBF results: a) Regression plot, b) Training, and c) Best validation performances

Response surface methodology (RSM)

From the previous results, it was concluded that the effect of pH in phosphate removal efficiency is minor. The effect of pH, dose, initial concentration, contact time and stirring rate in the removal efficiency was studied. As listed in Table 4, positive linear effect of the independent variables "stirring rate", "dose" and "initial concentration" on phosphate removal were observed to be significant ($p < 0.05$). Additionally, a significant effect ($p < 0.05$) was noticed for independent variable "time". However, insignificant effect ($p > 0.05$) was determined for the linear term of "pH". The coefficient of determination between measured data and simulated results (r^2) and adjusted r^2 were 0.763 and 0.732, respectively. This r^2-value suggested that about 76% of the variation is explained by the RSM model. Equation 5 showed the all regression models (significant and insignificant).

$$Y = 71.696 - 0.389\, x_1 + 22.358\, x_2 + 0.019\, x_3 + 0.171\, x_4 - 2.942\, x_5 \qquad \text{Eq. 5}$$

where, Y is the predicted response of heptachlor removal efficiency (%); x_1 is initial pH values (3 – 11); x_2 is adsorbent dose (0.1 – 1 g); x_3 is stirring rate (50 – 500 rpm); x_4 *is* contact time (5-90 min), and x_5 initial concentration (1 – 10 mg/L).

Table 4. *t*-statistics and *p*-values for coefficients of a linear regression model.

Term	Estimate	Standard Error	*t*-Ratio	Prob > |*t*|	Effect*
β_0	71.696	3.902	18.375	.000	Significant
β_1	-.389	.261	-1.488	.145	Insignificant
β_2	22.358	3.757	5.951	.000	Significant
β_3	.019	.008	2.472	.018	Significant
β_4	.171	.039	4.426	.000	Significant
β_5	-2.942	.376	-7.830	.000	Significant

Abbreviations: Standard Error, the error of the estimated difference between the means; t-Ratio, the t-ratio for the test of whether the estimated difference between the means is zero; Prob> |t|, the p-value for the test
*The significant levels at the 95% level ($p < 0.05$) were considered to have a greater impact on the response

Conclusions

The results of the batch studies indicated that iron carbide nanoparticles are effective in phosphate removal under specific conditions. A removal efficiency of 82% was achieved for 5 mg/L phosphate solution after 30 min of contact time using dose 0.6 g/L at pH 3 and stirring rate 250 rpm. A maximum removal efficiency of 92% was achieved at the same environmental conditions. The isothermal data showed that the Hill isothermal model was the best fitting model with the lowest Chi error 0.0637. Modeling results showed that ANN with a structure of 5 - 2 – 1 was more reliable than RSM in describing the adsorption of phosphate onto the iron carbide nanoparticles under the tested experimental conditions with a sum of squares error of 0.33 and 0.623 for testing and training, respectively. In this context, the proposed ANN model could be used to optimize the adsorption rate of phosphate by iron carbide nanoparticles under different environmental conditions in future studies.

References

CLEARY, J., SLATER, C. & DIAMOND, D. 2009. Analysis of phosphate in wastewater using an autonomous microfluidics-based analyser.

CORDELL, D., DRANGERT, J.-O. & WHITE, S. 2009. The story of phosphorus: global food security and food for thought. *Global environmental change,* 19, 292-305.

EL SALIBY, I., SHON, H., KANDASAMY, J. & VIGNESWARAN, S. 2008. Nanotechnology for Wastewater Treatment: in brief. *Encyclopedia of Life Support System (EOLSS).*

FEDERATION, W. E. & ASSOCIATION, A. P. H. 2005. Standard methods for the examination of water and wastewater. *American Public Health Association (APHA): Washington, DC, USA*.
HILL, A. V. 1910. The possible effects of the aggregation of the molecules of haemoglobin on its dissociation curves. *J Physiol (Lond),* 40, 4-7.
HO, Y., PORTER, J. & MCKAY, G. 2002. Equilibrium isotherm studies for the sorption of divalent metal ions onto peat: copper, nickel and lead single component systems. *Water, Air, and Soil Pollution,* 141, 1-33.
JARONIEC, M. 1975. Adsorption of gas mixtures on homogeneous surfaces Extension of Jovanovic equation on adsorption from gaseous mixtures. *Chem. zvesti,* 29, 512.
KANAMARLAPUDI, S., PRADGNA, K. & VANGALPATI, M. 2015. OPTIMIZATION OF PARAMETERS FOR REMOVAL OF PHOSPHATE FROM PHARMACEUTICAL EFFLUENT BY CHEMICAL PRECIPITATION USING CaCO3. *INTERNATIONAL JOURNAL OF ENGINEERING SCIENCES & RESEARCH TECHNOLOGY*
KHAN, A., AL-WAHEAB, I. & AL-HADDAD, A. 1996. A generalized equation for adsorption isotherms for multi-component organic pollutants in dilute aqueous solution. *Environmental technology,* 17, 13-23.
KHAN, A., ATAULLAH, R. & AL-HADDAD, A. 1997. Equilibrium adsorption studies of some aromatic pollutants from dilute aqueous solutions on activated carbon at different temperatures. *Journal of colloid and interface science,* 194, 154-165.
KOBLE, R. A. & CORRIGAN, T. E. 1952. Adsorption isotherms for pure hydrocarbons. *Industrial & Engineering Chemistry,* 44, 383-387.
KRÖGER, S. & LAW, R. J. 2005. Sensing the sea. *Trends in biotechnology,* 23, 250-256.
LEBOUKH, S. Development of Enzyme-Based Biosensor for Environmental Monitoring. International Nanotechnology Conference and Exhibition, Hammamet, Tunisia, April 24-26, 2014., 2014.
MALEK, A. & FAROOQ, S. 1996. Comparison of isotherm models for hydrocarbon adsorption on activated carbon. *AIChE Journal,* 42, 3191-3201.
MARQUARDT, D. W. 1963. An algorithm for least-squares estimation of nonlinear parameters. *Journal of the society for Industrial and Applied Mathematics,* 11, 431-441.
MEZENNER, N. Y. & BENSMAILI, A. 2009. Kinetics and thermodynamic study of phosphate adsorption on iron hydroxide-eggshell waste. *Chemical Engineering Journal,* 147, 87-96.
NAIDU, R., LAMB, D., BOLAN, N. & GAWANDAR, J. 2012. Recovery and reuse of phosphorus from wastewater sources. Occasional Report.
OECD 2011. *Fostering Nanotechnology to Address Global Challenges: Water.*
PHILANDER, S. G. 2008. *Encyclopedia of global warming and climate change*, Sage Publications.
REDLICH, O. & PETERSON, D. L. 1959. A useful adsorption isotherm. *Journal of Physical Chemistry,* 63, 1024-1024.
RINGOT, D., LERZY, B., CHAPLAIN, K., BONHOURE, J.-P., AUCLAIR, E. & LARONDELLE, Y. 2007. In vitro biosorption of ochratoxin A on the yeast industry by-products: Comparison of isotherm models. *Bioresource technology,* 98, 1812-1821.
TANBOONCHUY, V., GRISDANURAK, N. & LIAO, C.-H. 2012. Background species effect on aqueous arsenic removal by nano zero-valent iron using fractional factorial design. *Journal of Hazardous Materials,* 205–206, 40-46.
TOTH, J. 1971. State equations of the solid-gas interface layers. *Acta Chim Acad Sci Hungar,* 69, 311-328.
XI, Y., MALLAVARAPU, M. & NAIDU, R. 2010. Reduction and adsorption of Pb 2+ in aqueous solution by nano-zero-valent iron—a SEM, TEM and XPS study. *Materials Research Bulletin,* 45, 1361-1367.

539776 Extraction of Chlorophenols from Wastewater Using Ionic Liquids As Green Solvents

Abstract:

Chlorophenols (CPs) are synthetic chemicals which are generated and used widely by different industries such as textile, pharmaceutical, metallurgic, oil, pulp and paper manufacturing. CPs are considered as hazardous pollutants due to their negative impacts on ecosystem, resistance to biodegradation and persistence in the environment. Their toxic effects are severe including DNA damage, endocrine disruption, cytotoxicity, mutagenicity, and carcinogenicity. Thus, some members of them are considered as priority pollutants by United States Environmental Protection Agency (EPA) and potential carcinogenic chemicals by the International Agency for Research on Cancer (IARC). The water contaminated by phenols must be treated before releasing to the environment. Ionic liquids (ILs) are molten salts that are liquid below 100 ºC, which have been used as solvents for many processes and as electrolytes in electrochemical devices. ILs are characterized by negligible vapor pressure and non-flammability. These compounds are liquid over a wide range of temperatures (down to -81 ºC), possess high thermal and chemical stability as well as high density and electrical conductivity in comparison to other solvents. ILs as green solvents in water treatment have become research hotspot and solubility data of water pollutants in ILs is important for assessing the potential of the use of ILs in water treatment, but relevant data for ILs are scarce. No solubility data of chlorophenols in ILs were not reported in the literature. In addition, a limited number of publications on the extraction of CPs using ILs is available. Therefore, in this work, the solubility of 3-chlorophenol (3-CP), 2,5-Dichlorophenol (DCP), 2,4, 6-Trichlorophenol (TCP), and Pentachlorophenol (PCP) in six hydrophobic ILs were measured at different temperatures. Moreover, the extraction of these CPs from water using the same ILs was conducted at different conditions of temperature, IL/water mass ratio, and extraction time. In order to investigate the cross contamination, mutual solubility of water and ILs were measured at 25 °C.

The solubility experimental results showed that 3-CP is miscible with all tested ILs. As for DCP, TCP and PCP, the solubility in all ILs decreased as the number of chlorine atoms in chlorophenols increased. Additionally, the solubility increased with increasing temperature, but the degree of increase depended on the structure of both IL and chlorinated phenol. Through extraction experiment, when1-Ethyl-3-methylpyridinium bis(trifluoromethylsulfonyl)imide ([1,3emPY][Tf2N]) was used, 98.7% extraction efficiency for 3-CP, 99.4% extraction efficiency for DCP and 99.1% extraction efficiency for TCP were achieved. When 1-Butyl-3-methylimidazolium bis(trifluoromethylsulfonyl)imide ([bmim][Tf2N]) was used, 98.3% extraction efficiency for 3-CP and 99.8% extraction efficiency for DCP were achieved. When 1-butyl-1-methylpiperidimium bis(trifluoromethylsulfonyl)imide ([C4mPip][Tf2N]) was used, 99.9% extraction efficiency for TCP were achieved. From mutual solubility results, it was found that the solubility of ILs in water is very small and is much smaller than the solubility of water in ILs.

537831 Ecohydrological Management and Valuation Insights of Ecosystem Services in Salt Lakes through Advanced Dynamic Optimisation Strategies

Abstract:

In this work, we propose a mathematical model within a dynamic optimization framework to address the management and enhancement of ecosystem services of a salt lake in a semiarid region in Argentina (Chasicó Lake). Further, useful criteria for valuation of ecosystem services in continental salt water environments are provided. Ecosystem services are the direct and indirect contributions of ecosystems to human well-being (TEEB, 2010, Jorda-Capdevila et al., 2016). They support directly or indirectly the survival and quality of life. The salt lake basin under study provides multiple ecosystem services that include production of silverside (*Odonthestes bonariensis*), a fish species of regional importance and high commercial value; recreational space for boating; sports fishing; the appreciation of the wilderness; natural wetlands; birds sanctuary; etc.

We formulate dynamic mass balances for water and salt; evaporation is calculated based on energy and momentum balances on a daily basis. The model has been validated with collected data for the period 2008-2016 (Siniscalchi et al., 2018a). Based on two bathymetric studies, we include rigorous calculations for lake volume, as composed of several geometric forms. We formulate an optimal control problem to manage the water resource during wet and dry periods. In wet periods the water body is managed to prevent flooding of the nearby fields. In dry periods, management actions are carried out to address an important ecological and economic issue, the maintenance of optimal environmental conditions for silverside.

The objective is to keep lake salinity around a desired value considered as optimal for silverside reproduction, while fulfilling the differential algebraic equations system describing the hydrological model. Therefore, the objective function is integral of the square difference between a set point value and current salinity value at each time instant. There are two control variables: the diverted flowrate to a nearby artificial reservoir in a wet scenario and the required inflow from the artificial reservoir in dry periods. We consider a four-year wet period (mean annual rainfall 800 mm), followed by a two-year dry period (annual rainfall 408 mm). Numerical results show diverted river flowrate profiles in the wet period, which are between 0 and 85% of the river discharge. During the four-year wet period, a new artificial reservoir of 80 hm^3 volume (area: 11 km2) could be constructed fed with the diverted water flowrate. This reservoir, for which mass balances and evaporation calculation are performed, can later provide an input stream to Chasicó Lake throughout the two-year dry period. Stream profiles are obtained. In this way, salinity can be kept within desired values, as well as lake volume. The model is implemented within a control vector parameterization environment in gPROMS (PSEnterprise, 2017). The driving forces of the ecosystem services provided by the salt lake, identified through the proposed integrated model, represent useful tools towards their valuation.

References

SchrÃ¶ter,M., E. Zanden, Oudenhoven, Remme, Serna-Chavez, de Groot, Opdam, (2014) Ecosystem services as a contested concept: A synthesis of critique and counter-arguments. Conserv. Lett. 7(6),514â€"523.

Jorda-Capdevila, D., B. RodrÃguez-Labajos, M. Bardina (2016), An integrative modelling approach for linking environmental flow management, ecosystem service provision and inter-stakeholder conflicto, Environmental Modelling & Software 79 22-34

Siniscalchi, AG., Kopprio, G., Raniolo, LA., Gomez, EA., Diaz, MS., Lara, RJ. (2018a). Mathematical modelling for ecohydrological management of an endangered endorheic salt lake in the semiarid Pampean region, Argentina. J. Hydrol. (under review)

A.G. Siniscalchi, L. Fritz, C. Garcia Prieto, V. Estrada, M. Hoffmeyer, R.J. Lara, M.S. Diaz (2018b), Modelling and advanced dynamic optimisation strategies for hydrological and water quality management in continental water bodies, Proceedings of the 28th European Symposium on Computer Aided Process Engineering, June 10th to 13th, 2018, Graz, Austria.

TEEB (2010), The Economics of Ecosystems and Biodiversity Ecological and Economic Foundations. Edited by Pushpam Kumar. Earthscan, London and Washington

538551 From Thermopower Waves to Asymmetric Chemical Doping – New Concepts in Energy Storage and Generation Using Molecular Interactions with Single-Walled Carbon Nanotubes

Abstract:

There is a pressing need to find alternatives to conventional energy generation techniques, specifically those that rely on elements in finite global supply. Thermopower wave (TPW) devices, which convert chemical to electrical energy by means of self-propagating reaction waves guided along nanostructured thermal conduits, have the potential to address this demand.[1] We show that conversion efficiency can be increased significantly by selecting molecules such as sodium azide or sucrose with potassium nitrate to offset the inherent penalty in chemical potential imposed by strongly *p*-doping chemicals,[2] a validation of the predictions of Excess Thermopower theory.[3] Such chemical-potential-gradient-induced-electricity can be further exploited in a more direct manner, decoupled completely from the combustion reactions, affording another novel energy generation scheme using only molecular interactions, and subsequent charge transfers, with single-walled carbon nanotubes.[4] Specifically, we demonstrate that chemically-modified carbon nanotube fibers enable unique power sources driven entirely by a chemical potential gradient.[5] Short circuit electrical current (11.9 $\mu A\ mg^{-1}$) and open circuit potential (525 mV) are reversibly produced by localized acetonitrile doping under ambient conditions. An inverse length-scaling of the maximum power as $L^{-1.03}$ that creates specific powers as large as 30.0 kW kg^{-1} highlights the potential for microscale energy generation.

(1) Choi, W.; Strano, M. S. et. al. *Nat Mater* **2010**, *9*.

(2) Mahajan, S. G.; Liu, A. T.; Strano, M. S. et. al. *Energy & Environmental Science***2016**, *9*, 1290.

(3) Abrahamson, J. T.; Strano, M. S. et. al. *ACS Nano* **2013**, *7*, 6533.

(4) Kunai, Y.; Liu, A. T.; Strano, M. S. et. al. *J. Am. Chem. Soc.* **2017**, *139*, 15328.

(5) Liu, A. T.; Kunai, Y.; Strano, M. S. et. al. *Adv. Mater.* **2016**, *28*, 9752.

526989 Fabrication of Loose Inner-Selective Polyethersulfone (PES) Hollow Fibers By One-Step Spinning Process for Nanofiltration (NF) of Textile Dyes

Abstract:

The development of a loose inner-selective polyethersulfone (PES) hollow fiber by one-step spinning process for nanofiltration (NF) has been demonstrated in this study. It is found that the addition of a small amount of sulfonated polysulfone (SPSf) and the use of a specially designed bore fluid made of hydrophilic additives such as polyethylene glycol (PEG) and polyvinylpyrrolidone (PVP) are of great importance to fabricate this type of NF membranes. Not only are they able to reduce the pore size but also suppress the formation of finger-like macrovoids. After optimizing both dope solution and spinning conditions, the as-fabricated membrane can have a 70 % rejection to PEG 1000 and a pure water permeability (PWP) of 12.8 $Lm^{-2}h^{-1}bar^{-1}$. Its rejection of indigo carmine (MW: 466 $gmol^{-1}$) can reach 96.64 %. In addition, it shows good stability in a continuous 76-hour operation with good efficiency of dye removal. The newly developed membrane may have great potential in the treatment of textile wastewater.

530439 Design of a Molecular-Recognition Material for Capture and Release of Phosphate

Abstract:

As many resources are being unsustainably depleted as the global population exponentially rises, society is turning towards wastewater as a new source for reclaiming valuable nutrients. Novel materials need to be engineered to sequester and recycle these valuable resources for reuse, with phosphate as a key target for reuse as a fertilizer. This work specifically aims to design a new material based on biologically-inspired peptide amphiphiles (PAs) that can molecularly recognize, sequester and recycle phosphate out of water. To perform these functions, we are synthetically manipulating the natural ability of proteins to specifically bind to ions and small molecules based on their molecular sequence and secondary structure. Thus, we have incorporated phosphate-binding amino acid sequences into the peptide "headgroup" of PAs, which are conjugated to an alkyl "tail." These molecules then spontaneously self-assemble into supramolecular structures that display the binding sequence to the aqueous environment. We have also programed these PA micelles to entangle and physically bind to each other, creating a dense, gelled network of molecular-recognition structures that will retain phosphate at high pH and controllably release phosphate at low pH, as water is flowed through the hydrogel. In our work thus far, we used solid-phase-peptide synthesis to successfully synthesize two systems of PAs, incorporating an elucidated phosphate-binding amino acid sequence into the headgroup of the PAs. Fundamental self-assembly and structural properties were then characterized. We determined the critical micelle concentrations and used scattering and electron microscopy techniques to validate that these systems self-assemble into worm-like micelles at the binding pH, which is the desired architecture to form a crosslinked hydrogel that is able to capture and release phosphate.

535625 Reactive Membranes for the Degradation of Emerging Wastewater Contaminants

Abstract:

Membrane systems offer a highly effective means of water purification; however, some harmful contaminants cannot be removed by this technology. These emerging contaminants include carcinogenic compounds such as N-nitrosodimethyalmine (NDMA) and 1,4-dioxane. Traditionally, removal of such chemicals would require a separate process to allow for their oxidative degradation. Therefore, the objective of this project is to develop modified membranes capable of oxidizing such emerging contaminants at the membrane surface, allowing for wastewater treatment to be accomplished in a single stage.

A widely employed and broadly effective advanced oxidation process used in wastewater treatment is the Fenton process, which relies on the catalytic decomposition of hydrogen peroxide to generate reactive free radical species for the oxidation of organic molecules. The reactive membranes used here were modeled on previously developed anti-fouling reverse osmosis and nanofiltration membranes. A polydopamine coating was applied to the membrane surface, creating a layer capable of embedding copper oxide nanoparticles. These particles then served as a catalyst for the decomposition of hydrogen peroxide added at low concentrations to the influent stream, thereby generating the same reactive species employed in Fenton oxidation. Availability of these species near the membrane surface allowed oxidation of organic compounds to occur before they could pass through the membrane, while the polydopamine layer scavenged excess radicals to protect the membrane itself from degradation. The efficacy of the modified membranes was tested against NDMA and 1,4-dioxane as model contaminants.

537629 Electrochemically Mediated Regeneration of Ionic Liquids (EMRIL) for Heavy Metal Removal and Water Disinfection

Abstract:

The increased use of heavy metals and metalloids in industrial, agricultural and technological applications has led to their wide distribution and persistence in natural water bodies and soil. Elements such as lead, cadmium, nickel, mercury, arsenic and copper may cause multiple organ damage even at low exposure and are therefore of public health significance.

We report the use of novel, task-specific hydrophobic ionic liquids for liquid/liquid extraction of metal ions from an aqueous phase. Selective extraction of 99.95% of solute heavy metals was achieved directly from the aqueous solution of various metal nitrates (Cu, Pb, Ni, Co) at concentrations as low as tens of ppm. The ionic liquid was then regenerated for further extraction via the electroplating of the metal ions onto a steel surface. This allowed for the continuous extraction of such metal ion species from a contaminated stream and the recovery of precious metals via selective electroplating. This concept is of great utility to municipal water treatment as well as metal recovery for the mining industry. Furthermore, these ionic liquids also demonstrate antimicrobial activity, which makes them attractive for in-house water treatment units in the form of immobilized ionogels for both heavy metal removal and water disinfection

538132 A Polypyrrole-Based Asymmetric System for Electrochemically Mediated Separations of Organics from Water

Abstract:

Water contamination by neutral organic species, including industrial chemicals, pesticides, and pharmaceuticals, and personal care products is a long-standing issue globally. Adsorption is one of the most widely used technologies for mitigating uncharged organic contaminants in the aquatic system. However, conventional adsorbent materials such as activated carbons suffer from costly regeneration and high attrition rates. Polymeric adsorbents have shown great potential to overcome the challenges around regeneration and replace activated carbons for removing a wide range of toxic pollutants. We use electrochemical stimuli to modulate the affinity between target hydrophobic organic molecules and polymeric adsorbents as an energetically efficient and environmentally friendly means of regenerating the adsorbents. We choose polypyrrole (PPy), an intrinsically conducting polymer, as the primary component of the electrochemically regenerable adsorbent. We have developed two approaches to electrochemically modulate the hydrophobicity in PPy, by incorporating redox-responsive ferrocene moieties, and by doping PPy with amphiphilic surfactant anions. We functionalize the two types of PPy onto conductive carbon fibers and assemble the two electrodes into an asymmetric system for treating water contaminated with organics. Both electrodes have high affinities for organics when the asymmetric system is uncharged while their affinities can be lowered upon applying a mild potential (1.5V) to render the polymer hydrophilic. We can therefore apply an electrical potential to drive the desorption of organics from the electrodes when the polymer is saturated. Because the reactivation of the two electrodes in the asymmetric system is spontaneous, the asymmetric system improves the efficiencies of the separation by avoiding water splitting. We demonstrate experimentally that the asymmetric system can be used reversibly for removing model organic contaminants from water in a cyclic fashion with minimal decay in capacity over a number of cycles. We also design and simulate a multi-stage separation process using the asymmetric PPy-based system to further enhance the efficiency with careful choice of process parameters.

515519 Polyaniline Nanofiber Electrodes for Reversible Capture and Release of Mercury(II)

Abstract:

Highly effective removal of heavy metal ions especially mercury (II) from aqueous media is very important as release of those are only increasing. Traditional methods by adsorption mechanism include use of clay, activated carbons, zeolites. However, these method normally have low capacity and weak binding affinity. Metal organic frameworks and covalent organic frameworks have been explored as new type of adsorbents for mercury removal due to their easy synthesis, high surface area, and recyclability. However, they usually suffer from instability in water or aqueous solutions with a wide pH range and possess low adsorption capacity and weak affinity for Hg(II). Recently, beyond the traditional methods, electrochemical mediated capture and release of contaminants received the spotlight. Electrochemical separation and recycle method have distinctive advantages. It could be a general platform for selective ion recovery with high throughput and separation factors with advantages including fast kinetics, control by electrical potential, modularity, and reuse. Here we demonstrate that polyaniline nanofiber electrodes (PNE) can selectively adsorb mercury(II) from water and subsequent release of it can be controlled electrochemically through the redox processes. Our PNEs showed the fastest kinetics and the highest uptake capacity among known polyaniline adsorbents. This is the first demonstration of high performing redox active adsorbents for electrochemically controlled capture and release of mercury ions. Change of cyclic voltammetry is a simple and straightforward indicator for sensing mechanism.

522127 Cyanide Recovery from Barren Solution Using UV Photodissociation and Gas-Filled Membrane Technology

Abstract:

A process was developed to recover cyanide from barren solution containing simple and metal cyanide complex. The technology includes pretreatment of metals and oxyanions, UV-light-aided photodissociation of metal-cyanide complex, and Gas Filled Membrane (GFM). Using a 200w medium pressure UV light at 254nm wave length, more than 85% cyanide ions (CN^-) were recovered from soluble iron cyanide complex at pH between 10.0 and 10.5 within one hour of reaction. The effluent from the UV reactor was filtered and pH was adjusted between 5.0 and 5.5 in a leak proof container, prior to pumping to the GFM that separated hydrocyanic acid (HCN^0) from water. Liquid solutions were pumped on both sides of the membrane. One side contained hydrocyanic acid solution, and 1.0 N sodium hydroxide solution was recirculated on the other side of the membrane. Within one hour, more than 95% of the cyanide was recovered in the sodium hydroxide solution. The impacts of pH, UV dosage, sequestering agent dosage, and reaction time on the photodissociation efficiency will be presented. Furthermore, the impact of membrane flux rate on cyanide recovery will be discussed.

529887 Assessment of the EU Member Countries Electricity Mix Using DEA and Effmixf

Abstract:

The assessment of the electricity mixes sustainability level is a crucial step in the change towards a more sustainable energy process. Here we assess the three dimensions of sustainability: the economic (i.e., the levelized cost of electricity), environmental (considering six impact indicators) and social performance (i.e., the number of direct job-yr) of the electricity mix of 28 EU members in 2015 by combining life cycle assessment, data envelopment analysis (DEA) and mathematical programming tools. Our method encompasses three main steps: i) the DEA application to the electricity portfolios of the 28 EU countries ii) the optimization of inefficient countries portfolio solving the EffMixF optimization problem that we pose in this work, iii) the evaluation of the optimized electricity portfolios obtained in STEP 2. The EffMixF model proposed takes into account the main technical constraints modeling electricity generation within each country which complements standard DEA in order to extend previous research and obtain more realistic information for policy-making. We find that 12 countries are inefficient, being United Kingdom, Lithuania and Italy the ones with lower efficiency. In our main findings, we determined that the non-renewable electricity sources have to be reduced a 10% (with respect the total electricity production) in order to become efficient. On the other hand, hydropower and wind electricity technologies are the sources to be enhanced to improve the environment performance, whereas solar and coal are (respectively) the preferred ones to improve the social and economic sustainability dimensions.

539309 Assessment of Urban Sustainability: An Exploration Based on Two Metrics for the Chicago Metropolitan Area

Abstract:

Cities and urban agglomerations are undeniably becoming humanity's habitat as they embody the centers of economic, educational, and social activities; in fact, about 70% of the world's population is expected to live in urban areas by 2050. These urban areas significantly deplete natural and environmental resources. Consequently, it is essential that urban systems be sustainable to minimize the excessive depletion of such resources. In this regard, the assessment of urban sustainability becomes critical and it should be performed with metrics capable of elucidating the sustainability trends of urban systems rapidly. Herein, we deploy the Ecological Footprint (EF) and the Green Net Metropolitan Product (GNMP) for exploring the sustainability trends of the Chicago Metropolitan Area (CMA) between 1990 and 2015. The CMA is formed by seven counties in northern Illinois under the jurisdiction of the Chicago Metropolitan Agency for Planning (CMAP). Our exploration is in line with those research efforts at the U.S. Environmental Protection Agency (USEPA) concerning the quantification of regional sustainability based on multidisciplinary assessments. First, the EF aims at capturing the impacts (expressed in space units) of human consumption on the regenerative capacity of ecosystems via an ecological accounting system. Second, the GNMP is an economic measure of sustainability that adjusts the Gross Domestic Product (GDP), a classical indicator of economic well-being, for the degradation of natural and environmental resources. We compute the regional EF and GNMP with data collected from publicly available, secondary sources regarding variables such as population, land area, per-capita consumption of various items, Gross Metropolitan Product, net income, pollution emissions, solid waste, etc. The preliminary results of our analyses appear to indicate that there is no evidence the region is moving away from sustainability. We expect that the results of this exploration will serve as a basis for a refined computational framework that might assist CMAP in crafting policy aimed at maintaining the sustainability of the CMA.

530918 Dimensionality Reduction in Sustainability Assessment: A Combined Use of Mixed-Integer Programming and Data Envelopment Analysis

Abstract:

The sustainability assessment of industrial systems often requires the consideration of a wide range of indicators. Among the available tools to tackle this task, Data Envelopment Analysis (DEA) has recently emerged as an effective technique to measure the comparative efficiency of a set of homogeneous units considering multiple indicators (i.e. inputs and outputs in DEA notation) simultaneously whilst providing clear guidelines on how to improve inefficient alternatives. However, involving too many indicators in the analysis could inevitably turn the problem into larger dimensions, leading to difficulties in the visualization of results. In addition, this could deteriorate the discriminatory power of DEA and ultimately produce results that are less meaningful and hard to interpret. Hence, there is a clear need to retain in the DEA analysis only some indicators that are the most essential by eliminating some of those that might be redundant in the sense that they can be excluded from the analysis without changing the results.

In this paper, we proposed a systematic MIP-DEA model to enhance the DEA application in sustainability assessments where many indicators need to be considered. Our method poses the task of identifying metrics that can be omitted with minimum information loss as a bi-level programming model. The outer problem seeks to minimize the difference between the efficiency scores calculated considering all the inputs and outputs and those obtained in a reduced subset of them. On the other hand, the inner problem provides the efficiency scores that would be obtained for any potential combination of inputs and outputs proposed by the outer problem. Here, binary variables model the selection of inputs and outputs in the master outer problem, while continuous ones represent the DEA weights in the inner problem.

We explored the capabilities of our approach through the assessment of several electricity generation technologies according to multiple criteria, some of which are based on life cycle metrics that are modelled as inputs to be minimized. The results show that our systematic approach can effectively reduce the number of indicators from 10 to 5 without information loss. These results evidence that there are significant redundancies in sustainability indicators, which makes it possible to reduce the problem dimensionality with no information loss. The same approach can be used to reduce the number of efficient units, thereby improving the discriminatory power of DEA. However, the later approach would likely modify the efficiency scores, so a compromise should be attained between the number of indicators included in the analysis and the quality of the final results.

Furthermore, to gain further insight into why our approach decides to keep some specific inputs in the model and how the inputs correlate with each other, we calculated the Spearman correlation matrix of the inputs and applied a â€œk-meansâ€ clustering algorithm to identify groups of variables performing similarly. It turns out that the optimal combination of inputs selected by the MIP-DEA is consistent with the clustering results. This suggests that the proposed approach reduces the error by selecting proxies of each set of correlated variables. This opens up new research directions on combining these two tools effectively for dimensionality reduction. Overall, our approach for dimensionality reduction has the potential to greatly simplify sustainability studies from the viewpoints of visualization and interpretation of the results.

536186 Air Emission Reduction Benefits of Biogas Electricity Generation at Municipal Wastewater Treatment Plants

Abstract:

Municipal wastewater treatment is an energy and materially intensive process. Generating the electricity and manufacturing the chemicals that are used for these processes leads to criteria air pollutants (e.g. NO_x, SO_2, $PM_{2.5}$, and volatile organic compounds) and greenhouse gas emissions (e.g. CO_2, CH_4, and N_2O). In this talk, we quantify the implications of these criteria air pollutants and greenhouse gas emissions resulting from energy consumption, chemical use, and direct pollutant release at publicly operated treatment works (POTWs) across the United States. We also assess the potential to reduce air emission damages by generating electricity and heat resulting from upgrading POTWs through the installation of biogas produced during anaerobic sludge digestion. We find that the embedded and on-site health, environmental, and climate (HEC) damages from air emissions are on the order of $1.63 billion. The overwhelming majority of these damages are associated with electricity consumption. Retrofitting all existing POTWs with anaerobic sludge digestion and biogas-fueled heat and electricity can reduce the HEC damages from POTWs by 25%. This work emphasizes the need to account for use-phase embedded air emissions and use spatially resolved marginal damages in evaluating sustainable infrastructure systems.

536386 Integrating Market Effects into Sustainable Process Design – Application to Urea Production

Abstract:

Conventional engineering analysis and design approaches try to identify and develop production systems that have high efficiency, low cost, and low environmental impacts. The conventional approaches assume those production systems can be adopted by the market due to the above technological advances. However, the market does not always choose the "best" technology. The advanced technology may not be adopted as much as industry may like or assume because of market constraints. For example, the production of a key resource for the advanced technology may not be enough to satisfy the market demand. Other economic factors, such as labor, capital, land area, and regulations, along with competing alternatives may limit the use of the technology as well. In that case, the market selects among multiple technologies to satisfy its demand. In this sense, there are some gaps between engineering decisions based on technological excellence and market decisions based on the economy and human preferences.

The Rectangular Choice-of-Technology (RCOT) framework has been developed to account for those market constraints and multiple technology uses in analyzing economic systems [1]. In the traditional economic input-output model, economic transaction matrix that contain commodity exchanges between economic sectors is the square matrix. In the RCOT model, the transaction matrix becomes the rectangular matrix with additional columns. The additional columns represent multiple technologies that produce the same product. Also, by constraining the maximum available amounts of economic factors, multiple technologies are chosen together to satisfy the market demand in case the cheapest option is not able to satisfy the demand due to the constraints. The RCOT framework has also been applied to process-based life cycle assessment (LCA) model in a value chain scale [2]. However, none of the previous studies applies the RCOT framework to multiscale models and solves engineering design problems.

In this work, the RCOT framework is applied to a multiscale model based on the Process-to-Planet (P2P) framework [3] to account for the resource availability and environmental regulations in designing process engineering models. A methodology for the RCOT-P2P model and its toy example are described and the RCOT-P2P model is applied to the urea manufacturing process. The urea P2P model has been developed by including inputs from value chain and economy scales to the urea manufacturing engineering model. Those inputs include ammonia and carbon dioxide resources and electricity and natural gas uses. As for multiple technology options, different technologies that produce those inputs are considered. Specifically, different fuel and boiler types for electricity generation, natural gas extraction options between conventional NG and shale gas, different hydrogen sources for ammonia manufacturing, different sources of carbon dioxide are considered as multiple technology choices. The available amounts of relevant economic factors are constrained in limiting each of those technology uses. Also, to account for environmental regulations, environmental interventions, such as emissions and use of natural resources, are constrained as well [2].

Since this work reflects the market effect by considering economic factors in choosing technologies, this would inform better insights and understanding on how technologies are adopted in certain economic conditions and how they need to be developed. In particular, this work could be beneficial for modeling emerging technologies that need to substitute the conventional one, because emerging technologies will replace the conventional one gradually. Also, since the availability of natural resources can be considered as constraints, this work could be useful for the regional analysis of specific areas where the amounts of resources are limited and for solving spatial optimization problems.

References

[1] Duchin, F. & Levine, S. H. Sectors may use multiple technologies simultaneously: The rectangular choice-of-technology model with binding factor constraints. Econ. Syst. Res. 23, 281–302 (2011).

[2] Kätelhön, A., Bardow, A. & Suh, S. Stochastic Technology Choice Model for Consequential Life Cycle Assessment. Environ. Sci. Technol. 50, 12575–12583 (2016).

[3] Hanes, R. J. & Bakshi, B. R. Process to planet: A multiscale modeling framework toward sustainable engineering. AIChE J. 61, 3332–3352 (2015).

523448 Design of Residential Polygeneration Systems Applied to Isolated Communities

Abstract:

Water, energy and food are critical resources for meeting the social demands and socioeconomic development of communities and societies around the world. Nevertheless, there are communities that do not have access to energy and, consequently, there is a lack of access to water and food. Many rural communities or small towns still do not have access to these resources mainly due to their geographic location that makes the interconnection with the electrical grid difficult; moreover, the provision of water from the public network is a big problem. Distributed energy generation is a promising way to generate electricity in remote places by isolated systems due to the generation units are close to the consumers and the transmission energy losses are negligible. One of the schemes of Distributed Generation is polygeneration, wich can be defined as the simultaneous production of two or more energy services and products, seeking to take advantage of the maximum potential of the consumed resources. Even though, it is necessary to generate energy for the development of communities, it also represents an environmental problem caused by the use of fossil fuels. Because of this, in this project we propose to include the use of renewable energies that complement the polygeneration system as well as the implementation of waste-to-energy technologies that aim to recover energy in a form of heat, electricity or fuels from a waste source.

This work presents a mathematical programming model for the design of a residential polygeneration systems in an isolated community considering the water-energy-food nexus. The problem consists in determining the optimal and sustainable configuration to satisfy the energy, water and food demands of the inhabitants. To meet energy demands, the existence of different cogeneration units, renewable energies such as solar collectors, aerogenerators and solar panels were considered. Water demand was satisfied by rainwater collection systems and wells. Wastewater generated in the community was sent to blackwater and greywater treatment plants. The treated greywater was used in gardening, cattle and agriculture. Different types of crops and animal production were considered to satisfy the food demands of the inhabitants. The municipal solid wastes generated were separated in plastic, metal, paper, glass and non-recyclable, the latter were treated in process plants such as pelletization, incineration, gasification, pyrolysis and anaerobic digestion to obtain, pellets, steam and natural gas that can be used as an energy source.

The objective functions consist in maximizing the profit of the project that involves sales of crops, animals, pellets, biofuels and solid waste that can be reused and the total annual cost that is constituted by the fixed cost of the technologies or process and their operation cost. Additionally, other objective functions consist in minimizing the fresh water consumption and minimization of greenhouse gas emissions.

The proposed model is a Linear Programming problem and it was implemented in the software GAMS to solve a case study in the community of Cochoapa el Grande located in the state of Guerrero in Mexico, which is the municipality with the highest index of poverty and marginalization in the country, mainly due to its geographical location. The results show that there are economically, socially and environmentally attractive solutions with the implementation of the proposed approach.

525948 An Innovative Microalgae-Bacteria Symbiotic Process for in-Situ Secondary and Tertiary Treatment of Wastewater

Abstract:

Activated sludge process (ASP) is one of the aerobic bioprocesses widely used in secondary wastewater treatment. Despite substantially reducing the organic content of wastewater, the process cannot remove nitrogen and phosphorus significantly. Hence, a tertiary treatment needs to be implemented in addition to ASP for enhancing the effluent quality. Existing secondary and tertiary wastewater treatment methods are limited by high energy demand and waste generation. One of the approaches to alleviate these limitations is through co-culture of heterotrophic bacteria and autotrophic microalgae. Microalgae furnish oxygen through photosynthesis, which can be consumed by bacteria to oxidize the organic matter and produce carbon dioxide (facilitating further photosynthesis). Therefore, the co-cultivation of microalgae and bacteria obviates the need for mechanical aeration as oxygen produced by symbiotic interactions supports the organic carbon removal. Furthermore, microalgae assimilate nitrogen and phosphorus and subsequently enhance the effluent quality, without additional tertiary treatment. Carbon dioxide utilization during this process also assists in reducing the carbon footprint. While co-cultivation of microalgae and bacteria seems an attractive solution for wastewater treatment, its implementation faces several challenges. Bioreactors with mixed microalgal-bacterial consortium generally operate at longer hydraulic retention times (HRT) given the slower growth rate of microalgae. Besides, light availability for photosynthesis may also get impeded if bacteria outgrows the microalgae during co-cultivation.

In this research, microalgae and bacteria were cultivated in two separate bioreactors and a membrane-mediated symbiotic interaction was established between them. The gas transfer was achieved by recirculating oxygen and carbon dioxide-rich streams through submerged membranes. Effect of different recirculation rates on the organic carbon removal was investigated. At the optimum recirculation rate, complete removal of organic carbon was observed, without any external aeration. Nitrogen and phosphorus removal was optimized by varying operating parameters of the microalgae bioreactor. Microalgal biomass accumulation during the operation was monitored. It was observed that even in the absence of an external carbon source, high-density microalgae culture was obtained. The chemical composition of microalgal biomass was studied under different operating conditions to evaluate its potential application.

Overall, this research is first to show that collaboration between physically segregated microalgae and bacteria enables an efficient symbiotic process for removal of organic and inorganic pollutants from wastewater.

529146 Synergistic Cr(VI) Treatment with *S. Oneidensis*-reduced Metal-Organic Frameworks

Abstract:

Developing effective methods for Cr(VI) removal is a priority for sustainable wastewater treatment solutions. A variety of material and biological schemes have been developed, but they suffer from slow kinetics, low capacities, limited treatment condition ranges, or harsh regeneration schemes. Metal-organic frameworks are a class of highly porous materials that have garnered attention for catalysis, gas storage, and, notably, adsorption. Recently, a synergistic effect of pairing Fe(II) ions with a metal-organic framework was reported to dramatically improve Cr(VI) adsorption. We hypothesized that we could create a self-regenerative system for Cr(VI) adsorption that capitalizes on this synergy by pairing iron-based metal organic frameworks with the electroactive bacterium Shewanella oneidensis MR-1, which is capable of reducing Fe(III) based materials. We first confirm that water stable metal-organic frameworks (MIL-100, FeBTC, and MIL-88A) can serve as growth substrates for S. oneidensis in a similar manner to iron oxides. We then show that bacterially reduced frameworks promote a 10-fold increase in Cr(VI) adsorption (after a single challenge) relative to abiotic controls. Finally, we demonstrate that S. oneidensis growing on MIL-100 can survive multiple challenges of lethal Cr(VI) concentrations and that the metabolic action of this bacteria dramatically improves Cr(VI) adsorption kinetics and capacities by continuously regenerating the metal-organic framework. Overall, our system constitutes a recyclable, living process for Cr(VI) treatment, as well as expands the potential substrates for metal-reducing bacteria to include tunable frameworks.

531432 Humidification/Dehumidification for Low Cost, Energy Efficient, Zero Liquid Discharge Desalination Using Solar Thermal Sources

Abstract:

Humidification/Dehumidification (HDH) has developed into a robust technology for small-scale desalination. HDH has multiple advantages such as low capital and maintenance costs and compatibility with low grade heat sources such as non-tracking solar thermal collectors. Despite the substantial advantages offered by HDH, significant shortcomings hinder its application. A primary limitation of HDH is recovery ratio for a single cycle. Since energy for evaporation is generally provided by sensible heat of the feedwater itself, the maximum potential freshwater recovery is limited by the ratio of heat capacity times viable operating temperature difference and heat of vaporization. The ratio of recovered to feed water volume must be < 14% for a single cycle operating under normal conditions. Here we introduce and analyze novel approaches to increase recovery ratio of an HDH cycle via introduction of additional sensible heat to the system allowing independent control of the sensible to latent heat ratio. The decoupling of latent and sensible heat allows for operation with zero liquid discharge. We address material constraints to deal with precipitates, and we consider the coupling of this system with low cost solar thermal sources. The efficiency of HDH is primarily limited by the exchange of heat from condensing vapor to the feedwater in the dehumidification stage. There is generally a tradeoff between heat exchanger size and complexity and heat transfer efficiency. Thus, the overall efficiency of HDH is constrained for current economically viable heat exchanger implementations. We also consider approaches to enhance sensible heat recovery from condensation using low capital cost, high-efficiency heat exchangers.

535976 Improving Speed and Efficiency of Global Sensitivity Analysis Using Metamodeling-Based Approach: A Case Study on Wastewater Treatment Plant Modeling

Abstract:

Global sensitivity analysis (GSA) provides model developers with invaluable insights into the interactions between uncertain model parameters and model output uncertainty. Among its large class of applications are identifying relevant input variables for uncertainty quantification, model quality assessment, model selection, calibration, optimization, to name a few. In the wastewater treatment modelling community, several global sensitivity analysis techniques using Monte Carlo (MC) approach has been applied. (Sin et al., 2011; Cosenza et al., 2013; Ramin et al. 2014). Although the MC approach to sensitivity analysis is relatively well established, it requires a large number of model evaluations (typically in the order 10^4 for variance decomposition based techniques) to estimate sensitivity indices of each parameter accurately, making its use impractical for computationally intensive models such as wastewater treatment process models. In an alternative approach, the original model behavior is approximated using computationally less costly metamodels (also referred to as surrogate models), which are then used to calculate Sobol' sensitivity indices. Two of the most efficient type of metamodels used for this purpose are polynomial chaos expansions (PCE) (Sudret, 2008) and Gaussian process (GP) models (Marrel et al., 2009).

This study explores applications of these surrogate models for performing global sensitivity analysis of models used within the BSM2 framework. To this end, three different scenarios are framed. The first scenario looks into the case of uncertain influent fractionation (pollutant loads), the second into the uncertain stoichiometric and biokinetic parameters of Activated Sludge Model 1 (ASM1) whereas the third investigates sensitivity of secondary settler model to hydraulics and design related parameters. Experimental designs (i.e. learning sample) are generated using two different space-filling sampling algorithms, i.e., Latin hypercube sampling and Sobol sequence (quasi-Monte Carlo) and surrogate models of type PCA using least angle regression and GP model are constructed. Confidence intervals of the sensitivity indices are also estimated using the bootstrap method for PCE and the internal model structure of GP. Sobol sensitivity indices of metamodeling-based approach showed great similarity with those obtained from MC approach for each of the investigated scenarios, yet with much faster computation times. An experimental design size of 200 was found to be adequate to construct efficient surrogate models. Effect of using different sampling algorithms was found negligible. Further, the obtained results also suggest surrogate models can efficiently be used to replace complex mechanistic models used in wastewater treatment for complex modelling needs.

References

Sin, G., Gernaey, K. V., Neumann, M. B., van Loosdrecht, M. C. M., & Gujer, W. (2011). Global sensitivity analysis in wastewater treatment plant model applications: Prioritizing sources of uncertainty. *Water Research*, *45*(2), 639–651. https://doi.org/10.1016/j.watres.2010.08.025

Cosenza, A., Mannina, G., Vanrolleghem, P. A., & Neumann, M. B. (2013). Global sensitivity analysis in wastewater applications: A comprehensive comparison of different methods. *Environmental Modelling and Software*, *49*, 40–52. https://doi.org/10.1016/j.envsoft.2013.07.009

Ramin, E., Flores-Alsina, X., Sin, G., Gernaey, K. V., Jeppsson, U., Mikkelsen, P. S., & Plósz, B. G. (2014). Influence of selecting secondary settling tank sub-models on the calibration of WWTP models – A global sensitivity analysis using BSM2. *Chemical Engineering Journal*, *241*, 28–34. https://doi.org/10.1016/j.cej.2013.12.015

Sudret, B. (2008). Global sensitivity analysis using polynomial chaos expansions. *Reliability Engineering and System Safety*, *93*(7), 964–979. https://doi.org/10.1016/j.ress.2007.04.002

Marrel, A., Iooss, B., Laurent, B., & Roustant, O. (2009). Calculations of Sobol indices for the Gaussian process metamodel. *Reliability Engineering & System Safety*, *94*(3), 742–751. https://doi.org/10.1016/j.ress.2008.07.008

538897 Rational Design of Tailor-Made Threshold Scale Inhibitor Dendrimers and Dendrons

Abstract:

For years, water-based industries have been facing operational challenges as a result of sparingly soluble salt formation and adhesion to unit operations, referred to as scaling, increasing heat, mass, and momentum transfer resistances. Macromolecular additives, often decorated with negatively charged functional groups such as carboxylates, sulfates, and phosphates may interact with the early stages of precipitating inorganic polymorphs, thermodynamically destabilize and dissolve them back to their ionic constituents (threshold effect). Despite significant effort in developing new macromolecular platforms for scale inhibition, no mechanistic studies have been conducted on how the synergy between the functional groups and molecular conformation (backbone flexibility) influence the antiscaling properties of macromolecules. Here, we design and synthesize two new families of dendrimers and dendrons with precisely tailored structures to shed light on the structure-antiscaling property relationships. While phosphonate groups are the most efficient functional moieties to regulate calcium carbonate scaling, the backbone attains an equally imperative role in the macromolecule-scale interactions. We show how increasing the generation of dendrimers and dendrons, partial substitution of anionic functional groups with soluble inert moieties, and altering the molecular core directly affect the antiscaling properties. This work may set the stage for rational design of next generation threshold scale inhibitors, which will globally benefit water industry.

525798 Biochar-Alginate Novel Composite Adsorbent: Synthesis Characterisation and Application in Water and Wastewater Treatment

Abstract:

Biochar-Alginate novel composite adsorbent: synthesis characterization and application in water and wastewater treatment

Subrata Biswas[a,b], Tushar Kanti Sen[b], Bhim Charan Meikap[a,b]

[a]Department of Chemical Engineering, Indian Institute of Technology, Kharagpur, W.B, India, 721302.

[b]Chemical Engineering, Curtin University, GPO BOX U 1987, 6,845 WA, Australia.

Email: b4subrata@gmail.com, T.Sen@curtin.edu.au, bcmeikap@che.iitkgp.ernet.in

Abstract

Water contamination is a serious risk and of great apprehension in today's context due to its harmful effects on the environment. Currently, large attention has been made to industrial wastewater discharged into the land and water bodies. Effluents from various process industries often contain various hazardous metals, harmful gases, and several organic and inorganic toxic compounds. Contamination of natural resources such as water & soil by heavy metals is of great concern over the last few decades because of their toxicity, very low degradability and carcinogenic nature to the living organisms. Over the last few decades, various physio-chemical methods are in practice to minimize the heavy metal ion contamination in wastewater. Among all the treatment process, adsorption is one of the most effective processes for water and wastewater treatment. Commercial activated carbon (CAC) is already used as an effective adsorbent in water treatment. However, its use is costly because of its coal-based synthesis and difficulties in regeneration. Therefore many researchers have reported that cost-effective alternative adsorbents derived from natural biomass, industrial solid waste, agricultural by-products, and biosorbent. Biochar produced by pyrolysis from waste biomass was found a widespread application in wastewater treatment. Biochar is very effective in removing hydrocarbon and inorganic heavy metals from industrial wastewater. Along with biochar use of algae in living and nonliving form is also very effective and innovative alternative in heavy metal ions and organic pollutant removal.

In this work, an attempt has been made to synthesize and characterize of a novel Ca-alginate –biochar composite as adsorbent and tested its effectiveness in the removal of aqueous phase Zn^{+2} ions under various physiochemical process conditions. Figure 1a, 1b& 1c represents the synthesized composite beads and their morphological characteristics. Physical characteristics like FTIR, XRD is shown in Figure 2a & 2b which indicates promising properties for this composite to be a good adsorbent. Percentage removal of Zn^{+2} with a variation of initial concentration and Langmuir isotherm plot has been presented here in Figure 3a & 3b. The removal efficiency was studied by varying with initial Zn^{+2} ion concentration, adsorbent doses, solution pH, contact time, temperature. Between two kinetic models namely pseudo-first-order kinetic model and pseudo-second-order kinetic model, the later resulted in a better fit with the experimental data for this system. Mechanism of adsorption and Intra-particle-diffusion coefficient and boundary layer thickness was determined by using Webber –Morris (Intra-particle-diffusion) model. The nature of adsorption process is, confirmed by activation energy calculation as well as by diffusion coefficient value. Various thermodynamic parameters such as Gibbs free energy change (DG^0), enthalpy change (DH) and entropy change (DS) for which graphical representation is not present here, suggested that the adsorption process becomes spontaneous, feasible, endothermic and irreversible in nature. From the performance analysis, it has been found that adsorbent like this is very efficient compared to other reported Activated carbon from various sources. Novel adsorbent like this can be a good alternative for large-scale application in various process industries for removal of heavy metal from the process effluent.

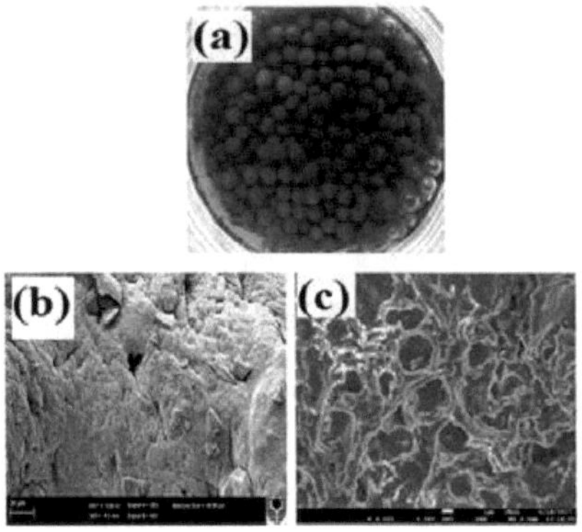

Fig 1. a) Synthesized Beads in water. b) outer and c) inner surface of beads

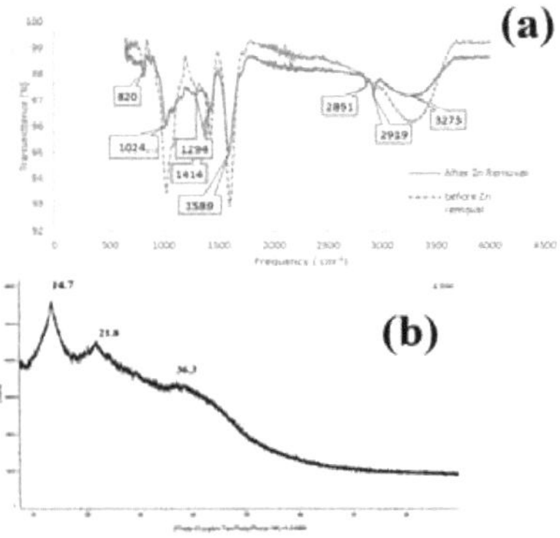

Fig. 2. a) FTIR spectra of treated and untreated composite adsorbent. b) X-ray diffraction spectra of composite adsorbent

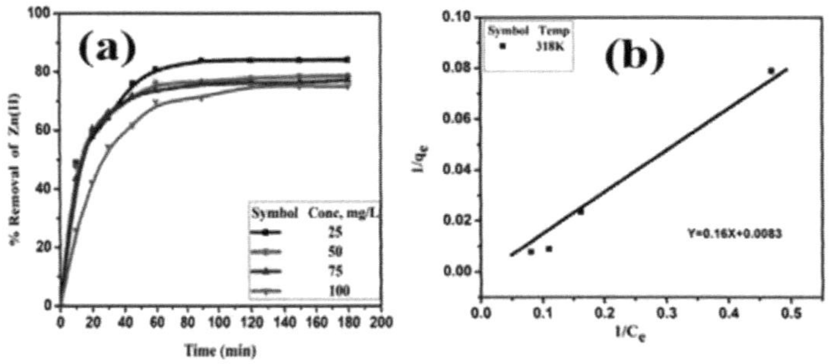

Fig 3. a)Variation of Percentage removal of Zn^{+2} with initial concentration b) Langmuir isotherm plot

Keywords: Biochar alginate composite, wastewater treatment, adsorbent characteristic.

Reference

S. Dawood, T.K. Sen, C. Phan, Synthesis and characterization of novel-activated carbon from waste biomass pine cone and its application in the removal of congo red dye from aqueous solution by adsorption, Water. Air. Soil Pollut. 225 (2014). doi:10.1007/s11270-013-1818-4.

524834 Process and Supply Chain Design: Sustainability Metrics and Structural Considerations

Abstract:

The design of sustainable manufacturing processes and supply chains is rapidly becoming a critical issue, because of the need to provide for the needs of a growing human population which is increasingly prosperous across the globe. To address this growing issue, we review the most important elements of sustainability science including relevant sustainability criteria and metrics that can be used to assess sustainability. It is notable that there approximately six critical processes that must be maintained for sustainability to be possible. These represent energy resources, land use, energy efficiency, economic strength, human well-being, and system stability. We also discuss the importance of process and supply structures which is an often-overlooked aspect of process and supply chain design, and we discuss the application of the P-graph framework. This can be used to generate easily a large number of feasible structures which can then be assessed for sustainability. We illustrate the application of these ideas with examples of energy generation processes and supply chains.[1]

[1] Cabezas, H., Argoti, A., Friedler, F., Mizsey, P, and J.P. Pimentel, â€œDesign and Engineering of Sustainable Process Systems and Supply Chains by the P-Graph Framework,â€ Env. Prog. Sust. Energy, in press. DOI: 10.1002/ep.12763.

531539 Sustainability Assessment of Nanocoating Manufacturing

Abstract:

Nanopaint is a very promising coating material in the coating industry, as the coating layer using this type of material could demonstrate significantly improved coating performance or even a number of new functional capacities. However, there have been a number of serious environmental concerns, as during the nanopaint manufacturing and application phases, nanoparticles in paint and coating layers could enter the working environment through emission into air and water. A further concern is the nanoparticle release into the environment during the product use phase. The unintended environmental and health risks, as well as the economic and social challenges associated with the use of nanocoating products, have not been systematically estimated. Over the past few years, we have explored the feasibility of using a unique life cycle-based sustainability assessment method to investigate the nanocoating sustainability issues.

In this paper, we present our recent progress on nanocoating sustainability by focusing on the nanocoating manufacturing process and product performance. We propose a sustainability metrics system that could be used to provide a more comprehensive assessment of the environmental, economic, and social performances of the manufacturing process and the product. We will also introduce a hybrid approach by combining both subjective and objective methods to determine weighting factors associated with the metrics in different sustainability categories. A comprehensive case study on sustainable manufacturing of automotive nanocoating will be illustrated to show methodological efficacy.

533370 Comparative Life Cycle Assessment of Methanol Production Processes Based on Externalities Under Uncertainty

Abstract:

Carbon dioxide is considered as the main contributor to climate change and thus to global warming. Several technologies have been proposed to reduce atmospheric CO_2, such as CO_2 capture and storage (CCS), CO_2 capture and utilization (CCU) and enhanced oil recovery (EOR). From these proposals, converting the captured CO_2 into valuable products was seen as a promising area to explore. This research focuses on methanol production using captured carbon dioxide and hydrogen and compare it to the business as usual (BAU) process that is based on steam reforming of natural gas (SRNG) according to sustainability criteria. To achieve this objective, a systematic approach has been followed. Starting with a technoeconomic analysis of the process using simulation software Aspen-HYSYS. Then, a â€œcradle-Âto-gateâ€ life cycle assessment has been carried out for the methanol production processes from different CO_2 and H_2 sources that will inform on the impact caused in several damage categories. ReCipe 2008 LCA method was followed in this work that quantifies the environmental impacts per 1 kg of methanol. Later, the obtained impacts were monetised using external costs and combined with the economic performance of each alternatives. Finally, the analysis was repeated for different scenarios generated using Monte Carlo sampling in order to understand how various uncertainty sources affect the results. The holistic approach demonstrated by the life cycle assessment (LCA) combined with the externalities does not only enable decision makers to assess the proposed technologies compared to the conventional production technologies based on their economic value but also gives a better insight about the overall performance of each process based on techno economic and environmental measures.

539172 Determining Chemical Release Profiles By Rapid Estimation Methods

Abstract:

Rapid estimation methods can be used to determine the chemical release profiles of a chemical of interest. This chemical release profile provides information needed for chemical screening and contributing to more complete risk assessments. Looking over the life-cycle of where a chemical could be released, methods are needed to estimate emissions and water discharges for the extraction, manufacturing, transport, processing, use, and end-of-life phases. Paying attention to both accuracy and the level of required effort, the estimation methods to use can depend on circumstances. An approach will be described that combines four efforts developed in the U.S. Environmental Protection Agency, focusing on those advanced by the Office of Research and Development. The first effort uses EPA data sources and ways to use top-down data mining to extract and subsequently manipulate the data into release profiles. The second effort applies EPAâ€™s ChemSTEER software, generic scenarios, and bottom-up simulation methods, which define material flows and allow for the estimation of emissions. The third effort estimates releases through proxies based on properties of similar chemicals with respect to release profiles. Finally, a reconciliation effort will provide appropriate methods for determining which methods to use.

The views expressed in this presentation are those of the authors and do not necessarily represent the views or policies of the U.S. Environmental Protection Agency.

538646 Design and Manufacture of a Torus Microreactor for the Removal of Azo Dyes By Laccase Immobilized on Magnetite Nanoparticles

Abstract:

Water pollution is one of the main environmental problems that mankind is facing over the coming few decades. Oceans, aquifers, rivers, lakes, and groundwater are being polluted at an unprecedented rate. This directly affects not only the health of an increasingly higher number of human communities but ecosystems of other living organisms [1]. Industrial and agricultural processes are responsible due to the continuous discharge of large volumes of wastewaters to these aquatic environments [2][3]. Some of the most common discharged pollutants include pathogens, excess nutrients, suspended solids and sediments, pesticides, plastics, fertilizers, acids, detergents, phenols, minerals, and heavy metals [4][5].

The annual production of synthetic dyes or azo dyes approaches 70 million tons [6]. These xenobiotic chemicals stand as one of the major sources of water pollution, they are relatively simple to synthesize, soluble in water and their natural occurrence is rare [7][8][9]. Azo dyes find applications in various industries including textiles, leather goods, paper, plastics, foodstuffs, cosmetics and candles [10][11]. From them, the vivid colors develop, most notably yellow, orange and red [11]. In recent years, the growing demand of these products has in turn increased the production of dyes and consequently the amount of waste [9].

Despite the industry's efforts to couple wastewater treatment processes to their manufacturing plants, 90% of reactive textile dyes entering activated sludge sewage treatment plants will pass through unchanged and will be eventually discharged to rivers [12][8]. As a result, between 30 to 150 thousand tons of dyes are discharged into water bodies, soil and aquatic ecosystems annually [13][6]. The continuous exposure to azo dyes is suspected to be harmful for living organisms [6]. Water sources contaminated by these compounds have low penetration of sunlight and oxygen, which is essential for the survival of various aquatic organisms [13]. Moreover, the anaerobic degradation entails sub-products of high biological toxicity that may end up in the food chain. An important number of dyes are made of known carcinogens and toxics, such as benzidine and other aromatic compounds, which main effects to humans include damage to ADN and proteins [8][13][11]. Current regulations in Europe, China, Taiwan, Korea and Japan limit the discharge of synthetic dyes to approximately 30 parts per million when used in textiles and leather items that could have prolonged contact with the skin [11]. Accordingly, there has been a recent effort to implement more stringent effluent treatment regulations that enforce industry to find methods to lower the color level in their wastewater prior to discharging into surface waters [11][13].

The primary methods to remove azo compounds are coagulation/flocculation, adsorption, precipitation, flotation, membrane filtration, bioflocculants treatments, ion pair extraction, ultrasonic mineralization, electrolysis, ion exchange, advanced oxidation processes, sonication, photocatalysis, and ozonation [6][7][13]. A major limitation of this processes is however, their simultaneous dependence on both anaerobic and aerobic conditions. This leads to costly processes and/or extremely toxic byproducts [10][20]. Additionally, the biocatalytic degradation of azo dyes and wastewater treatment via laccase has been previously described in different contributions [9][12][7]. Laccase is an oxidoreductase enzyme, capable of oxidizing phenolic compounds into phenoxyl radicals, with the aid of 4 copper electrons in its structure [9][7]. The enzyme promotes decolorization in azo dyes, due to the presence of aromatic polyphenolic moieties in their structures [7].

Here we explore the design and manufacture of a microreactor to enable the enzyme-based degradation of Azo dyes. Laccase molecules for the biodegradation of azo compounds were covalently immobilized on amino-terminated silanized magnetite (Fe_3O_4). Silanization was conducted with 3-Aminopropyltriethoxysilane (APTES) while laccase molecules were conjugated via imine bonds with the aid of glutaraldehyde as crosslinker. We hypothesize that a torus microreactor is suitable for maximizing biodegradation processes due to the absence of dead volume, efficient mixture of reagents and the continuous reaction loop allowed by the toroidal shape [21]. Additionally, the strong magnetic response of magnetite allows the application of magnetic fields to maintain the nanoparticles suspended during the treatment process and maximize contact between the components. A first attempt for find an optimal configuration for the micro-torus reactor was explored *in silico* with the aid of Comsol Multiphysics® by analyzing mixing patterns and fluid dynamics. This was accomplished by coupling the CFD and particle tracing modules. Device prototyping was conducted in polymethyl methacrylate using a laser cutter system and commercially available fittings for the assemblage and subsequent testing. Currently, a spectrophotometry assay is performed before and after to evaluate the degradation achieved with the process. Finally, to monitor the extent degradation of azo molecules in an in-line and real-time manner, the device was instrumented with an *in situ* spectrophotometry system. Functionalization of nanoparticles surfaces in conjunction with torus microreactors appear well suited for degradation and/or capture of other pollutants of concern.

REFERENCES

[1] M. A. Atieh, «Removal of Phenol from Water Different Types of Carbon – A Comparative Analysis», *APCBEE Procedia*, vol. 10, pp. 136-141, 2014.

[2] W. W. Anku, M. A. Mamo, y P. P. Govender, «Phenolic Compounds in Water: Sources, Reactivity, Toxicity and Treatment Methods», en *Phenolic Compounds - Natural Sources, Importance and Applications*, InTech, 2017.

[3] Z. M. Wang, *Nanotechnology for Water Treatment and Purification*, vol. 22. 2014.

[4] N. Iit y K. Web, «Classification Of Water Pollutants And Effects On Environment», *NPTEL IIT Kharagpur Web Courses*, pp. 1-7.

[5] R. Helmer y I. Hespanhol, *Water Pollution Control - A Guide to the Use of Water Quality Management Principles*. London: E & FN Spon, 1997.

[6] H. Zou y Y. Wang, «Azo dyes wastewater treatment and simultaneous electricity generation in a novel process of electrolysis cell combined with microbial fuel cell», *Bioresour. Technol.*, vol. 235, pp. 167-175, 2017.

[7] J. Kanagaraj, T. Senthilvelan, y R. C. Panda, «Degradation of azo dyes by laccase: Biological method to reduce pollution load in dye wastewater», *Clean Technol. Environ. Policy*, vol. 17, n.º 6, pp. 1443-1456, 2015.

[8] S. Seshadri, P. L. Bishop, y A. M. Agha, «Anaerobic/aerobic treatment of selected Azo dyes in wastewater», *Waste Manag.*, vol. 14, n.º 2, pp. 127-137, 1994.

[9] A. Zille, B. Górnacka, A. Rehorek, B. Go, y A. Cavaco-paulo, «Degradation of Azo Dyes by Trametes

villosa Laccase over Long Periods of Oxidative Conditions», *Appl. Environ. Microbiol.*, vol. 71, n.º 11, pp. 6711-6718, 2005.

[10] F. I. Vacchi *et al.*, «Chlorine disinfection of dye wastewater: Implications for a commercial azo dye mixture», *Sci. Total Environ.*, vol. 442, pp. 302-309, 2013.

[11] B. J. Brüschweiler y C. Merlot, «Azo dyes in clothing textiles can be cleaved into a series of mutagenic aromatic amines which are not regulated yet», *Regul. Toxicol. Pharmacol.*, vol. 88, pp. 214-226, 2017.

[12] A. Cavaco-paulo y G. M. Gu, «Decolorization and Detoxi cation of Textile Dyes with a Laccase from», *Society*, vol. 66, n.º 8, pp. 3357-3362, 2000.

[13] S. P. Buthelezi, A. O. Olaniran, y B. Pillay, «Textile dye removal from wastewater effluents using bioflocculants produced by indigenous bacterial isolates», *Molecules*, vol. 17, n.º 12, pp. 14260-14274, 2012.

[14] B. Bertrand, F. Martínez-Morales, y M. R. Trejo-Hernández, «Fungal laccases: induction and production», *Rev. Mex. Ing. Química*, vol. 10, n.º 1, pp. 17-28, 2011.

[15] J. F. Osma, J. L. Toca-Herrera, y S. Rodríguez-Couto, «Uses of Laccases in the Food Industry», *Enzyme Res.*, vol. 2010, pp. 1-8, sep. 2010.

[16] N. Durán y E. Esposito, «Potential applications of oxidative enzymes and phenoloxidase-like compounds in wastewater and soil treatment: A review», *Appl. Catal. B Environ.*, vol. 28, n.º 2, pp. 83-99, nov. 2000.

[17] A. Tsioulpas, D. Dimou, D. Iconomou, y G. Aggelis, «Phenolic removal in olive oil mill wastewater by strains of Pleurotus spp. in respect to their phenol oxidase (laccase) activity», *Bioresour. Technol.*, vol. 84, n.º 3, pp. 251-257, sep. 2002.

[18] A. Jaouani, F. Guillén, M. J. Penninckx, A. T. Martínez, y M. J. Martínez, «Role of Pycnoporus coccineus laccase in the degradation of aromatic compounds in olive oil mill wastewater», *Enzyme Microb. Technol.*, vol. 36, n.º 4, pp. 478-486, mar. 2005.

[19] E. Abadulla, T. Tzanov, S. Costa, K. H. Robra, A. Cavaco-Paulo, y G. M. Gubitz, «Decolorization and detoxification of textile dyes with a laccase from Trametes hirsuta», *Appl. Environ. Microbiol.*, vol. 66, n.º 8, pp. 3357-3362, ago. 2000.

[20] F. P. Van Der Zee y S. Villaverde, «Combined anaerobic-aerobic treatment of azo dyes - A short review of bioreactor studies», *Water Res.*, vol. 39, n.º 8, pp. 1425-1440, 2005.

[21] L. Pramparo, F. Stüber, J. Font, A. Fortuny, A. Fabregat, y C. Bengoa, «Immobilisation of horseradish peroxidase on Eupergit®C for the enzymatic elimination of phenol», *J. Hazard. Mater.*, vol. 177, n.º 1-3, pp. 990-1000, 2010.

538417 Gold Recovery from Electronic Waste By Nanoporous Polymers

Abstract:

Urban mining of precious metals from electronic waste is not yet feasible because of its lengthy process, health risks and environmental impact. We recently developed a nanoporous porphyrin polymer that captures precious metals quantitatively even in the presence of over 35 metals from the periodic table. The nanoporous polymer is synthesized in two steps from widely available monomers and can be scaled up without loss of activity. Through an unprecedented reductive capture mechanism, gold is particularly recovered with ten times the theoretical limit, reaching to a record 1,637 g/g. With 99 % uptake taking place in the first 30 minutes, the porous polymer can be rapidly desorbed and reused for repetitive batches. We also demonstrated that gold can be readily recovered from actual printed circuit boards.

524665 Bio-Inspired Immobilization of Casein-Coated Silver Nanoparticles on Cellulose Acetate Ultrafiltration Membranes for Biofouling Control

Abstract:

The focus of this project was on immobilizing Ag-NPs on ultrafiltration membranes to provide the membranes with antimicrobial properties while preventing nanoparticle leaching and membrane degradation. To this end, low-biofouling nanocomposite membranes were loaded with casein-coated silver nanoparticles (casein-Ag-NPs) using two approaches, physical blending of Ag-NPs in the dope solution (^{P}Ag-NP/CA membranes) and chemical attachment of Ag-NPs to cast membranes (^{C}Ag-NP/CA membranes). The functionalization of Ag-NPs onto the CA membranes was achieved via attachment with functionalized thiol groups with the use of glycidyl methacrylate (GMA) and cysteamine chemistries, which was inspired by the affinity of silver to the thiol groups of cysteine proteins in bacteria. The performance of two different membranes were characterized via structural, morphological, permeation, and biofouling studies. The ^{P}Ag-NP/CA membranes displayed more pore channels in their cross-sections, exhibited higher average flux values, and had the highest average flux decline during short-term filtration studies. There were minor differences in pore channel morphology between CA and ^{C}Ag-NP/CA membranes, but the latter displayed a tighter pore network compared to the other membranes due to GMA crosslinking, which contributed to its lower average flux. While the ^{C}Ag-NP/CA membranes had a similar average flux decline compared to the CA membranes, the chemically attached Ag-NPs expressed the highest salt rejections. Lastly, ^{C}Ag-NP/CA membranes appeared to possess better biofouling control as compared to CA membranes, based on the reduction of observable cells. There was minimum leaching of silver during cross-flow filtration of ^{C}Ag-NP/CA membranes, and the leached water could still meet drinking water regulations.

530720 Concurrent Desalination and Boron Removal Via Reverse Osmosis

Abstract:

Desalination of saline water is a critical technology for supplying the potable and irrigation water needed for the rapidly increasing world population. The boron content of seawater averages 4.6 ppm and is as high as 9.6 ppm in the Eastern Mediterranean. A small amount of boron is required for all plants; however, it affects the yield and can become toxic to plants at levels ranging between 0.5 and 20 ppm. The World Health Organization (WHO) recommends that the boron concentration in drinking water and irrigation water be 0.5 ppm or less. Since boron is present in water as boric acid, its concentration cannot be reduced adequately by single-stage reverse osmosis (SSRO) using commercially available reverse osmosis (RO) membranes. This can be done using commercially available RO membranes by using a two-pass SSRO at the cost of reduced water recovery. Increased water recovery is possible by raising the pH that causes boric acid to dissociate into hydrated borate ions for which commercially available RO membranes have a good rejection. However, this requires adding chemicals and an additional process step to reduce the pH again. Desalination followed by ion exchange or dialysis can also reduce the boron concentration at the cost of increased process complexity. None of these processes can achieve desalination and boron removal with a high water recovery using just commercially available membranes. This paper describes the Concurrent Desalination and Boron Removal (CDBR) process for water desalination and reduction of the boron concentration to 0.5 ppm or less using commercially available membranes. (Krantz, 2016; Kürklü, 2017)

In the CDBR process, the saline water is fed to one or more SSRO stages in series whose retentate is fed to a countercurrent membrane cascade with recycle (CMCR) and whose permeate is fed to a low pressure membrane stage (LPMS) for further boron removal. The permeate from the LPMS is blended with that from the CMCR to obtain the desired product water. This process configuration employs countercurrent flow of the retentate and permeate, permeate recycle, and retentate reflux in the CMCR to reduce the required osmotic pressure differential (OPD) significantly. This reduces the pumping, maintenance, pretreatment and brine disposal costs. The performance metrics of the process have been predicted for operation at the thermodynamic limit.

The performance metrics for this novel CDBR process are compared with those for using two SSRO stages in series for desalination and boron removal. The assumptions are a seawater feed containing 35,000 ppm salt and 10 ppm boron producing water containing 0.5 ppm boron and less than 100 ppm salt and, pump and energy recovery device (ERD) efficiencies of 85% and 90%, respectively. For these conditions the CDBR process employing commercially available RO membranes with a 90% boron rejection can reduce the OPD relative to two SSRO stages in series by 35% at a water recovery of 55% and 37% at a recovery of 75%, and can reduce the net specific energy consumption (SEC_{net}) by 0.87% at 55% recovery and 39% at 75% recovery. The SEC_{net} and OPD can be further reduced by recycling and blending the retentate of the LPMS stage with the feed to the SSRO.

A total cost analysis of water production for the CDBR configurations is also carried out in terms of the fixed and operating costs, relative to a two-pass SSRO process. In the comparison, both processes are designed using commercial membranes to yield a water product with less than 0.5 ppm boron concentration from a saltwater feed containing 10 ppm boron and 35000 ppm TDS. While the two-pass RO process yields an overall water recovery of 46%, the yield of the CDBR process is 61%. The gross specific energy consumption of the CDBR process 4.44 kWh/m^3 in comparison to 4.76 kWh/m^3 for the two-pass RO process. If energy recovery devices are used, then the net specific energy consumption of the CDBR process is reduced to 3.51 kWh/m^3 in comparison to 3.19 kWh/m^3 for the two-pass RO process. The cost comparison is based on a water production capacity of 10000 m^3/day.

A novel CDBR process for achieving concurrent desalination and boron removal has been developed. The CDBR process uses only membrane technology without the need for increasing the pH or using other technologies such as ion exchange that are required for current commercial processes for desalination and boron removal. As such, the CDBR process permits concurrent desalination and boron removal rather than requiring sequential processes.

This work is supported by the Scientific and Technological Research Council of Turkey (TUBITAK) through Project No. 114Y165.

References

Krantz, W.B., S. Velioğlu, S. Kűrklű, M.G. Ahunbay, S.B. Ersolmaz, "Concurrent Desalination and Boron Removal (CDBR) Process," Patent application no. PCT/TR2016/050387, October 19, 2016.

Kürklü, S.; Velioglu, S.; Ahunbay, M. G.; Tantekin-Ersolmaz, S. B.; Krantz, W. B., Concurrent Desalination and Boron Removal (CDBR) Process, *Desalination*, 423,79-94, 2017.

532764 Surface Nano-Structuring with Hydrophilic Polymer Brush Layers for Tailored Performance of Fouling Resistant RO and UF Membranes

Abstract:

Surface nano-structuring with hydrophilic polymer brush layers is an effective approach for imparting fouling resistance to membrane surfaces due to improved surface wettability as well as partial chain mobility. In the present approach, a base membrane of a suitable level of rejection and permeability is selected, and its surface is activated via treatment with atmospheric pressure plasma (APP) followed by free radical graft polymerization in an aqueous vinyl monomer solution. The resulting polymer brush layer grafted onto the membrane surface alters the base membrane separation performance (i.e., permeability and solute rejection), which can be tailored by the plasma and graft polymerization conditions. Therefore, selection of the base membrane and the conditions at which surface nano-structuring is performed is critical to achieve a fouling resistant membrane with the target membrane performance. Accordingly, the present study focused on evaluating the impact of plasma surface activation (i.e., APP type and treatment time) and graft polymerization (i.e., monomer type and concentration, grafting temperature and reaction time) conditions on the performance of surface nano-structured (SNS) polyamide (PA) RO and polysulfone (PSf) UF membranes with respect to water permeability and solute rejection. Polymer surface grafting was confirmed via X-ray photoelectron spectroscopy (XPS), and surface wettability and topography of the SNS membranes were evaluated via contact angle measurement and AFM, respectively. Permeability and solute rejection of the SNS-PA and SNS-PSf membranes were evaluated using laboratory scale plate-and-frame and stirred cells, respectively. Fouling propensity of the SNS-PA and SNS-PSf membranes was evaluated using model foulants (i.e., BSA and alginic acid), and membrane cleaning effectiveness was evaluated with DI water flushing at the end of foulant filtration. Results of the present study indicate that the present membrane surface nano-structuring approach is suitable for tailoring the performance of both RO and UF membranes in addition to mitigating membrane fouling and improving membrane cleaning effectiveness.

528629 Dual-Functionalized Nanofiltration Membranes Exhibit Multifaceted Anti-Fouling and Ion Rejection Performance

Abstract:

As interest in decentralized water treatment units that enable localized water reuse grows, the development of nanofiltration (NF) membranes as a complementary technology to reverse osmosis membranes has gained traction. Furthermore, the use of advanced materials (e.g., graft copolymers), which form uniform nanoscale pores that are decorated with tunable chemical moieties can provide a ready means for post-assembly functionalization. For example, the incorporation of charged functionalities increases ion rejection by NF membranes due to repulsive electrostatic interactions. However, the incorporation of charge functionality tends to exacerbate the process of fouling, which is a continued hurdle in the development and application of NF membranes. Fouling is the process whereby attractive dispersive, electrostatic, and hydrodynamic interactions between dissolved solutes and the membrane leads to the attachment and accumulation of material on the surface of the membrane. This process reduces the throughput and performance of the membrane. This study evaluates novel NF membranes that exhibit both high ion rejection and high fouling resistance through the use of a post-fabrication multi-functionalization process performed on the membrane template.

Membranes cast from a poly(acrylonitrile-*co*-oligo-(ethylene glycol) methyl ether methacrylate-*co*-(3-azido-2-hydroxypropyl methacrylate)) P(AN-OEGMA-AHPMA) copolymer are the parent substrates utilized in this study. The azido groups lining the pore wall of the parent substrate lead to multiple functionalization avenues. Here, a dual functionalized membrane containing both ion-selective and anti-fouling functionality is attained through a controlled reaction between propiolic acid and the membrane. This copper-catalyzed azide-alkyne cycloaddition (CuAAC) reaction occurs rapidly and by varying the concentration and the exposure time the penetration depth of the initial reaction can be controlled. This control is confirmed through a decrease in the signal intensity associated with the azido group in the Fourier transform infrared spectroscopy (FTIR) spectrum. This layer achieves an anti-fouling character through a 1-ethyl-3-(3-dimethylaminopropyl)-carbodiimide (EDC) coupling reaction of 1H,1H-perfluorooctylamine with the propiolic acid. The azido moieties not exposed to propiolic acid in the initial stage are then charge-functionalized through a CuAAC reaction of propargyl amine. When challenged with a 0.5 g/L Bovine Serum Albumin (BSA) solution, these membranes express increased anti-fouling properties compared to a fully charged membrane, while maintaining similar ion rejection performance.

This study begins the analysis of incorporating multiple, useful functionalities into a single membrane template. The rapid exposure of a reactive solution works to prove that through readily controllable reactions, the required depth for inhibiting foulant interactions could be reduced down to the minimal distance in which the attractive forces between the foulants and the surface is shielded. Minimizing this layer incorporates more charged-groups on the pore wall, increasing the overall resistance to ion-transport. This membrane fabrication process can be further extended to ion-exchange membranes, in which the inclusion of a top surface anti-fouling layer could increase the lifespan of the membrane while maintaining a high ionic throughput.

530611 Inorganic Microfiltration Membranes Modified with Hydrophilic Silica Nanoparticles for Oil-in-Water Emulsion Separation

Abstract:

On a global scale, approximately three barrels of produced water are produced for every barrel of oil extracted. The existence of oil droplets can result in the formation of stable oil-in-water emulsions. In order to meet the discharge requirements, there is a significant need for efficient, cost-effective strategies to remove hydrocarbons from produced water. Our work aims at developing new membrane strategies that result in high water flux and low membrane fouling for produced water applications. To accomplish this, alumina microfiltration membranes were modified by incorporating silica nanoparticles into the membrane structure. The modified alumina tubular membranes were tested using cyclohexane-in-water emulsions. The experimental results showed that the alumina membrane treated with a 0.5 wt.% silica nanoparticle solution yielded a water flux of 350 L/m^2-h and oil rejection efficiency of 93%. This is a 21% enhancement in the flux and a 6% enhancement in the oil rejection efficiency compared with the unmodified membrane. The cyclohexane concentration of the permeate was less than 40 ppm when the cyclohexane concentration of the feed was 500 ppm at 40 â„*f* with a feed pressure of 43.7 psia.

532326 Effect of the Surface Charge of Monodisperse Particulate Foulants on Cake Formation

Abstract:

In microfiltration and ultrafiltration, particulate foulants are inevitably deposited on the membrane surface, forming a cake whose structure and behavior play crucial roles in the subsequent filterability of the suspensions. This study investigated the impact of fouling by three types of latex particulate foulants, which were of the same size (3 Î¼m) but with different surface charges. Surprisingly, although the positively charged aminated latex was expected to perform the worst in the flux-decline experiments due to attractive electrostatic interaction with the negatively charged membrane, this latex displayed the best performance relative to the two negatively charged latex. To understand these counter-intuitive results, a novel network model and three-dimensional (3D) optical coherence tomography (OCT) image analysis were employed to reveal the underlying reasons for the different fouling behaviors. Two mechanisms were found to contribute to the worse performance of the negatively charged latex. Firstly, these particles tended to deposit on the pore rather than non-pore region of the membrane due to the repulsive particle-membrane electrostatic interactions, which led to a more complete pore blockage and thereby greater initial cake resistance. Secondly, these particles had a greater tendency to cluster and deposit on other deposited latex due to similarly repulsive particle-membrane and particle-particle interactions, which led to a more inhomogeneous cake and thereby greater specific cake resistance.

532835 Zwitterion-Substituted Polysulfones As Fouling-Resistant Desalination Membranes

Abstract:

We have demonstrate a novel synthesis method to prepare a zwitterion-modified copolymer: poly(arylene ether sulfone-*co*-sulfobetaine arylene ether sulfone) (PAES-*co*-SBAES), which was blended with unmodified polysulfone (PSf) to prepare asymmetric, free standing membranes. The polymer's chemical structure was analyzed by ^{1}H-NMR spectroscopy and the molecular weight of polymer was determined using size exclusion chromatography. Then, asymmetric PSf/PAES-co-SBAES blend membranes with various zwitterionic SBAES segment contents were fabricated via the non-solvent induced phase separation process. The cross-section morphology, surface hydrophilicity, and surface roughness of the membranes were analyzed by scanning electron microscopy, water contact angle measurements, and atomic force microscopy. The results indicated that both the porosity of the support layer and surface hydrophilicity increased drastically due to the incorporation of hydrophilic zwitterionic (or SBAES) segments. Surface roughness remained low for all membranes. The water permeance and antifouling ability of the PSf/PAES-co-SBAES blend membranes were both remarkably improved, with permeance values significantly higher than the native PSf and at or above levels of commercial polyamide membranes. Also, salt rejection was maintained at a high level, despite increases in hydrophilicity. The flux recovery when challenged with protein-rich solutions was ~93%, and complementary fluorescence microscopy experiments revealing very small levels of protein adhesion. Finally, no decrease in performance was observed when the membranes were challenged with chlorinated solutions. In summary, the strategy outlined to prepare zwitterion-modified polysulfones yielded asymmetric membranes with a high water permeance and salt rejection that also exhibit fouling resistance and tolerance toward chlorinated solutions. This will talk will discuss these developments and ongoing experiments to further improve the performance and scalability.

530245 Economic Assessment of Novel Process Turning Industrial Waste Gases (mixed CO/CO_2 streams) into Intermediates for Polyurethanes for Rigid Foams and Coatings

Abstract:

The chemical and manufacturing industries in particular is an extremely large contributor to greenhouse gas emissions and the global warming effect[1]. The steel industry alone accounts for around 4.5% of global CO_2 emissions[2]. Reduction of emissions and a decrease in the usage of fossil fuels must happen in order to fight this. Industrial symbiosis is one way in which this can be done. The EU funded Carbon4PUR project develops a novel process for the manufacture of polyol intermediates for polyurethane plastics from steel production flue gases (mixed CO/CO_2 streams). The project consortium comprises 15 industrial partners and academic institutions throughout Europe.

The project aims to make polyurethane production less dependent on fossil fuels. The goal is to decrease the carbon footprint of production of polyurethane intermediates by approximately 20-60%[1] and reduce the cost by around 25%. The project also contributes to creation of new employment, reducing the dependency of industry on imports and provide a secure supply of carbon feedstock.

The process begins with hydrogen removal from the flue gas streams, shown in Figure 1 as eFlue Gas Conditioningf. The CO and CO_2 in these streams are transformed to lactones and cyclic carbonates, respectively, which are used as building blocks for the different kinds of polyols to be produced in the process. These novel polyols are reacted with isocyanates to produce specific kinds of polyurethanes, particularly rigid foams for building insulation and wood coatings. A diagram of this process is shown in Figure 1.

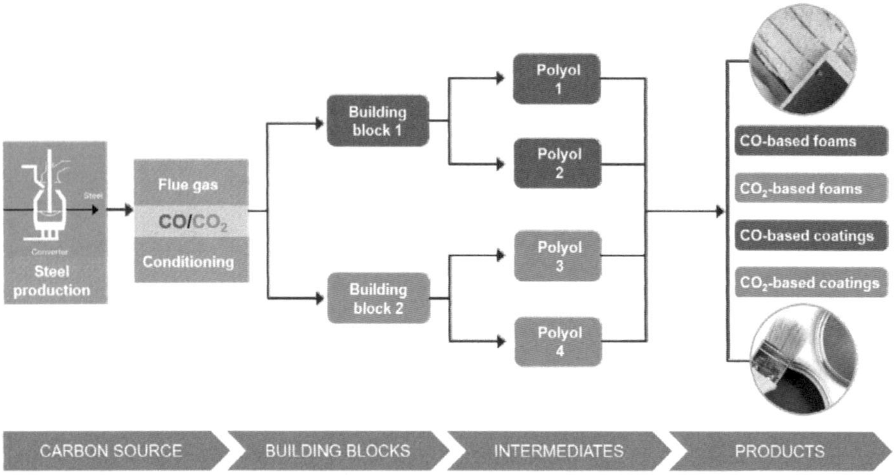

Figure 1. Basic schematic of the Carbon4PUR process.[1]

It is very useful to be able to determine the cost of the flue gas taken from steel plants, as it is the chief novel feed stock and the most crucial to this process. A method is also presented for the determination of the production cost of the flue gas, depending on its usage on a temporal basis and therefore its economic value to the steel producer. This is necessary information when attempting to assess the economic viability of a novel industrial symbiosis type project as it can be easily compared to current production methods.

Acknowledgments

This project has received funding from the European Unionfs Horizon 2020 research and innovation programme under grant agreement No 768919.

References

[1] Carbon4PUR (2018). *The project – Concept*. Carbon4PUR. Retrieved from https://www.carbon4pur.eu/about/the-project/

[2] D.E. Wiley, T.M. Ho, A. Bustamante (2011). *Assessment of Opportunities for CO2 Capture at Iron and Steel Mills: An Australian Perspective.* Elsevier, Energy Procedia 4, pp 2654-2661.

533211 Techno-Economic AssessmentÂ of CO_2 and SO_x Capture Process By Dilute Aqueous Ammonia

Abstract:

Techno-economic Assessment of CO_2 and SO_x Capture Process by Dilute Aqueous Ammonia Hoan Le Quoc Nguyen, David Shan-Hill Wong* Department of Chemical Engineering, National Tsing Hua University, Hsinchu 30013, Taiwan *Corresponding Author's E-mail: dshwong@che.nthu.edu.tw **Keywords:** Economic analysis; CO_2 capture; Co-capture; Aqueous ammonia.

Dilute aqueous ammonia (NH_3) solvent is one of the promising for post-combustion CO_2 capture. This technology provides many advantages such as low solvent cost, negligible solvent degradation and the simultaneous capture of multiple acid pollutants. For example, Li et al. (2015) suggested an advanced NH_3-based CO_2 capture process that can be simultaneously achieved SO_2 and CO_2 removal. However, the economics of this co-capture was not been analyzed in details.

In this work, we evaluate the techno-economic feasibility of this concept using a 500 MWe gross coal-fired power station with this advanced CO_2 capture process. Cost estimation of Power plant was performed using the Integrated Environmental Control Model (IECM) v9.5 and the capture plant was cost by ASPEN Plus v8.4. Figure 1 shows a brief description of two case studies including base case and co-capture case. In the base case, the power plant has all the treatment steps including selective catalytic reduction (SCR), electrostatic precipitator (ESP), flue gas desulfurization (FGD) and post-combustion capture (PCC) by aqueous ammonia process. In the co-capture case, FGD was eliminated and additional ammonia was used in the PCC to remove the SO_x as ammonia sulfate and ammonia sulfite. Besides, the reaction for absorbing SO_3 by aqueous ammonia is also included to the chemistry system in order to fully describe the characteristic of the SO_x absorption in the capture process, instead of that of SO_2 absorption only in literature. Figure 2 shows the capital cost (CAPEX) and operating cost (OPEX) of two cases. It can be seen that the total capital cost for a whole power plant with full treatment steps in the base case is about 1300 million dollars based on the U.S. cost reported in 2014, which is included in the IECM framework. In that, the capital cost of CO_2 capture unit is significant almost equal to 27% the capital cost of total other units. By applying the advanced CO_2 capture unit to remove SO_x and CO_2 simultaneously, the FGD unit can be completely eliminated in the co-capture case. This leads to a CAPEX savings of about 124 million dollars. The increase of the total operational cost due to the increase of the amount of ammonia solvent make-up can also be nearly offset by the elimination of operation cost by elimination of operation cost of the FGD. Although there is an increase in operating cost of about 5.6 million dollars due to the high recirculation rate required for the co-capture process, the saving capital cost can cover this increase in 22 years. Moreover, without consumption from FGD unit, the increase in net electrical output for sale leads to raising the benefit of the power plant significantly. In addition to considering the simple technique and operation of a single process, the combined SO_x and CO_2 capture plant with the power plant is preferred rather than the use of separate capture systems.

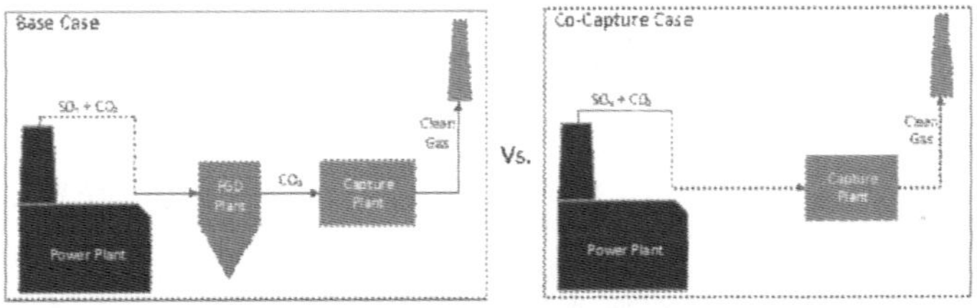

Figure 1: A brief description of two case studies: Base case and Co-capture case.

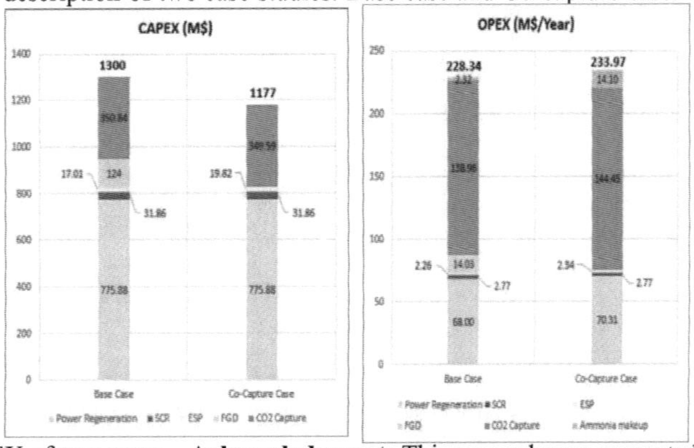

Figure 2: CAPEX and OPEX of two cases. **Acknowledgment:** This research was supported by Grant MOST 107-3113-E-007-002 from the Ministry of Science and Technology of the Republic of China. **References**

1. Li, K., Yu, H., Qi, G., Feron, P., Tade, M., Yu, J., and Wang, S., Rate-based modelling of combined SO_2 removal and NH_3 recycling integrated with an aqueous NH_3-based CO_2 capture process, *Applied Energy,* **148**, 66-77, 2015.

535714 Worldâ€™s Largest Commercial CO_2 to Methanol Demonstration Plant

Abstract:

The George Olah CO_2 to methanol plant, commissioned in April 2012, currently produces ~ 5 million liters/year renewable methanol and captures up to ~ 5600 ton CO_2/year in the process. [Lim 2016, *Nature*, 526(630)] This Carbon Recycling International (CRI) plant is located in Svartsengi, near Grindavik, Iceland.

The process was originally developed by a small CRI team in Reykjavik, and has undergone several iterations to arrive at the present state of technology and functionality. Taking the process from pilot scale to industrial scale was not trivial. Several difficulties encountered along the way were resolved to arrive at the current robust version of the technology. The high purity renewable methanol currently produced is sold as gasoline additive, similar to ethanol in the USA.

There exist two viable catalytic routes to convert CO_2 to methanol. Perhaps the most familiar option is to first reduce CO_2 to CO through the RWGS reaction and then reduce CO with H_2 to methanol in a second step. The CRI process instead implements the direct hydrogenation of CO_2 with H_2 over a mixed metal oxide catalyst.

The presentation will include a brief history of the R&D and early development of the process, followed by a discussion of selected process features and performance metrics.

The presentation will conclude with a brief discussion of world-wide implementation opportunities, for example the transformation of stranded H_2 into a liquid commodity, or as a combined CO_2 and energy sink for intermittent renewables.

541010 Electrochemical CO_2 Conversion

Abstract:

The utilization of CO2 as a C1 carbon source through electrochemical reduction could improve industrial sustainability, as the process is readily powered by renewable energy sources. However, limited solubility of CO2 at atmospheric pressure starves electrocatalysts of substrate and results in low conversion rates. We are investigating electrochemical CO2 reduction under elevated CO2 pressures to increase the CO2 solubility and reaction rates. This talk will provide details on some of our recent work in this area.

522967 Direct Carbonation of $Ca(OH)_2$ Using Super Critical CO_2 at Different Temperatures Along with the Introduction of SiO_2 Aggregate

Abstract:

The production of portland cement is responsible for about 7% of the global carbon dioxide (CO2) emissions. Roughly one ton of CO2 is produced for every ton of cement. The direct carbonation of calcium hydroxide Ca(OH)2 has recently been suggested as a possible CO2-neutral replacement for portland cement in some applications, e.g. prefabricated cement-based construction components. The art of burning limestone ($CaCO_3$) to form lime (CaO), the formation of slaked lime (calcium hydroxide) and subsequent carbonation has been practiced since at least Roman times as a form of cement. The carbonation process, however, is generally too slow to be practical for most modern applications. This study examines the use of supercritical carbon dioxide at three different temperatures at a constant pressure. The different temperatures were chosen to study how temperature effects the rate of carbonation while in the supercritical state. Discs of Ca(OH)2 were compacted at various pressures and thicknesses (different mass) to produce samples of different density and size. The induction of an aggregate, sand (SiO_2), was also studied to test the effectiveness of formed CaCO3 as a binder. The aggregate was mixed and compacted along with the mass of the $Ca(OH)_2$ into pellets. The main goals of this study were to determine the practical limits of carbonation of Ca(OH)2 at different temperatures within the supercritical state, with the introduction of an aggregate, and to collect relevant morphological (microstructural) information. X-ray diffraction (XRD), thermal gravimetric analysis (TGA), and scanning electron microscopy (SEM) was used to characterize the carbonated samples. A full-scale heat and material balance for a pilot plant that utilizes this technology was also conducted in order to determine its feasibility on the industrial scale.

523744 Catalytic CO_2 Desorption in CO_2-Loaded Aqueous MEA Solution over $SO_4^{2-}/ZrO_2/\gamma-Al_2O_3$ Catalysts

Abstract:

Catalytic CO_2 desorption in CO_2-loaded aqueous MEA solution over $SO_4^{2-}/ZrO_2/gamma-Al_2O_3$ catalysts

Xiaowen Zhang [a], Helei Liu [b], Jieling Hong [a], Paitoon Tontiwachwuthikul [a, b], and Zhiwu Liang [a*]

[a] Joint International Center for CO_2 Capture and Storage (iCCS), Provincial Hunan Key Laboratory for Cost-effective Utilization of Fossil Fuel Aimed at Reducing CO_2 Emissions, College of Chemistry and Chemical Engineering, Hunan University, Changsha 410082, PR China

[b] Clean Energy Technologies Research Institute (CETRI), Faculty of Engineering and Applied Science, University of Regina, Regina, Saskatchewan, S4S 0A2, Canada

Keywords: CO_2 capture; solid acid catalyst; Br nsted acid site; heat duty reduction.

* Author for correspondence: Dr. Zhiwu Liang Email: zwliang@hnu.edu.cn

Abstract

Greenhouse gases, especially carbon dioxide (CO_2), have attracted significant focus as one of the predominant contributors to climate change. Chemical absorption of CO_2 using amine solvents is the most commercial mature technology that can be adopted for CO_2 capture. 5 M monoethanolamine (MEA) solvent is the baseline solvent and has been the most frequently investigated. However, this solvent suffers from high energy requirement for the rich solution regeneration process, resulting in its immensely high operating costs of CO_2 capture.

A potential approach to greatly reduce energy requirement was proposed by Idem et al. [1]. They demonstrated that the addition of solid acid catalyst into the rich amine solution for CO_2 desorption is an effective method to decrease the heat duty. Several zeolite molecular sieve catalysts such as, H-ZSM5 and gamma-Al_2O_3 have been studied and showed higher catalytic activity for this purpose. To further develop this promising method, in the present work, a composite catalyst SO_4^{2-}/ZrO_2 (SZ) supported on gamma-Al_2O_3 (SZA) was prepared by the precipitation-impregnation method and used for the first time as a catalyst in a rich anime solvent regeneration process. SZ is a representative solid superacid catalyst, which possesses the superior Br nsted acid sites (BAS), but encounters small surface area and rapid deactivation. gamma-Al_2O_3 is an acid catalyst and the most extensively employed support materials, and has large physical properties, strong Lewis acid sites (LAS), and high thermal stability. Potentially, the composite catalyst SZA may show a superior catalytic CO_2 desorption performance for a CO_2-loaded amine solvent regeneration process.

In this work, SZA was prepared and applied for the rich 5 M MEA solvent regeneration process to increase the CO_2 desorption performance. Three SZA catalysts with different ZrO_2 and gamma-Al_2O_3 mass ratios were synthesized. The procedure of catalyst preparation is based on Hua et al. [2], with a modification. The SZA with ZrO_2/gamma-Al_2O_3 mass ratio of 0.5, 1, and 2 was marked as SZA1/2, SZA1/1, and SZA2/1, respectively. All the catalysts were measured with XRD, FT-IR, N_2 absorption, Py-IR, and NH_3-TPD techniques.

The regeneration experiment was conducted using a batch reactor, which is displayed in Figure 1. Typically, 1 L rich MEA solutions with the CO_2 loading of 0.5 mol CO_2/mol amine were introduced into a 2 L four-necked flask, which was immersed in the oil bath. 12.5 g catalyst was put into the MEA solution. The time was recorded until the amine solvent temperature reached at 338 K. The final temperature of this system was kept at 371K for 540 minutes. The CO_2 loadings of solution at time of at 15, 30, 45, 60, 90, 120, 240, 360 and 540 minutes were measured by Chittick apparatus.

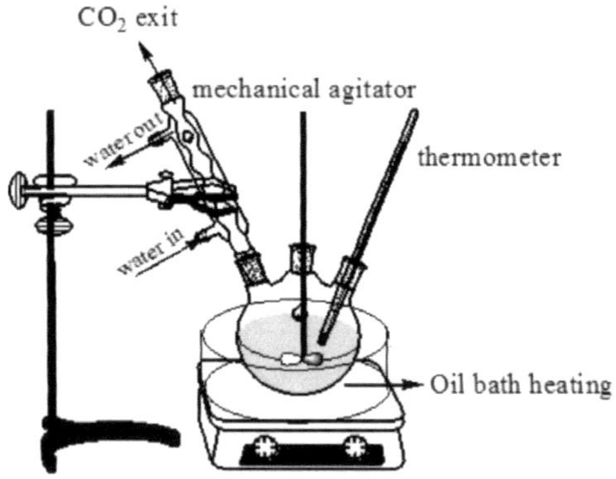

Figure 1. Schematic of the batch reactor.

The heat duty (HD, kJ/mol) for the CO_2 desorption process was monitored by an electric energy meter, which was calculated by equation (1).

$$HD = \frac{\text{Heat input / time}}{\text{Amount } CO_2 \text{ / time}} = \frac{\text{Electricity(kJ)}}{\text{Amount of desorbed } CO_2\text{(mol)}} \quad (1)$$

The relative heat duty (RH, %) was calculated by equation (2).

$$RH = \frac{HD_i}{HD_{baseline}} \times 100\% \quad (2)$$

where $HD_{baseline}$ is the HD of MEA without catalyst regeneration in first 120 minutes, while HD_i is the HD of MEA with different catalysts regeneration in first 120minutes.

The CO_2 desorption rate (DR, mol/min) was calculated by equation (3).

$$(3) \quad DR = \frac{Amount\ of\ desorbed\ CO_2(mol)}{120\,\text{min}}$$

The XRD patterns and FT-IR spectra of these catalysts are shown in Figure 2, which certify that the SZA catalysts were successfully synthesized. The N_2 physisorption isotherms for all catalysts are illustrated in Figure 2 Panel C. All the isotherms are type IV with a specific hysteresis loop, representing typical mesoporous material characteristics. The structural and acid properties of the five catalysts are summarized in Table 1. It can be seen that the introduction of gamma-Al_2O_3 as a carrier contributes to the increase of textural properties for SZ. For the acid sites, the addition of gamma-Al_2O_3 enhanced both of weak acid sites and strong acid sites of SZA catalysts, and the total acidity of SZA catalysts increased with the increasing gamma-Al_2O_3 content. SZA1/1 possesses the greatest strong acid sites and total acid sites as 1.105 and 2.317 mmol/g. Also, the BAS was greatly increased after loading gamma-Al_2O_3 support.

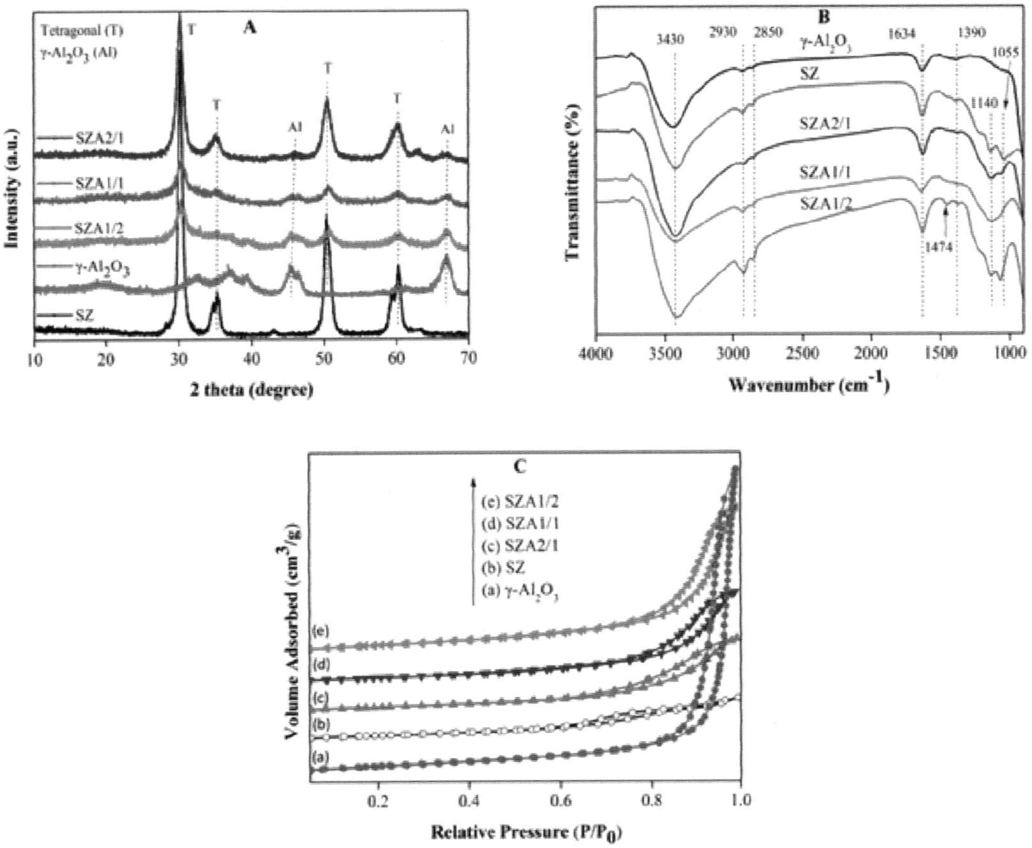

Figure 2. Characterization results of various catalysts.

Table 1 Physicochemical properties of various catalysts.

Catalysts	surface area (m^2/g)		Acidity by strength (mmol/g) [a]			Acidity by type [b]		
	MSA	BET	weak	strong	Total	BAS (mmol/g)	LAS (mmol/g)	B/L ratio
gamma-Al_2O_3	163	180	0.354	0.894	1.248	0.044	0.838	0.05
SZ	34	59	0.582	0.801	1.383	0.228	0.073	3.12
SZA2/1	114	124	1.149	0.942	2.091	0.302	0.292	1.04
SZA1/1	115	128	1.212	1.105	2.317	0.493	0.450	1.10
SZA1/2	158	166	1.441	0.951	2.392	0.316	0.061	5.18

a. Measured by NH_3-TPD.

b. Measured by Py-IR.

The CO_2 desorption curves of the 5 M MEA solution with different catalyst are displayed in Figure 3. These curves manifest the catalytic performance of five catalysts in MEA solution for the whole 540 min duration conducted at 371K. It is noticed that the CO_2 loading decreased rapidly in the first 120 min after which the extent of CO_2 loading change was smaller. Thus, all analyses for the catalytic performance of each MEA-catalyst system were performed based on the first 120 min of the desorption processes. As shown in Figure 3 Panel A£¬all the catalysts quickened the CO_2 desorption rate as compared with the blank run. Among these catalysts, three SZA catalysts demonstrated better catalytic activity than the parent catalysts SZ or gamma-Al_2O_3. It is important to highlight that the lean CO_2 loading of all the different MEA-catalyst systems tended to the same value at 0.354 mol CO_2/mol amine, implying that the catalyst only accelerated the desorption rate but did not change the thermodynamic equilibrium. As shown in Figure 3 Panel B, with regard to the RH, SZA1/1 offered the best catalytic activity, which decreased the heat duty by 36.9% as compared with the blank run. Moreover, the reusability of the SZA was studied, and the result indicates that SZA presented good catalytic stability. Therefore, the SZA catalysts can be considered as a potential catalyst for reducing the energy consumption for CO_2 desorption reaction and for use in industrial CO_2 capture processes.

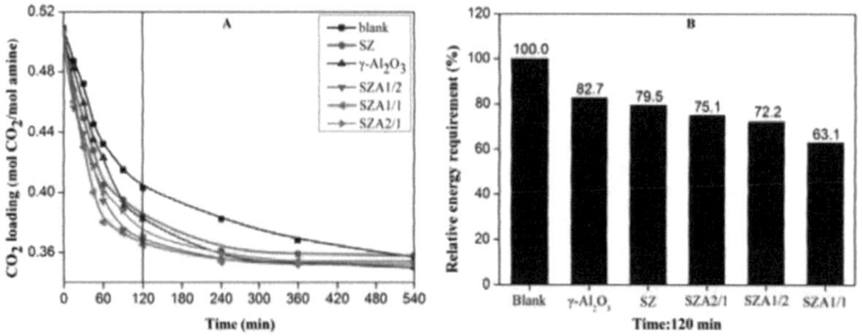

Figure 3. Catalytic CO_2 desorption performance.

The structure-activity correlation of the catalyst is investigated. The relationship between the BAS, strong acid sites and the catalytic activity were explored, and the result is shown in Figure 4. It can be seen that the catalytic performance increased with increasing BAS. The same trend is observed for strong acid sites as well. This confirms that the BAS and strong acid sites of a catalyst play a significant role in its catalytic performance. Moreover, because the mesopore surface area (MSA) of a catalyst affects the fraction of the active sites available for reaction, the combinations effect of the product of MSA and BAS (BAS*MSA) is studied. As shown in Figure 4 Panel C, an increase of BAS*MSA results in an increase in CO_2 desorption rate as well as a decrease in the heat duty. This reveals that the more BAS distributed on the catalyst surface, which are potentially provided by the larger mesopore surface area, the better catalytic performance for the MEA solvent regeneration process.

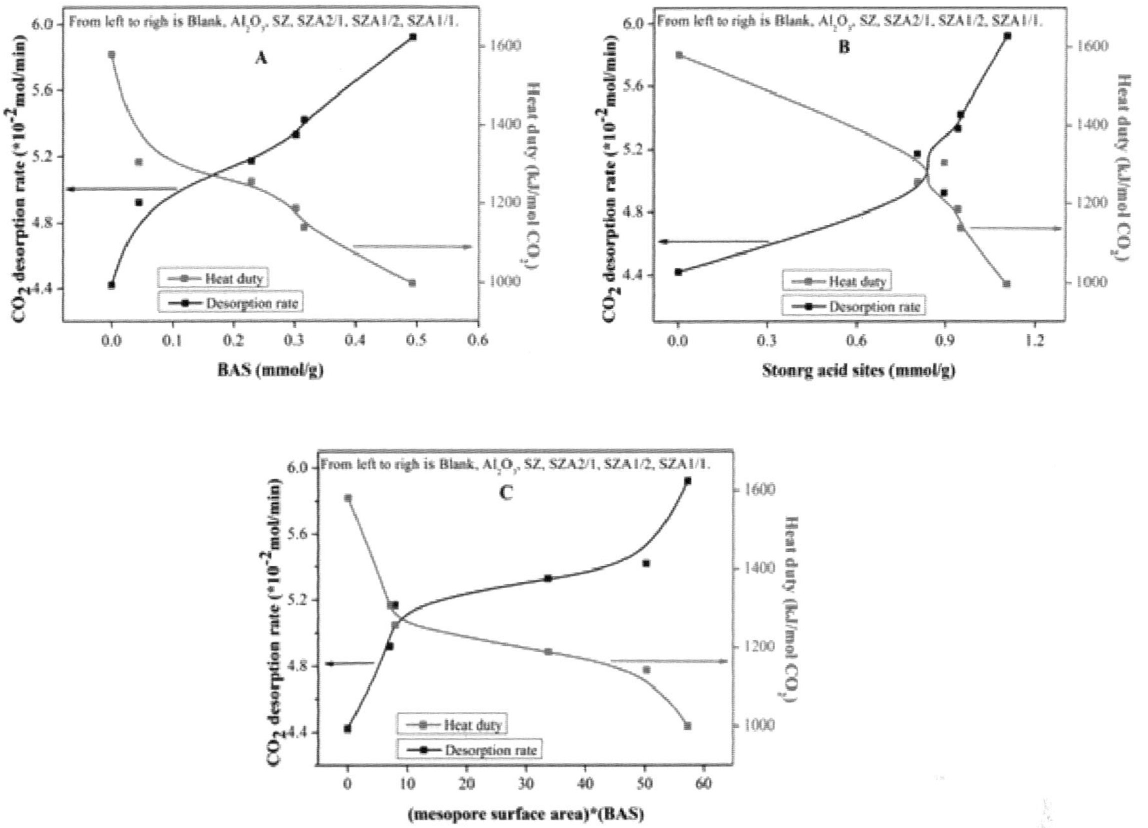

Figure 4. The structure-activity correlation of the catalyst

Acknowledgment

The financial support from the National Natural Science Foundation of China (NSFC-Nos. 21536003, 21776065, 21476064, and 51521006), China Scholarship Council (201606135004), and China Outstanding Engineer Training Plan for Students of Chemical Engineering & Technology in Hunan University (MOE-No.2011-40) are gratefully acknowledged. The authors also acknowledge support from the NPRP Grant # 7-1154-2-433 from the Qatar National Research Fund (a member of Qatar Foundation). Furthermore, the authors gratefully acknowledge the support of K.C WONG education foundation.

References

1. Shi H et al., *International Journal of Greenhouse Gas Control.* 2014;26:39-50.

2. Hua W et al., *Journal of Catalysis.* 2001;197(2):406-413.

541189 Carboxylation of Propylene Oxide to Propylene Carbonate

Abstract:

Polycarbonates, a class of thermoplastic polymer, have wide ranging applications in the transportation, construction, and packaging industries. Non-phosgene route for making polycarbonates involves transesterification of bis-Phenol-A and diphenyl carbonate, made from cyclic carbonates. Cyclic carbonates are produced from the cycloaddition of $CO_2$and epoxides employing either homogenous or heterogeneous catalysts. The main challenges of these systems are low catalyst activity and selectivity, and high CO_2pressures. In this study, we systematically investigate both homogeneous (TBAB) and heterogeneous catalysts (Ion exchange resins) (which show promise among the catalysts evaluated) aimed at enhancing catalyst performance with respect to activity and selectivity. We measured the liquid-phase solubility of CO_2in various solvents to better delineate solvent effect on carboxylation of propylene oxide to propylene carbonate (PC). The experiments were performed in a 100 mL Parr reactor with provision for temperature & pressure monitoring and in-line sampling of liquids. At the end of each experiment, the liquid products are analyzed for concentrations of PO, PC and solvent to assess the material balance. Various commercial solvents were investigated, and the order of activity is: Methanol > Ethanol > Isopropyl alcohol > Dimethyl formamide > N-methyl-2-pyrrolidone > Propylene carbonate > Dimethyl carbonate. The polar protic solvents (such aslinear alcohols) are strong nucleophiles, strong hydrogen-bond donors and acceptors and interact strongly with the electron-deficient electrophiles, epoxides and CO_2. In contrast, polar aprotic solvents are weak nucleophiles compared to the polar protic solvents resulting in low activity. These conclusions, based on CO2 solubility measurements in methanol and PC solvents along with kinetic studies of PO carboxylation, will be presented and discussed.

536942 Optimization of Sustainable Processes Incorporating Data Envelopment Analysis

Abstract:

Developing effective tools for process design is of paramount importance in the transition towards sustainable systems. In this work, we present a framework to incorporate sustainability principles in the design of chemical processes which combines a palette of tools, including life cycle assessment, surrogate modeling, objective reduction, multi-objective optimization, and multi-criteria decision analysis tools. In particular, we propose the use of Data Envelopment Analysis (DEA) to facilitate the post-optimal analysis of the Pareto front generated during the optimization. The use of DEA in the framework allows to narrow down the number of designs and rank them without the need to define weights in an explicit manner. Additionally, DEA also provides improvement targets for the sub-optimal alternatives that if attained would make them optimal. This characteristic of DEA can help to guide retrofit efforts when benchmark processes are compared against the best technologies available. The application of all the stages of the framework is demonstrated in the hydrogenation of CO_2 to produce methanol, where the process was initially modeled using a commercial process simulator to be further replaced by a surrogate model applying Artificial Neural Networks. The sustainability of the process was measured with the environmental categories included in the CML-2001 methodology and the total annualized cost. The objective-reduction stage narrowed down the number of total objectives to evaluate from 11 to five and the optimization was performed using a multi-objective Genetic Algorithm. The final application of DEA to analyze the solutions obtained from the optimization presented only 9 designs as the best options out of the 126 solutions that conformed the Pareto front. Finally, these 9 solutions were ranked and the best design identified. Overall, the methodology allowed an entire assessment of the process without compromising the performance of any sustainable indicator while reducing the computational resources used in the evaluation. The development of contributions such as the one presented in this work aim to facilitate the sustainable development of the chemical industry, not only by assessing new processes and technologies but also by identifying improvement targets for current sub-optimal ones.

535113 A Novel Model Predictive Control Scheme for Sustainability: Application to Biomass/Coal Co-Gasification System

Abstract:

Industrial, business and government communities have begun to shift from economic stand-alone focus to inclusion of sustainability in the decision-making process. This shift is due to the adverse environmental impact and unsustainable development caused by human activities, including chemical industry releases. As a result, process systems engineering (PSE) approaches have been developed, predominantly for incorporating sustainability into chemical process design and optimization [1]. However, the development of sustainability-oriented control schemes is still scarce, especially when compared to the number of steady-state design and optimization studies available [2]. Process control plays a critical role in realizing the efficient, sustainable and safe operation of the chemical and energy systems defined by the process design and optimization studies. The main barriers that have been prevented the integration of sustainability into process control are: 1) complexity of integrated large-scale chemical and energy processes; and 2) existence of multiple objectives for optimization involving social, economic and environmental issues.

To fill this gap, a novel model predictive control (MPC) scheme is proposed to drive the system to a sustainable operating point that is defined using a multi-objective optimization algorithm. In particular, the sustainable MPC is formulated based on dimensionless sustainability performance indicators from the U.S. EPA's GREENSCOPE (Gauging Reaction Effectiveness for the ENvironmental Sustainability of Chemistries with a Multi-Objective Process Evaluator) tool [3]. These established indicators are associated with the process state variables, releases, resource consumption, and products and can capture the sustainability information of the current process condition, including economic, environmental and social aspects. Such sustainability indicators are employed as hard/soft constraints in the implemented controller in order to maintain the process operation within a pre-defined sustainable zone, where the performance indicator values are higher than desired thresholds. To explicitly visualize the multidimensional sustainability indicators during transient, a multivariate plotting method is developed using dynamic radar diagrams.

The developed method is illustrated via a biomass/coal co-gasification process for syngas production with the end goal of methanol manufacturing. For this application, the whole process model is developed in Aspen Hysys based on existing literature information [4-5]. With the established models in Aspen and a link for data communication between Aspen and MATLAB, a multi-objective optimization problem is solved to maximize profit and optimize the process sustainability performance (e.g., environmental release and resource use minimization), by employing a genetic algorithm-based approach developed in MATLAB. In this presentation, the details on the application results of this novel framework for improving sustainability performance are discussed, focusing on the tradeoffs between using coal and biomass for the sustainable production of chemicals. The results show that the proposed sustainable MPC scheme can effectively drive the process to the optimal operating point, while maintaining the process within sustainable zones during transients.

References:

1. Daoutidis P, Zachar M, Jogwar SS. Sustainability and process control: A survey and perspective. J. Process Control. 2016;44:184–206.
2. Li S, Mirlekar G, Ruiz-Mercado GJ, Lima FV. Development of chemical process design and control for sustainability. Processes. 2016;4(3):23.
3. Ruiz-Mercado GJ, Gonzalez MA, Smith RL. Sustainability indicators for chemical processes: III. biodiesel case study. Ind. Eng. Chem. Res. 2013;52(20):6747–60.
4. Li S, Feliachi Y, Agbleze S, Ruiz-Mercado GJ, Smith RL, Meyer DE, Gonzalez MA, Lima FV. A process systems framework for rapid generation of life cycle inventories for pollution control and sustainability evaluation. Clean Technol. Environ. Policy. 2018. DOI: 10.1007/s10098-018-1530-6.
5. Robinson PJ, Luyben WL. Simple dynamic gasifier model that runs in Aspen Dynamics. Ind. Eng. Chem. Res. 2008;7784–92.

536285 Sustainability Dynamics and Control for Distributed Manufacturing of Renewable Energy

Abstract:

Prediction of a long-term sustainability performance of a large-scale system is always very challenging largely due to the existence of various types of uncertainties. Such a system usually consists of a number of interacted sub-systems, each of which could be structurally and operationally sophisticated. Distributed biodiesel manufacturing in a large geographical region could be a good example. A further challenge is the development of effective strategies, frequently for implementation in multiple stages, to improve regional sustainability performance in a time horizon of interest. Since sustainability is an act of balance among economic, environmental, and social domains, thus from system and control science point of view, sustainability assessment, performance prediction, and decision making is a non-conventional multi-objective, multi-input-multi-output control problem [1].

Recently, we developed a hierarchical control method to enhance manufacturing sustainability, which was successfully applied to a biodiesel manufacturing process [2]. In that framework, a two-layered hierarchical control scheme is introduced to implement strategic and tactical control for dynamic sustainability control. In this paper, we introduce an extended hierarchical control method, which is capable of conducting an integrated analysis of subsystems' dynamics and control strategies and developing strategies for coordinated control of the entire system to achieve a long-term sustainability goal under various uncertainties. The introduced methodology is tested through a case study on the sustainability performance prediction and improvement of a geographically distributed manufacturing of biodiesel using different feedstocks and different manufacturing technologies.

References:

[1] Moradi-Aliabadi, M. and Y. Huang, "Multistage Optimization for Chemical Process Sustainability Enhancement under Uncertainty," ACS Sustainable Chemistry and Engineering, 4(11), 6133-6143, 2016.

[2] Moradi-Aliabadi, M. and Y. Huang, "Decision Support for Enhancement of Manufacturing Sustainability: A Hierarchical Control Approach," ACS Sustainable Chem. Eng., 6(4), 4809–4820, 2018.

530944 An Analysis of Socioeconomic Impacts of Aviation Biofuel Development in Brazil

Abstract:

Aviation biofuel has been recognized by its potential to reduce greenhouse gas emissions of the aviation industry. Derived from renewable feedstocks, aviation biofuel is generally perceived sustainable. However, some of the sustainability issues such as land use change, food insecurity and biodiversity loss, are closely related to different phases of biofuel supply chains. Thus the overall impacts of biofuel production require a full investigation considering all dimensions of sustainability. While many studies have evaluated the environmental impacts or the techno-economic feasibility of aviation biofuel, very few have addressed the social or socioeconomic aspects in a systematic manner. The objective of this study is, therefore, to evaluate socioeconomic impacts of aviation biofuel production with a supply chain perspective.

A scenarios-based input-output (IO) analysis is used to evaluate socioeconomic effects including employment, GDP and trade balance (represented by imports) of aviation biofuel supply chains. This approach is exemplified for the local production of aviation biofuel in Brazil. Given that the commercialization of aviation biofuel is still at an early stage, available knowledge and data on the deployment of aviation biofuel are limited. A scenario analysis is useful to explore how possible futures of aviation biofuel would develop. The timeframe of the designed scenarios is set for 2050. The storyline of each scenario is elaborated on the market of aviation biofuel, conversion technologies, selection of feedstocks, and potential competition for biomass. Furthermore, the future demand of aviation biofuel under each scenario is estimated, which can subsequently be used in IO analysis to further determine the effects on employment, GDP and imports. In addition, we propose a stochastic simulation to address the uncertainty of IO analysis, by performing a Monte Carlo (MC) simulation for the technical coefficients in the IO matrix. The uncertainty analysis provides insights into the robustness and reliability of the assessment results.

Four scenarios are developed based on the diverging trends of two key driving forces: i) biofuel policies (conservative or proactive) and ii) technological advancement (gradual or breakthrough). Different demands of aviation biofuel are projected for each scenario, ranging from 3% to 15% of the total fuel demand. Three combinations of technologies and feedstocks are considered for producing aviation biofuel in Brazil: alcohol to jet (ATJ) with sugarcane, hydro-processed esters and fatty acids (HEFA) with macauba, and Fischer-Tropsch (FT) with eucalyptus.

Results of IO analysis show that the scales of socioeconomic effects on employment, GDP and imports are proportional to the estimated demand of aviation biofuel in each scenario. Specially, Scenario 3 (where proactive biofuel policies go hand in hand with advanced technology) leads to the highest employment, GDP and imports potential, whereas the lowest socioeconomic effects are expected in Scenario 1 (in which biofuel policies are conservative and biofuel technologies progress slowly). Furthermore, under the same scenario, the macauba-HEFA chain performs best in terms of employment creation, followed by the sugarcane-ATJ chain and at last the eucalyptus-FT chain. With regard to GDP, the macauba-HEFA chain leads to higher effects than the other two chains. On the other hand, the most significant effects on imports are estimated in the eucalyptus-FT chain, followed by sugarcane-ATJ chain, while the macauba-HEFA chain sees the lowest imports. Additionally, the uncertainty analysis suggests that socioeconomic effects calculated with deterministic method are generally higher than the stochastically simulated results. The variances range from 1% to 28%.

To conclude, it is clear that enabling policy environment, high level of technology readiness, as well as availability of sustainable biomass play a key role in maximizing socioeconomic benefits. The feedstock sectors are the key sectors responsible for employment and GDP generation. While this reveals substantial potential of job opportunities and social development, the cultivation of feedstocks is associated with many uncertainties such as climate, land availability, and agricultural practices. Only when the production of aviation biofuel are developed in a way without threatening food security or deteriorating land conditions, the positive socioeconomic impacts do not contradict sustainability. In addition, the proposed stochastic simulation approach has been useful to address uncertainty in IO analysis. This contributes to an informed decision-making process regarding aviation biofuel development, from the perspective of social sustainability. To the best of the authorsâ€™ knowledge, this is an innovative attempt to address uncertainty of IO analysis in the context of ex-ante socioeconomic assessment by using stochastic simulation to capture the uncertainty in the input-output matrix.

521791 A Coordinated Multi-Product Market for Organic Waste Management

Abstract:

One of the major challenges associated with organic waste management is determining an economic value for products. For instance, in current organic waste management systems, gate/tipping fees are normally required for processing, indicating that the waste supplier is willing to pay someone else for the waste disposal [1]. On the other hand, organic waste is a potential source for deriving value-added products such as biofuels, electricity, and fertilizers [2,3,4]. Thus, this suggests that waste suppliers can, in fact, earn income by letting others use their waste. Unfortunately, systematic waste exchange frameworks that can determine prices for waste streams are currently lacking. This is currently introducing significant levels of uncertainty on the market potential for derived products, technologies, and hindering investment.

We propose a coordinated market framework to exchange waste streams, processing, transportation services, and derived products. The framework is motivated and designed after coordinated electricity markets that currently exist in operation and that have led to high economic efficiency and technology innovation [5]. Also, the suppliers and consumers of waste and derived products, as well as technology and transportation providers bid into the market to offer their services. Prices for waste and derived products are obtained by solving a market clearing problem that balances products across a supply chain network that connects all the market participants. The market clearing problem provides prices for all products and services across the network.

We prove that the market clearing problem provides prices under which the profits of all market players are non-negative and provides a Pareto optimal solution (thus representing a true trade-off between the profits of all players). We provide illustrative examples to show the capability and properties of this coordinated market framework and demonstrate that the framework is flexible and can accommodate more sophisticated market players such as environmental remediation costs, waste storage, and policymakers and local/state/federal governments (which provide incentives/penalties to drive the market). The market can be cleared on a regular basis to adjust waste flows and prices due to seasonal effects (e.g., nutrient run-off during spring or poor air quality in particular areas). Moreover, the market naturally captures geographical effects (by providing space-dependent prices). This property can be used to redistribute waste streams across geographical regions (e.g., push nutrients away from endangered areas).

We use the proposed framework in a real case study of the Rock River basin in the State of Wisconsin. We consider (i) multiple organic waste streams including cow manure, wastewater sludge, and household food waste; (ii) multiple waste processing technologies that include nutrient recovery, energy production, and natural environmental degradation; (iii) regulatory drivers, such as RECs (Renewable Energy Certificates) and RINs (Renewable Identification Numbers) [6,7]; as well as (iv) environmental and social constraints on key factors such as TMDL (Total Maximum Daily Loads) [8].

References:

[1] R. Raven and K. Gregersen, "Biogas plants in Denmark: successes and setbacks," Renewable and sustainable energy reviews, vol. 11, no. 1, pp. 116–132, 2007.

[2] K. Krich, D. Augenstein, J. Batmale, J. Benemann, B. Rutledge, and D. Salour, "Biomethane from dairy waste," A Sourcebook for the Production and Use of Renewable Natural Gas in California, pp. 147–162, 2005.

[3] L. Young and C. C. Pian, "High-temperature, air-blown gasification of dairy-farm wastes for energy production," Energy, vol. 28, no. 7, pp. 655–672, 2003.

[4] Z. Ye, Y. Shen, X. Ye, Z. Zhang, S. Chen, and J. Shi, "Phosphorus recovery from wastewater by struvite crystallization: Property of aggregates," Journal of Environmental Sciences, vol. 26, no. 5, pp. 991–1000, 2014.

[5] V. Zavala, K. Kim, M. Anitescu, and J. Birge, "A stochastic electricity market clearing formulation with consistent pricing properties," Operations Research, 2017, 65(3): 557-576.

[6] "Renewable Portfolio Standard," Database of State Incentives for Renewables & Efficiency (DSIRE) available at http://programs.dsireusa.org/system/program/detail/564. [Online; accessed 06-December-2017]

[7] L. McPhail, P. Westcott, and H. Lutman, "The renewable identification number system and us biofuel mandates," ERS Report BIO-03, USDA, November 2011.

[8] D. Columbia, F L. Dodge, G L. Green, "Total Maximum Daily Loads for Total Phosphorus and Total Suspended Solids in the Rock River Basin," 2011.

536446 Vulnerability of United States Industrial Sectors Dependent on Insect-Mediated Pollination Service

Abstract:

Of the many critically important ecosystem goods and services that sustain human and industrial activity, one crucial service with high economic value is pollination by insects. The agricultural sector of the U.S. economy as well as many other industrial sectors (fiber, drugs, and fuel) are directly and indirectly dependent on insect-mediated pollination by both commercially-managed and wild species of bees and other insects. In forty-five years, production of pollination-dependent crops has increased more rapidly (50-62%) than the global population of managed bees (45%). Simultaneously, wild insects in all regions have declined up to and exceeding 75 percent over only the last twenty-seven years, both in biomass and with decreasing biodiversity. This disparity creates a concern to humans nutritionally and economically as many stress factors continue to influence the fitness of pollinators around the world. Most existing methods focused on environmental sustainability assessments across engineering including traditional life cycle assessment (LCA) ignore the critical role of ecosystem goods and services in product life cycles.

The present work utilizes field study data available in an extensive literature review to quantify the economic dependence of the United States agricultural and related sectors on pollination service mediated by insects, updating existing pollination dependence coefficients when possible and bounding uncertainty of the estimates with Monte Carlo or bootstrapping methods. Integration of InVEST model data and public price and production information available through the United States Department of Agriculture (USDA) National Agricultural Statistic Survey (NASS) and Census of Agriculture compiled at a national, state, and county level considers the spatial relationship between crop-specific economic dependence of pollination and landscape-specific pollinator forage suitability. Specifically, our results highlight regions of the U.S. most susceptible to pollinator loss or decline, having the highest demand for insect pollination services while simultaneously lacking appropriate pollinator habitat. Downstream, cascading dependencies of non-agricultural industrial sectors are assessed using input-output analysis framework. The implications of these findings include expanding understanding of pollination services and pollinator health and identifying dependence and vulnerabilities of agricultural and at risk industrial sectors. These findings also direct conservation and revitalization efforts, advise future policy development, and support the incorporation of valuable ecosystem services as a component of LCA through the creation of an impact category for insect-mediated pollination service.

532935 Membrane Mineral Scaling in Semi-Batch and Steady State Reverse Osmosis Desalination - a Comparative Study

Abstract:

RO membrane desalination has proven to be a robust technology for desalting both brackish water and seawater. However, despite the popularity of RO technology, membrane mineral scaling remains a major impediment to achieving high recovery, particularly in brackish water desalination, and also leads to increasing operational cost (due to required membrane cleaning cleaning) and shortening membrane lifetime. Recent studies have proposed the use of semi-batch RO as an alternative operational approach that could enable high recovery operation at reduced level of antiscalant dosing which would provide more effective means of mitigating mineral scaling. In semi-batch RO operation, continuous concentrate recycling is adopted in a filtration period to enhance the system water recovery. During the filtration period salt concentration (including the concentration of sparingly soluble salts) rapidly increases in the feed flow channel as salt accumulates in the feed flow channel. Once the osmotic pressure in the feed channel reaches the upper allowed limit, the concentrate holdup is discharged and can be flushed with the raw-feed water. Such intermittent flushing after each filtration period has been said to reduce membrane mineral scaling by resetting the nucleation induction period (or â€œinduction clockâ€) for the onset of mineral scaling on the membrane surface. Current studies have argued that the above operational mode would lead to more effective RO operation with respect to scale mitigation relative to conventional steady state RO. In order to assess the above hypothesis, an investigation was undertaken to evaluate the impact of repeated filtration-flushing cycles in semi-batch RO operation on mineral scaling. In order to unambiguously assess mineral scaling under semi-batch operation, a direct real-time membrane surface imaging system was developed to detect the onset of membrane scaling, with gypsum (calcium sulfate dihydride) selected as the model mineral scalant. The surface monitoring system was interfaced with a spiral-wound pilot scale RO system, capable of achieving up to 90% water recovery. RO desalination tests were then carried out in both semi-batch and steady-state operational modes to compare the above two operational modes with respect to resetting of the scaling induction period and quantify mineral scale kinetics at various water recoveries. Real-time membrane surface image analysis enabled quantification of gypsum scaling kinetics in terms of both the number density of gypsum crystals and the extent of membrane surface scale coverage. The above information enabled evaluation of: (a) impact of intermittent raw-feed water flushing for semi-batch RO operation on the formation of nuclei and rate of crystal growth, (b) effect of convective residence time in the RO system on the scale induction period, (c) mineral scaling at different levels of raw feed supersaturation with respect to gypsum, and (d) mineral scaling at semi-batch RO ad steady state RO at comparable levels of mineral saturation in the RO feed channel. Given the ability for early detection of mineral scaling, via the current direct and real-time membrane surface imaging, the present work provides an unambiguous demonstration that, contrary to hopes raised regarding the superiority of semi-batch RO, mineral scaling can be significantly exacerbated in this operational mode relative to steady state RO operation. However, there are certain limited operational domains in which semi-batch RO may be advantageous. The fundamental results of the current study provide a basis for assessing the recovery limits that are imposed by mineral scaling and possible operational approach to scale mitigation for semi-batch and steady-state RO.

521862 Water Desalination and Purification through Mixed-Matrix Membranes: An Atomistic Simulation Study

Abstract:

Development of technologies for water desalination is critical to cope with the global inadequate supply of clean water. Membrane-based separation is an efficient method to produce fresh water suitable for human consumption and irrigation. In particular, reverse osmosis (RO) has become the most widely used for water desalination. Currently, RO comprises 60% of the capacity of desalination and represents 75-85% new desalination projects worldwide. RO is also significant in purifying polluted water due to its low energy cost and simple operational process, especially in mining areas. In RO process, a selective membrane with high water ï¬‚ux is of vital importance to its performance. Compared with conventional polymeric membranes and zeolites, mixed-matrix membranes (MMMs) formed by incorporating organic/inorganic fillers into a polymer matrix have attracted considerable interest since they combine the merits of each type of material.

In this work, an atomistic simulation study is conducted for water desalination and purification through MMMs constructed by PB-1A organic cage and a polymer of intrinsic microporosity (PIM-1). PB-1A is a newly reported analogue of porphyrin box PB-1 with intrinsic porosity. The swelling of the MMMs in water and its effect on water permeation are examined. For desalination of seawater, the MMMs exhibit high water permeability and 100% salt rejection. For purification of wastewater (containing Pb^{2+}, Cu^{2+} and Cd^{2+} ions), high water permeation and ion rejection are also achieved at various conditions. For both applications, the dynamic and structural properties of water molecules in the membranes are investigated. This simulation study provides microscopic insights into water transport in PB-1A/PIM-1 MMMs and suggests their potential use for water desalination and purification.

538824 Metal Oxide Functionalized Graphene Oxide Membranes for Advanced Oxidation of Pollutants

Abstract:

Fast water transport, ease of membrane fabrication, separation on nanoscale, mechanical and thermal stability and stability in solvent media has made Graphene Oxide (GO)-based membranes a promising candidate for nanofiltration. Additionally, their ability to catalyze redox reactions have attracted much attention in environmental remediation of toxic organics. Aim of this study is to combine the ability of GO membranes for nanoscale separations with pursulfate mediated removal of smaller chlorinated organics from water. A novel membrane platform consisting of partially reduced GO and polyacrylic acid was fabricated using layer by layer assembly approach. Iron nanoparticles were embedded in the membrane domain by in-situ synthesis from ferrous ion precursors immobilized on the introduced carboxylic groups of polyacrylic acid. Extensive exclusion studies with various salts and dyes showed that the fabricated membranes retained separation abilities of the conventional GO membranes. Catalytic property of the GO membranes were investigated in a cross flow mode of operations. Extended study over 20 hours showed effective decomposition of persulfate on passing its aqueous solution (500 ppm) through the fabricated membranes for an operating flux of 2 LMH, pH 6.5 and 23°C. Evaluation of effectiveness of the fabricated membranes for removal of trichloroethylene (TCE) under same operating conditions showed near 100% removal of TCE (from 30 ppm feed). The GO-based platform with its capabilities of separation on nanoscale and catalyzing oxidative removal of toxic organics can further help in achieving more efficient water purification. This research is supported by the NIEHS-SRP (Grant no: P42ES007380) and NSF KY EPSCoR (Grant no: 1355438).

535985 Enhanced Biocidal and Antifouling Properties of Thin Film Composite Membranes Via Active Layer Modification with Polyrhodanine Nanoparticles

Abstract:

The demand for fresh water has been increasing and is expected to rise continuously for the foreseeable future. A freshwater scarcity can negatively affect the dynamics of ecosystems[1], food security[2], and so on. Forward osmosis (FO) is a high-water-recovery and low-cost membrane-based technology that has potential applications in desalination and wastewater treatment[3-5]. However, the lack of FO membranes with effective anti-fouling and antibacterial properties is still a challenge[3]. The incorporation of hydrophilic and antimicrobial nanomaterials into the active layers of membranes has been found to be a promising strategy to improve anti-fouling and anti-biofouling properties of nanocomposite membranes[6].

In this work, polyrhodanine (PRh) nanoparticles are synthesized via oxidization polymerization and incorporated into the active layers of thin film composite (TFC) membranes. High compatibility of the PRh with polyamide (PA) chains is corroborated using several analysis methods and molecular dynamics (MD) simulations. TFC-PRh membrane containing 0.01 wt% PRh displays enhanced water flux and sieving ability when tested in a FO process. Compared to the pristine TFC, smoother and more hydrophilic surfaces of TFC-PRh membranes provide more fouling resistance against a sodium alginate solution. Furthermore, outstanding biocidal activity of the TFC-PRh-0.01 membrane decreases bacteria adhesion hindering biofilm formation and leading to an enhancement in membrane anti-biofouling properties.

Keywords: Polyrhodanine, Thin-Film Composite Membrane, Forward Osmosis, Antimicrobial, Antifouling

References:

1. Barkaoui, K.; Navas, M. L.; Roumet, C.; Cruz, P.; Volaire, F., Does water shortage generate water stress? An ecohydrological approach across Mediterranean plant communities. *Functional Ecology* **2017,** *31* (6), 1325-1335.
2. Kummu, M.; Guillaume, J.; De Moel, H.; Eisner, S.; FlÃ¶rke, M.; Porkka, M.; Siebert, S.; Veldkamp, T.; Ward, P., The worldâ€™s road to water scarcity: shortage and stress in the 20th century and pathways towards sustainability. *Scientific reports* **2016,** *6*, 38495.
3. Lutchmiah, K.; Verliefde, A.; Roest, K.; Rietveld, L. C.; Cornelissen, E. R., Forward osmosis for application in wastewater treatment: a review. *Water research* **2014,** *58*, 179-197.
4. Shaffer, D. L.; Yip, N. Y.; Gilron, J.; Elimelech, M., Seawater desalination for agriculture by integrated forward and reverse osmosis: Improved product water quality for potentially less energy. *Journal of membrane science* **2012,** *415*, 1-8.
5. Chung, T.-S.; Zhang, S.; Wang, K. Y.; Su, J.; Ling, M. M., Forward osmosis processes: yesterday, today and tomorrow. *Desalination* **2012,** *287*, 78-81.
6. Zirehpour, A.; Rahimpour, A.; Arabi Shamsabadi, A.; Sharifian Gh, M.; Soroush, M., Mitigation of thin-film composite membrane biofouling via immobilizing nano-sized biocidal reservoirs in the membrane active layer. *Environmental Science & Technology* **2017,** *51* (10), 5511-5522.

538249 Modifying TiO_2 MagnÃ©li Phase Reactive Electrochemical Membranes with Pyrogenic Carbonaceous Materials for Adsorption and Electrochemical Removal of Water Contaminants

Abstract:

This study was focused on the synthesis, characterization, and performance evaluation of Ti_4O_7 reactive electrochemical membranes (REMs) amended with pyrogenic carbonaceous materials (PCMs). Among the Magnéli phases of TiO_2 (Ti_nO_{2n-1} n= 4 to 10), Ti_4O_7 (n=4) is the most conductive phase and produces the highest yield of hydroxyl radicals. Ti_4O_7 possesses an electrical conductivity in the range of 1000-1500 S/m, but has limited adsorption capacity for organic contaminants. In this work, a method was investigated to improve the conductivity and adsorption capacity of Ti_4O_7 REMs and their performance for removal of different classes of water contaminants.

Two types of PCMs were added to REMs, 10 wt% multi-walled carbon nanotube (MWCNT) and 10 wt% powdered activated carbon (PAC). The composite PCM/Ti_4O_7 REMs were synthesized and characterized by a series of techniques (e.g., X-ray diffraction, thermal gravimetry analysis, Raman spectroscopy). Conductivity analysis showed that by adding MWCNTs and PAC, REM conductivity increased 3.3-fold and 2-fold respectively. Preliminary results indicated that a 10 wt% PAC and 10 wt% MWCNT loading to a Ti_4O_7 REM increased the adsorption capacity towards N-nitrosodimethylamine (NDMA)) by > 4.7-fold and > 7.1-fold, respectively. These results were determined by analyzing the time for permeate breakthrough curves to reach 50% of feed concentration.

By employing two different types of REMs, simultaneous chemical reduction/oxidation reactions and adsorption were used to separate and destroy different groups of dissolved water contaminants. For example, results showed for the Ti_4O_7 REM that steady state NDMA reduction was 60% during the three day experiment. By contrast, using 10 wt% PAC/Ti_4O_7 and 10 wt% MWCNT/Ti_4O_7 REMs, NDMA was no longer detected in the permeate during the three day experiment. These results indicated that the PCM/Ti_4O_7 composite REMs under cathodic potentials have potential for long-term removal of contaminants of concern.

Keywords: Separation, Water treatment, Reactive electrochemical membranes, pyrogenic carbonaceous materials.

534972 Quantification of Thermal Energy Delivery to Water-Membrane Interface in Membrane Distillation

Abstract:

Membrane Distillation (MD) is an attractive technology for treatment of high salinity waters (e.g. waste brines, produced water, saline ground waters) that can take advantage of low grade heat or solar thermal energy, reducing its operational costs. However, the MD process performance (i.e. flux, water recovery, and thermal efficiency) which is governed by the heat transfer to the membrane-water interface cannot be directly quantified in experimental systems. Past efforts to quantify energy transport to the water-membrane interface have used thermal transport models that rely on bulk temperature and flow rate measurements, and the inaccuracies from this approach have led to high uncertainty in translating process performance from the bench to commercial scale. Herein, we demonstrate through analysis of literature and experimental data the effect of system size on MD performance, and current limitations in our ability to predict its performance. We present a novel approach that leverages our fundamental understanding of vapor transport to correct existing heat transport correlations and significantly reduce the uncertainty in estimated MD process performance.

532965 Ceramic Membranes for Desalination By a Vacuum Flow-through Evaporation

Abstract:

Purification of salty water is a huge challenge in the present times. Several membranes like GO, zeolite, and GOF membranes have been used for the purification of salt water. But limitation in terms of water flux and ion rejection demands for better methods for water purification. The atomic layer deposition (ALD) process is one type of chemical vapor deposition process based upon surface limiting chemical reaction between precursors and substrate. ALD uses alternating, self-limiting chemical reactions between gaseous precursor and a solid surface to deposit thin films in an atomic layer by layer fashion. In this work, we deposited ZnO, Al_2O_3 and TiO_2 using ALD on macroporous Î±-alumina disk membranes having 1 in. diameter, 1 mm thickness, and 25% porosity for the purification of salt water. We used these membranes for salty water desalination by flow-through membrane evaporation. The interaction between ALD layer and water molecules enhanced the water flux and ion rejection values. A high water flux (~100 L. m^{-2}. h^{-1}) and ion rejection rate (~99%) were obtained. Moreover, performance of ZnO, Al_2O_3 and TiO_2 ALD disk were also compared and it was found ZnO showed better performance than Al_2O_3 and TiO_2

527041 Two Stage Stochastic Programming Modeling Approach for Energy Storage System Operation Under Uncertainty

Abstract:

Employing renewable energy sources has become an irreversible trend in the current energy systems of various sectors including process industries. Their decision-makers should cope with often occurring mismatches between varying energy supply and demand over multiple time periods. Considering the inherent variability of renewable output, it would be often witnessed that surplus energies from renewable sources is simply thrown away without being used. It is thus understandable to see the introduction of energy storage system (ESS) for the sake of the economic benefit of using the surplus again later. We can increase the penetration of renewable energies by installing the ESS. On the other hand, it is also right to point out that there is much to be explored in maximizing the potential of ESS with other entities in the energy system. In order to transform the expectation into reality, the corresponding decision-making framework should be rigorously designed. As an alternative, this paper investigates the impact of variation on the design and operation of ESS. Specifically, the design of ESS under uncertainty is transformed into a two-stage stochastic programming model. The stochastic model was formulated in terms of a mixed integer linear programming (MILP) problem. As illustrated in numerical examples, the proposed modeling approach allows us to gain the insight on renewable energy systems. More similar works are expected to establish the robust energy system in process industries.

527226 Theoretical and Experimental Investigation of the Microbial Degradation of Solitary Oil Microdroplets

Abstract:

Natural seeps and accidental releases of crude oil in the sea result in clouds of droplets that are carried away by underwater sea currents. The droplets may be created either at the sea surface during the breakup of an oil slick by sea waves, or at the seafloor during the atomization of live crude oil extruding at sufficiently high speed from a natural crack or a broken wellhead. Drifting droplet clouds disturb the established ecosystem dynamics and pose a high risk of toxic effects to many marine species. It is therefore of crucial importance to understand and quantify the factors that determine the retention of the dispersed droplets within the seawater column. Dissolution, biodegradation and aggregation with marine snow are the main physical processes that rule the fate of dispersed oil droplets in marine waters.

With regard to the biodegradation process, microbes have developed three fundamental strategies for accessing and assimilating oily substrates. Depending on their affinity for the oily phase and ability to proliferate in multicellular structures, microbes might either attach to the oil surface and directly uptake compounds from the oily phase, or grow suspended in the aqueous phase consuming solubilized oil, or form three-dimensional biofilms over the oil-water interface (Kapellos, 2017).

We have recently developed a compound particle model for the biodegradation of solitary oil microdroplets moving through a water column (Kapellos et al., 2018). The compound particle is of the core-shell type and consists of an oily core that is successively surrounded by a bioreactive skin of negligible thickness and another bioreactive shell of finite thickness. The bioreactive skin represents a thin layer of microbes that uptake oil directly from the oily phase, whereas the bioreactive shell represents a distinct biofilm phase. The new model accounts for all three modes of biodegradation: direct interfacial uptake, bioreaction in the bulk aqueous phase, and bioreaction in a biofilm formed around the droplet. A closed-form correlation has been established for the overall dissolution rate for a given particle configuration and two coupled ordinary differential equations have been derived for the evolution of the dimensions of the compound particle. Numerical analysis is currently used to extend the domain of validity of the model by relaxing key hypotheses. In this presentation, major findings of the theoretical analysis will be discussed in detail and in conjunction with preliminary results from the biodegradation of hexadecane droplets by Marinobacter sp. in a microfluidic setting.

Acknowledgements: This work has received funding from the European Unionâ€™s Horizon 2020 research and innovation programme under the Marie Sklodowska-Curie grant agreement No 741799 (project "OILY MICROCOSM").

References:

Kapellos, G.E. "Microbial strategies for oil biodegradation", In *Modeling of Microscale Transport in Biological Processes*; Becker, S.M., Ed.; Academic Press, pp. 19-39 (2017).

Kapellos, G.E., Paraskeva, C.A., Kalogerakis, N. and Doyle, P.S. "Theoretical insight into the biodegradation of solitary oil microdroplets moving through a water column", *Bioengineering*, 5:15, (2018).

530591 Modeling Faradaic Capacitive Deionization with Redox Active Porous Electrodes

Abstract:

Capacitive deionization (CDI) is an attractive technology for brackish water treatment, due to the low energy consumption at low source water salinity level compared to conventional water treatment technologies such as reverse osmosis or distillation.

It has been shown that the surface charge on the porous carbon electrodes has significant impact on CDI performance. By introducing immobilized chemical groups, the point of zero charge of carbon materials can be tuned, thus enhanced or inversed CDI performance could be obtained. Redox active electrodes have been investigated in CDI applications recently. The Faradaically induced chemical charges add to electronic charges within the electric double layers formed at the electrode surfaces to provide a much higher adsorption capacity than attained in purely capacitive processes, which is denoted as Faradaic capacitive deionization (FaCDI) here. The redox active ligand can be considered as variable chemical charges that depend on the cell voltage, as opposed to the fixed charges introduced by chemical modification.

Here, we developed a model which incorporates the effect of variable chemical charge due to general faradaic reactions of redox species immobilized on the electrode surfaces. The theory describes the variable chemical charge by the generalized Frumkin-Butler-Volmer equation across the stern layer as an extension of current electrosorption theory. As a first approximation, the model assumes the spacer channel is well-mixed with uniform electrosorption in each electrode. The model is able to predict the experimentally observed enhanced and inverted performance of CDI cells. Furthermore, the deionization performance of FaCDI cells is shown to be superior to that of CDI and enhanced-CDI systems with equilibrium adsorption capacities 50-100% higher than attained with CDI systems. By properly choosing the equilibrium potentials of the redox active material, FaCDI is able to achieve the same ion adsorption capacity at smaller voltages compared to conventional CDI, which indicates the promise of obtaining higher salt adsorption capacity with lower energy input through the use of FaCDI materials.

The model enables us to optimize the operation parameters of the process, such as flow rate, applied voltage as well as the design of CDI cells, such as the surface to volume ratio of the device, in order to obtain the optimal deionization performance.

532901 Finding the Ideal Feedstock for Biodiesel Production Via Generalized Kinetic Model for Transesterification and Saponification

Abstract:

The hunt for an assortment of energy sources is just about as old as humanity's need for energy. This search has been escalated by an expanding industrialization since the 19th century. The internal combustion engine is a prime example. Even though fuels obtained from fossil sources have been a prime choice since the onset of these engines, biofuels especially biodiesel have also found some interest. This interest has particularly gained momentum in the past few decades due to rapidly increasing energy demand and depletion in fossil fuel resources.

Biodiesel is a non-toxic, biodegradable, and environmentally benign fuel that can be used in diesel engines. The most commonly used method for the production of biodiesel is transesterification. It involves the reaction of oil with an alcohol in the presence of a catalyst to form fatty acid alkyl ester (biodiesel) and glycerol. Although, biodiesel is a promising alternative fuel, its production poses significant challenges due to its higher costs compared to that of petroleum-based diesel. Typically one of the major contributors to this issue is the cost of the feedstock, which accounts for approximately 60 to 80% of the total production cost.

Even though the literature is replete with the studies about modelling and optimization of transesterification [1, 2, 3, 4], very few researchers have focused their attention on the selection of feedstock for biodiesel production. Although, a few comparative studies have been reported in the literature [2, 5], all of them focus on static optimization, thereby neglecting the kinetics of biodiesel production. Therefore, in order to address these issues, we aim to achieve the following objectives:

1. Find optimal operating conditions of the reactor for biodiesel production using different vegetable oils as feedstock.
2. Develop meta-models depicting the effect of the different vegetable oils/their blends on the properties of the biodiesel produced using these feedstocks.
3. Use the developed meta-models to find the best possible feedstock for a particular country.

The purpose of this study is to find optimal feedstock (vegetable oil blend) for a particular country based on the standard property specifications and the availability of oils in that country. To this end, we use the generalized kinetic model, developed in our previous work [6] to find optimal conditions with 4,500 different blends of six major oils *viz.*, Jatropha, Linseed, Olive, Palm, Rapeseed, and Sunflower oil as feedstock. We employ a property based objective function (considering density, viscosity, cold point, pour point, cetane number, higher heating value, lower calorific value, and cold filter plugging point of biodiesel) to solve these optimization problems. We then employ correlations from the literature [3] to calculate properties of biodiesel formed using a particular feedstock and at corresponding optimal conditions. This is followed by developing meta-models with the computed properties of biodiesel as output variables and different characteristics of the feedstocks as the input variables. We use five different approaches to characterize the feedstock *viz.* a) average chain length and the unsaturation degree of the fatty acids that constitute the oil (adopted from [7], number of dimensions (N) = 2); b) average chain length, weighted unsaturation degree calculated based on the triglyceride composition of oil (proposed in this work, N = 2); c) triglyceride composition (proposed in this work, N = 34); d) weights on the considered oils for the different blends (proposed in this work, N = 6); and e) properties of oils, *i.e.* viscosity, density, heat capacity, enthalpy of vaporization, and vapour pressure calculated based on the triglyceride composition of oil (proposed in this work, N = 5). Finally, we use the best performing meta-models to obtain optimal feedstock for production of biodiesel in a particular country.

The predictions made using the developed meta-models are in accord with those made using the original models. We quantify this agreement, by computing the Percentage Average Absolute Error (PAAE) as follows,

PAAE = (Average Absolute Error/Average Response)x100

The error values are below 1% for most of the cases with the maximum value at 2.55%, which shows excellent prediction performance. Moreover, we use a normalized error metric to compare the performance of the meta-models developed using different data sets in this study. We observed that the meta-models developed using triglyceride composition and weight distribution of oils in the blend as input variables performed superior to the other meta-models. However, owing to the large number of dimensions in the former case, it is apt to use the latter. Finally, we used the meta-models to obtain optimal feedstock blend for a particular country. We analysed different scenarios based on the standard specification imposed on the biodiesel quality, availability and type of feedstock (*i.e.* edible or non-edible). We observed that the quality of biodiesel produced can change significantly based on the considered scenarios, thereby highlighting the importance of such a study. Besides, we also observed that it is possible to produce premium quality biodiesel even with just non-edible oils. Such a result can definitely bolster the efforts for commercialization of biodiesel production as the use of non-edible oils can significantly reduce the costs associated with the biodiesel production process.

REFERENCES

[1] V. Singh, K. Solanki, M. N. Gupta, Process optimization for biodiesel production, Recent patents on biotechnology 2 (2) (2008) 130{143.

[2] J.Yang, T. Astatkie, Q. S. He, A comparative study on the effect of unsaturation degree of camelina and canola oils on the optimization of bio-diesel production, Energy Reports 2 (2016) 211-217.

[3] S. Pinzi, D. Leiva, G. Arzamendi, L. Gandia, M. Dorado, Multiple response optimization of vegetable oils fatty acid composition to improve biodiesel physical properties, Bioresource technology 102 (15) (2011) 7280-7288.

[4] R. S. Kumar, K. Sureshkumar, R. Velraj, Optimization of biodiesel production

from manilkara zapota (l.) seed oil using taguchi method, Fuel 140 (2015) 90-96.

[5] T. Issariyakul, A. K. Dalai, Comparative kinetics of transesterication for biodiesel production from palm oil and mustard oil, The Canadian Journal of Chemical Engineering 90 (2) (2012) 342-350.

[6] P. Chhabra, S. Mosbach, M. Kraft, I. A. Karimi, A generalized kinetic model for transesterication and saponification (2017).

URL https://aiche.confex.com/aiche/2017/meetingapp.cgi/Paper/500597

[7] E. G. Giakoumis, A statistical investigation of biodiesel physical and chemical properties, and their correlation with the degree of unsaturation, Renewable Energy 50 (2013) 858-878.

530255 Accelerating the Generation of Coal Power Plant Property Models

Abstract:

Power plants are complex systems that are represented by first principles or empirical models. As power plants are developed with new technologies outside of well-studied thermodynamic regions, data-driven approaches have been used to model these complex systems [5]. For example, thermophysical properties or kinetic reaction models can be explained to an arbitrary level of complexity with an empirical model. The challenge is that these process models are not well understood, and it is not always intuitive what functional forms can be utilized to represent these systems. When optimizing proposed model systems, the efficiency of power plants relies on having accurate models. To maximize efficiency as well as to maintain safe operating conditions, it is essential that complex kinetic and thermophysical relationships are represented accurately, even though determining such a representation may be computationally expensive.

In this work we address the problem of speeding up the model generation process. In particular, we look at accelerating best subset selection algorithms with the use of GPU parallel computing. Best subset selection is a regression problem that seeks to determine a small set of features that best relates a set of inputs to an output [7]. There are many exact and heuristic approaches that can be used to solve the best subset selection problem, such as branch-and-bound [4], forward and backward selection [3], exhaustive search, the lasso [7], and mixed-integer optimization [6, 2, 1]. The proposed approach speeds up the model building process, allowing us to consider a larger feature space and generate more realistic models. The proposed methodology is compared against best subset selection techniques as well as the Automated Learning of Algebraic MOdels (ALAMO) methodology [2].

References

[1] D. Bertsimas, A. King, and R. Mazumder. Best subset selection via a modern optimization lens. The Annals of Statistics, 44:813–852, 2016.

[2] A. Cozad, N. V. Sahinidis, and D. C. Miller. Automatic learning of algebraic models for optimization. AIChE Journal, 60:2211–2227, 2014.

[3] M.A. Efroymson. Multiple regression analysis. Mathematical methods for digital computers, pages 191–203, 1960.

[4] C. Gatu and EJ Kontoghiorghes. Branch-and-bound algorithms for computing the best-subset regression models. Computational and Graphical Statistics, 15:139–156, 2006.

[5] S Lu and BW Hogg. Dynamic nonlinear modelling of power plant by physical principles and neural networks. International Journal of Electrical Power & Energy Systems, 22.

[6] A. Miller. Subset selection in regression. CRC Press, 2002.

[7] R. Tibshirani. Regression shrinkage and selection via the lasso. Royal Statistical Society, Methodological:267–288, 1996.

524382 Dynamic Optimisation of Water-Injection Wells Operation for Enhanced Oil Production from a Mature Oil and Gas Field

Abstract:

Field engineers in the oil and gas industry are constantly faced with the challenge of maintaining profitability amidst several operational constraints whose effects span a wide spectrum of timescales. Given the inevitably high environmental and financial stakes associated with the exploration and production of an oil and gas prospect, there is a strong incentive to enhance hydrocarbon recovery and production via systematic, model-based optimisation approaches [1-3]. In further response to this challenge, the application of sophisticated simulation methodologies to comprehensively ocapture the reservoir behaviour, multiphase flows (in wellbores and flowlines) [4-5] and gas-oil-water separation in the processing facilities is of exceptional industrial as well as academic interest. Liquid loading in gas wells and artificial lift design considerations, reservoir pressure maintenance via water injection, gas/water coning during production from deviated wells, pressure drop and liquid holdup of multiphase mixtures in highly deviated flowlines are some of the specific features of the problem which are frequently encountered and known to affect its complexity.

Simulation of these prevalent subsurface and surface phenomena does not always guarantee an accurate prediction of the onset of these problems, let alone the trouble-free operation of an existing petroleum field. In order to tackle the shortcomings and thus improve the fidelity of the currently employed models, it is necessary to also combine robust optimisation methods with these oil and gas CFD simulations [6-7]. This combination of simulation and optimisation algorithms increases the order of complexity due to model nonlinearity, non-convexity and the presence of discrete variables. There has been an increasing number of contributions in the application of optimisation techniques to field production, and this can be attributed to the advances in the development of specialised algorithms and accompanying computational power. To date, most of these attempts have focused on fields undergoing primary production through predominantly vertical wells [1-3].

This paper advances the state of the art in this field by formulating a production optimisation problem, via simultaneous consideration of production and injection wells with both vertical and deviated (inclined/horizontal) well geometries. A field undergoing secondary production (Enhanced Oil Recovery, EOR) is considered in this study. We use multiphase flow and reservoir simulation software in order to compute flowrates and pressure drops in wellbores and flowlines of the considered production network; extra measures are taken to ensure all flowlines are free from hydrate and wax at the prevalent temperature and pressure conditions. The problem is solved as a nonlinear program (NLP), comprising an economic objective function and several constraints to ensure operational feasibility. The adopted optimisation technique yields a reasonable increase in oil production compared to the usual and direct application of a black-box simulator. This also translates to an increased NPV with the current oil price, thus demonstrating the efficiency of the proposed method as a value addition tool.

REFERENCES

[1] Gunnerud, V., Conn, A. and Foss, B., 2013. Embedding structural information in simulation-based optimization. *Comput. Chem. Eng.* **53**: 35–43.
[2] Gerogiorgis, D.I., Georgiadis, M., Bowen, G., Pantelides, C.C. and Pistikopoulos, E.N., 2006. Dynamic oil and gas production optimization via explicit reservoir simulation. *Comput. Aided Chem. Eng.*, **21**: 179–184.
[3] Tavallali, M.S., Karimi, I.A., Teo, K.M., Baxendale, D. and Ayatollahi, S., 2013. Optimal producer well placement and production planning in an oil reservoir. *Comput. Chem. Eng.*, **55**: 109–125.
[4] Epelle, E.I. and Gerogiorgis, D.I., 2017. A multiparametric CFD analysis of multiphase annular flows for oil and gas drilling applications. *Comput. Chem. Eng.*, **106**: 645–661.
[5] Epelle, E.I. and Gerogiorgis, D.I., 2018. Transient and steady state analysis of drill cuttings transport phenomena under turbulent conditions. *Chem. Eng. Res. Des.* **131**: 520-544.
[6] Gunnerud, V. and Foss, B., 2010. Oil production optimization—A piecewise linear model, solved with two decomposition strategies. *Comput. Chem. Eng.*, **34**(11): 1803–1812.
[7] Kosmidis, V.D., Perkins, J.D. and Pistikopoulos, E.N., 2004. Optimization of well oil rate allocations in petroleum fields. *Ind. Eng. Chem. Res.* **43**(14): 3513–3527.

537394 Voluntary Oil Well Ignition As a Blowout Response: Analysis of Factors Influencing Viability

Abstract:

While most off-shore drilling use conventional steel rigs, for large reservoirs in shallow waters, artificial islands make an attractive alternative for offshore drilling because it offers more space, lower environmental impact, greater personnel safety, and lower cost. Drilling from an off-shore island may allow voluntary ignition as a blowout response if the oil contains sufficient gas. For voluntary ignition to be a viable blowout response, wellhead burning â€“ that depends on a number of factors â€“ needs to be nearly 100% efficient.

In this work, we investigate wellhead burning as a planned response to oilwell blowouts. The work presents a case study of proposed wellhead burning of 880 GOR, 27° API oil, should a blowout occur. The work takes into account some of the concerns regarding possible significant oil spill on Beaufort Sea as a result of incomplete combustion of the blowout effluent.

Burning of two-phase mixtures of oil and gas is a complex phenomenon that depends on the fraction of gas in the mixture, percentage of oil being entrained as liquid droplets in the gas stream, the size of the droplets, their residence time in the flame, among other parameters. Available models and correlations for the influence of these parameters on burning efficiency are examined in view of their relevance to wellhead burning.

This work will demonstrate that for some oil wells with high enough GOR, wellhead burning could be efficient enough for to be a reasonable response to a well blowout. In the studied case, a GOR of 880 combined with high total flowrate of the mixture, strongly implies nearly complete combustion of the well effluent during a blowout. The case study helped us formulate an algorithm to examine whether any particular well could be a possible candidate for wellhead burning as a planned response to a blowout. The approach suggests determining a wellâ€™s worst case discharge (WCD), estimating fluid velocities at well exit, estimating adiabatic flame temperature, and performing an energy balance to determine if all oil drops will evaporate.

AUTHOR INDEX

AUTHOR INDEX

AUTHOR INDEX

AUTHOR INDEX

AUTHOR INDEX

AUTHOR INDEX

AUTHOR INDEX